Advances in Experimental Medicine and Biology

Volume 983

More information about this series at http://www.springer.com/series/5584

Long-Cheng Li
Editor

RNA Activation

Springer

Editor
Long-Cheng Li
Laboratory of Molecular Medicine
Peking Union Medical College Hospital,
Chinese Academy of Medical Sciences
Beijing, China

ISSN 0065-2598 ISSN 2214-8019 (electronic)
Advances in Experimental Medicine and Biology
ISBN 978-981-10-4309-3 ISBN 978-981-10-4310-9 (eBook)
DOI 10.1007/978-981-10-4310-9

Library of Congress Control Number: 2017938618

Printed on acid-free paper

This Springer imprint is published by Springer Nature
The registered company is Springer Nature Singapore Pte Ltd.
The registered company address is: 152 Beach Road, #21-01/04 Gateway East, Singapore 189721, Singapore

To the memory of my beloved father
and to my family

Preface

RNA is a class of versatile nucleic acid molecules with a myriad of functions, and being able to base pair with both DNA and RNA, it is naturally used by organisms as a guide tool to specifically target proteins to other nucleic acid sequences as exemplified by the RNAi (RNA interference) and CRISPR (clustered regularly interspaced short palindromic repeats) systems. The serendipitous discovery made a decade ago that promoter-targeted short double-stranded RNAs (dsRNAs) are able to guide Argonaute proteins to promoters to activate gene expression has led to the development of a new research field known as RNA activation (RNAa). The past 10 years have been a long and rough journey for the RNAa field that I and a few others started. Since its birth, RNAa has met with hidden skepticism. While some believed RNAa adds another dimension to gene regulation, yet others thought RNAa is simply RNAi in disguise. Now mounting evidence indicates that RNAa is an endogenous mechanism conserved from *Caenorhabditis elegans* to human with important cellular and physiological functions which we are just beginning to appreciate. In just 10 years, a number of preclinical studies have reported the therapeutic efficacy of RNAa therapies for diseases from erectile dysfunction to cancer, and the first RNAa-based drug has advanced into clinical trial for treating liver cancer. It is high time that a book devoted to RNAa should be published. Contributed by world-recognized experts in RNA research, this book provides comprehensive coverage of different RNA-guided gene activation mechanisms and timely update of recent advances in RNAa research.

The chapters have been organized into four major parts: exogenous RNAa (RNAa triggered by artificially designed and promoter-targeted short dsRNAs), endogenous RNAa (RNAa induced by naturally occurring miRNA), RNA activation guided by other small RNAs (including translational upregulation by antisense oligonucleotides targeting upstream open reading frames and CRISPR-mediated transcriptional activation [CRISPRa]), and developing RNAa-based therapeutics.

Finally, I thank all the authors for their valuable contributions. It is my sincere hope that this book will stimulate interest among scientists to further explore this fascinating field and pharmaceutical companies to harness the potential of RNAa for managing diseases.

Beijing, China Long-Cheng Li

Contents

Contributors

Huiqing Cao Institute of Molecular Medicine, Peking University, Beijing, China

Pradeep Chaluvally-Raghavan Department of Obstetrics and Gynecology, Medical College of Wisconsin, Milwaukee, WI, USA

Department of Physiology, Medical College of Wisconsin, Milwaukee, WI, USA

Meng Chen Department of Bioengineering, Stanford University, Stanford, CA, USA

Gladstone Institute of Cardiovascular Disease, San Francisco, CA, USA

David R. Corey Departments of Pharmacology & Biochemistry, University of Texas Southwestern Medical Center, Dallas, TX, USA

Stanley T. Crooke Department of Core Antisense Research, Ionis Pharmaceutics, Inc., Carlsbad, CA, USA

Stephanos Ghobrial Department of Surgery, Hammersmith Hospital, Imperial College London, London, UK

Nagy A. Habib Department of Surgery, Hammersmith Hospital, Imperial College London, London, UK

Yunzhang Hu Institute of Medical Biology, Chinese Academy of Medical Sciences and Peking Union Medical College, Kunming, China

Vera Huang Molecular Stethoscope, Inc., San Diego, CA, USA

Moo Rim Kang Ractigen Therapeutics, Nantong, China

Benyi Li Department of Urology, University of Kansas Medical Center, Kansas City, KS, USA

Changlin Li Institute of Precision Medicine, Jining Medical University, Jining, China

Gongcheng Li Department of Urology, Wuhan General Hospital, Guangzhou Command PLA, Wuhan, China

Long-Cheng Li Department of Urology, The First Affiliated Hospital of Xiamen University, Xiamen, China

Laboratory of Molecular Medicine, Peking Union Medical College Hospital, Chinese Academy of Medical Sciences, Beijing, China

Xue-Hai Liang Department of Core Antisense Research, Ionis Pharmaceutics, Inc., Carlsbad, CA, USA

Ying Liang Laboratory of RNA Epigenetics, Institutes of Biomedical Sciences & Department of General Surgery, Huashan Hospital, Cancer Metastasis Institute, Fudan University, Shanghai, China

Department of Biochemistry and Molecular Biology, Shanghai Medical College, Key Laboratory of Ministry of Education, Department of Molecular Biology, Fudan University, Shanghai, China

Zicai Liang Institute of Molecular Medicine, Peking University, Beijing, China

Huaibing Luo Laboratory of RNA Epigenetics, Institutes of Biomedical Sciences & Department of General Surgery, Huashan Hospital, Cancer Metastasis Institute, Fudan University, Shanghai, China

Department of Biochemistry and Molecular Biology, Shanghai Medical College, Key Laboratory of Ministry of Education, Department of Molecular Biology, Fudan University, Shanghai, China

Antonello Mallamaci Lab of Cerebral Cortex Development, Scuola Internazionale Superiore di Studi Avanzati (SISSA), Trieste, Italy

Xing Meng Institute of Molecular Medicine, Peking University, Beijing, China

Tiejun Pan Department of Urology, Wuhan General Hospital, Guangzhou Command PLA, Wuhan, China

Lei Stanley Qi Department of Bioengineering, Stanford University, Stanford, CA, USA

Department of Chemical and Systems Biology, Stanford University, Stanford, CA, USA

ChEM-H, Stanford University, Stanford, CA, USA

Ramani Ramchandran Department of Obstetrics and Gynecology, Medical College of Wisconsin, Milwaukee, WI, USA

Department of Pediatrics, Medical College of Wisconsin, Milwaukee, WI, USA

Vikash Reebye Department of Surgery, Imperial College London, London, UK

John J. Rossi Department of Molecular and Cellular Biology, Beckman Research Institute of City of Hope, Duarte, CA, USA

Irell and Manella Graduate School of Biological Sciences, Beckman Research Institute of City of Hope, Duarte, CA, USA

Wen Shen Department of Core Antisense Research, Ionis Pharmaceutics, Inc., Carlsbad, CA, USA

Mikko P. Turunen A.I. Virtanen Institute for Molecular Sciences, University of Eastern Finland, Kuopio, Finland

Tiia A. Turunen A.I. Virtanen Institute for Molecular Sciences, University of Eastern Finland, Kuopio, Finland

Jon Voutila MiNA Therapeutics Ltd, London, UK

Bin Wang Department of Biology, Kunming University, Kunming, China

Xiaoxia Wang Institute of Molecular Medicine, Peking University, Beijing, China

Jin-Chun Xing Department of Urology, The First Affiliated Hospital of Xiamen University, Xiamen, China

Seppo Ylä-Herttuala A.I. Virtanen Institute for Molecular Sciences, University of Eastern Finland, Kuopio, Finland

Sorah Yoon Department of Molecular and Cellular Biology, Beckman Research Institute of City of Hope, Duarte, CA, USA

Wenqiang Yu Laboratory of RNA Epigenetics, Institutes of Biomedical Sciences & Department of General Surgery, Huashan Hospital, Cancer Metastasis Institute, Fudan University, Shanghai, China

Department of Biochemistry and Molecular Biology, Shanghai Medical College, Key Laboratory of Ministry of Education, Department of Molecular Biology, Fudan University, Shanghai, China

Hui Zhang Institute of Human Virology, Zhongshan School of Medicine, Sun Yat-sen University, Guangzhou, Guangdong, China

Key Laboratory of Tropical Disease Control of Ministry of Education, Zhongshan School of Medicine, Sun Yat-sen University, Guangzhou, Guangdong, China

Yijun Zhang Institute of Human Virology, Zhongshan School of Medicine, Sun Yat-sen University, Guangzhou, Guangdong, China

Key Laboratory of Tropical Disease Control of Ministry of Education, Zhongshan School of Medicine, Sun Yat-sen University, Guangzhou, Guangdong, China

Section of Infectious Diseases, Department of Internal Medicine, Yale University School of Medicine, New Haven, CT, USA

Xiaoyang Zhao Department of Surgery, Hammersmith Hospital, Imperial College London, London, UK

Qingping Zou Laboratory of RNA Epigenetics, Institutes of Biomedical Sciences & Department of General Surgery, Huashan Hospital, Cancer Metastasis Institute, Fudan University, Shanghai, China

Department of Biochemistry and Molecular Biology, Shanghai Medical College, Key Laboratory of Ministry of Education, Department of Molecular Biology, Fudan University, Shanghai, China

Chapter 1
Small RNA-Guided Transcriptional Gene Activation (RNAa) in Mammalian Cells

Long-Cheng Li

Abstract Small RNA partnering with Argonaute (Ago) proteins plays important roles in diverse biological processes mainly by suppressing the expression of cognate target sequences. Mounting evidence reveals that the small RNA-Ago pathway can also positively regulate gene expression, a phenomenon termed as RNA activation (RNAa), which is evolutionarily conserved from *Caenorhabditis elegans* to human. In this chapter, I provide a general overview of mammalian RNAa phenomena and their basic characteristics and discuss recent advances toward understanding the nature of the molecular machinery responsible for RNAa and the development of RNAa-based research tools and therapeutics.

Keywords RNA activation • RNAa • Small activating RNA • saRNA • Transcription • Argonaute

1.1 Introduction

Small RNA molecules of 20–26 nucleotides (nt) generated within the cells through multiple steps of processing can serve as a versatile and sequence-specific regulatory signal to affect the expression of longer nucleic acid sequences. Such regulation by small RNAs is largely achieved by partnering with Argonautes (Agos) proteins [32]. Agos belong to a highly evolutionarily conserved family of proteins, and once programmed by small RNAs, can function as a homology search engine which uses the sequence information of the small RNA as a querying keyword [21]. Besides matching small RNAs with cognate sequences, Agos assume additional tasks such as participating in biogenesis of small RNAs, executing catalytic activity on cognate sequences, serving as a platform for recruiting other proteins. The pathways formed by various types of small RNAs and Agos have mainly been associated with a function of suppressing or eliminating cognate nucleic acid sequences including RNA and DNA [41], and are mainly utilized by organisms

L.-C. Li (✉)
Laboratory of Molecular Medicine, Peking Union Medical College Hospital, Chinese Academy of Medical Sciences, Beijing, China
e-mail: long@urogene.org

L.-C. Li (ed.), *RNA Activation*, Advances in Experimental Medicine and Biology 983, DOI 10.1007/978-981-10-4310-9_1

as a defense mechanism against the invasion of foreign sequences and harmful effects of mobile genetic elements. However, in mammalian cells such defense mechanisms have become obsolete during evolution and the pathway has been repurposed as a device for tuning the output of endogenous genes.

1.2 A Historical View of RNAa

Back in 2006, in an attempt to address whether aberrant promoter hypermethylation occurring in cancer cells could be triggered by small RNAs, we designed short double-stranded RNAs (dsRNAs) to target the promoter of E-cadherin, a gene frequently silenced by promoter hypermethylation in human cancer cells. Unexpectedly, we found that dsRNAs could in fact induce the expression of E-cadherin. After reproducing this phenomenon with two additional genes, we were prompted to examine it in detail. Subsequent experiments revealed that (1) gene activation by small RNAs depended on Ago2 protein for it was abolished in the absence of Ago2; (2) gene activation resulted neither from demethylation of the target promoter DNA nor from increased methylation; (3) the 3′ end of the small RNA duplex was critical for activity; (4) gene activation had kinetics that differed from RNA interference (RNAi); and (5) epigenetic changes represented by a loss of methylation at lysine 9 of histone 3 (H3K9me) were associated with gene activation. Based on those observations, we concluded that the small RNA-induced gene activation could be a novel regulatory mechanism for the small RNA-Ago pathway and named it as RNA-induced transcriptional gene activation (RNAa, Fig. 1.1) [25]. To distinguish these promoter-targeting "activating dsRNAs" from siRNAs which silence gene expression by targeting mRNA sequences, we named them as small activating RNA (saRNA). Although the observation of RNAa was initially met with skepticism, it was soon corroborated by a study from the Corey Laboratory. Also conducted in human cells, Janowski and colleagues showed that the progesterone receptor (PR) and major vault protein (MVP) genes could be activated by promoter-targeting dsRNAs which they termed as agRNA (antigene RNA) [20]. The same group further showed that agRNAs targeting 3′ terminus of PR gene could also induce PR expression [70].

Subsequent studies identified endogenous triggers for RNAa. In 2008, our group further showed that a naturally occurring microRNA (miRNA) could serve as saRNA to induce gene expression by targeting promoter sequences [38]. By introducing synthetic miR-373 into human cells, we found E-cadherin and CSDC2 (cold shock domain-containing protein C2) gene expression were induced. Similar to miRNA targeting 3′UTR (untranslated region) of an mRNA, for a miRNA to activate gene expression, it does not need to have perfect sequence homology with its cognate promoter targets [38].

To search for evidence that RNAa is an endogenous mechanism functioning in cellular physiology, in 2010, our group performed gene expression profiling in cells with miRNA biogenesis genes perturbed [16]. It was found that along with many other genes, a subset of genes that promote cell-cycle progression depends on

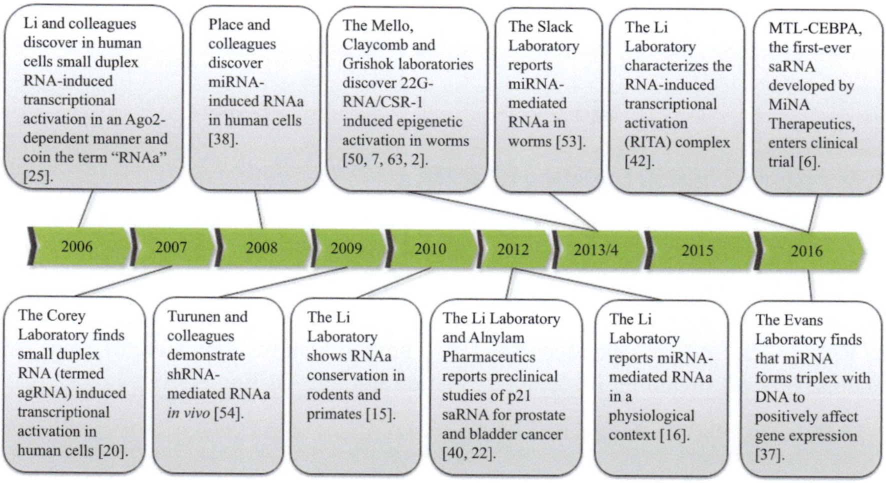

Fig. 1.1 RNAa timeline

miRNA for constitutive expression. For instance, the expression of Ccnb1 (cyclin B1) was downregulated when Drosha, Exportin-5, and Ago1 were depleted and was restored by overexpression of Ago1, suggesting that certain endogenous miRNAs are required to maintain Ccnb1 level. By validating several miRNAs predicted to target Ccnb1 promoter, miR-744 and miR-1186 were confirmed to be the "activator miRNA" of Ccnb1. Through activating Ccnb1, these miRNAs were found to have a pro-proliferative effect in the short term and cause genetic alternations (e.g., changes in chromosome number and structure) when they were stably overexpressed for a few dozens of cell divisions. This study provided strong evidence that miRNAs act as an endogenous trigger of RNAa, thereby participating in potentially fundamental cellular processes including carcinogenesis.

RNAa is conserved from *Caenorhabditis elegans* to human. In a 2010 study from our group, RNAa was demonstrated in several mammalian species including nonhuman primates (NHPs), mouse, and rat [15]. In 2013, Turunen et al. further demonstrated in vivo activation of Vegfa gene in mice by saRNA stably expressed as shRNA (short-hairpin RNA) [54]. Further conservation of RNAa traces back to *C. elegans*. In late 2013 and early 2014, several groups reported that, in *C. elegans*, a pathway composed of the 22G–RNAs (secondary RNAs derived from piwi-interacting RNAs or piRNAs) and CSR-1 (a *C. elegans* Ago protein) activates/maintains the expression of endogenous mRNAs in germline cells and also in whole worm to protect them from the silencing effect of RNAe (RNA-induced epigenetic silencing) [2, 7, 50, 63]. Mammalian and *C. elegans* RNAa differ in that the worm 22G–RNAs engage nascent mRNA transcripts on the chromatin of their target genes, instead of noncoding sequences such as promoters as in mammals. Furthermore, in 2014, the *C. elegans* lin-4 miRNA, the first-ever discovered miRNA, was found to sustain its own transcription by binding to its own promoter, thereby forming a positive-feedback loop [53].

1.3 Basic Characteristics of RNAa

RNAa shares with RNAi early steps in the small-RNA-Ago pathway but diverges from RNAi after an Ago is programmed by a guide RNA which is targeted to a nuclear noncoding sequence. Therefore, RNAa possesses characteristics common to RNAi such as Ago dependence and also unique features indicative of being an unclear process including its delayed kinetics and long-lasting effects.

1.3.1 Indispensable Role of Ago in RNAa

Similar to the role of Ago proteins in RNAi, in RNAa, Agos serve the role of a navigator and a recruiting platform on which an RNAa effector complex is assembled. The requirement of Ago proteins in RNAa was first demonstrated by loss of function analysis in our initial work reporting RNAa [25]. By depleting individual Agos (Ago1-Ago4) using siRNA, it was found that the knockdown of Ago2 almost abolished RNAa of p21 gene while depleting Ago1, Ago3, and Ago4 did not significantly affect p21 activation. Our findings were corroborated by subsequent studies from other groups [5, 13, 33, 54]. The involvement of Ago proteins in RNAa is supported by additional lines of evidence. First, nuclear presence of Ago proteins is evident by immunostaining and biochemical fractionation; second, Ago2 loading of saRNAs can be visualized by biochemistry analysis; third, enrichment of Ago2 can be observed at promoter targeted by saRNA; and fourth, strong interaction of Ago proteins with RNA polymerase II (RNAP II) has been demonstrated [17, 42, 75].

There are, however, inconsistent reports with regard to which Ago protein(s) is required for RNAa. Duplex saRNAs which have perfect complementarity appear to exclusively rely on Ago2 [25, 5, 54], whereas duplexes containing bulges and/or mismatches (i.e., miRNAs) have been shown to require Ago1 [3, 16, 17], Ago2 [30, 65], or both [75].

It is believed that Agos load small RNAs in the cytoplasm to form a complex, which then enters the nucleus via mechanisms that are not completely understood. This view is supported by the observation that Ago2 loading of small RNA is deficient in the nucleus [10].

1.3.2 Effective Concentrations of saRNA in RNAa

RNAa is a nuclear process and, for this reason, requires higher transfection concentrations of its trigger saRNA in cultured cells compared to the concentration of an siRNA used to elicit cytoplasmic RNAi. The same feature is shared by nuclear RNAi to knock down nuclear localized long noncoding RNAs (lncRNAs) [18]. The higher concentration needed for both nuclear RNAa and RNAi is necessary to compensate for the inherent nuclear exclusion properties attributed to duplex RNAs [67]. The effective dose for saRNA is generally in the nanomolar range (1–50 nM) with an EC_{50} in the range of 1–10 nM [9, 25, 26, 29, 59, 62], whereas that for cytoplasmic RNAi the EC_{50} can be as low as in the picomolar or femtomolar range [56].

1.3.3 Kinetics of RNAa

RNAa has a unique in vitro kinetics which distinguishes itself from RNAi. By transfecting siRNA into cultured cells, gene knockdown can be detected within a few hours [39]. In RNAa, however, induction of gene expression occurs after a 24–48 h delay and peaks on days 4–5 following saRNA transfection [20, 25, 39]. The mechanism behind this delayed kinetics is not entirely clear. Perhaps gaining access by saRNA to its nuclear target and subsequently gene induction involving histone modifications represent the rate-limiting steps. Another feature characteristic of RNAa is its prolonged effect following a single transfection of saRNA which may last for over 10 days [20, 25, 39, 59]. Presumably, this long-lasting effect can be attributed to epigenetic modulation of target promoters which can be passed onto daughter cells across cell divisions. Regardless, sustained gene activation is a key feature of RNAa that could hold significant therapeutic benefits.

1.3.4 Activation of Endogenous Promoters Leads to Relevant and Predictable Functional Consequences

RNAa generally upregulates genes by a magnitude of twofold to fivefold as assessed by the levels of steady-state mRNA and the change is considered to be within the physiological range [12]. It has been known that subtle change in gene dosage can have overt biological outcomes [11]. For instance, subtle reduction in the dosage of tumor suppressor genes can promote tumorigenesis [1, 66]. Therefore, despite the small addition to the dosage of endogenous gene by RNAa, it may have practical applications in studying gene function and treating disease. In a study comparing directly RNAa and vector-based overexpression of KLF4, a member in the Krüpple-like transcription factor family, RNAa induction of KLF4 by several fold was found to have predictable functional consequences equivalent to that resulting from hundreds fold of overexpression by a retroviral vector. These consequences included the modulation of KLF4-regulated downstream genes and phenotypic changes of treated cells. The seemingly unparalleled scales in gene induction and functional consequences can be attributed to the natural approach of turning on an endogenous promoter by RNAa. The activated promoter can then drive the transcription of the target gene with intact UTRs and introns, as well as the expression of potential splicing variants [20, 25, 27, 33, 55, 61].

1.4 Design Rules for Exogenous RNAa

Currently, designing effective saRNAs on gene promoters is largely a hit-or-miss process due to a lack of full understanding of RNAa mechanism. We have published a set of saRNA design rules which allowed us to achieve a success rate of 10–20% [15, 62]. Those rules do not take into consideration the existence or absence of known noncoding transcripts overlapping the promoter to be targeted, but focus more on the sequence features of a target per se and its sequence context especially for its 3′ terminus. These rules include: (1) use the sense DNA sequence of the promoter as the template for saRNA design; (2) targets can be selected within a promoter region between at −100 and −1000 bp upstream of the transcription start site (TSS); (3) targets should be 19-nt in size; (4) targets should have a GC content of 40–60%, GC-rich regions or CpG islands should be avoided; (5) the corresponding saRNA duplexes should have lower thermodynamic stability at the 3′ end than the 5′ end; (6) the 18th and 19th positions counted from the 5′ end of targets should be "A/T"s, preferably "A"s; (7) avoid sequences that have five or more consecutive nucleotides; (8) avoid simple repeat sequences such as di- or tri-nucleotide repeats; and (9) the 20–23rd nucleotides (flanking the 3′ end of a target) should preferably be "A/T"s.

By focusing on regions (~30 bp) surrounding TATA-box-like sequences which usually sit ~30 bp upstream of the TSS, Fan et al. derived a set of rules which overlap with ours, such as the first two nucleotides of the antisense strand should be "A/U"s [9]. Further, the authors found that a target sequence more than 19 nt (i.e., 21 nt and 23 nt) yields more potent RNAa activity for at least one saRNA tested. This feature, however, needs to be verified in saRNAs targeting additional promoters.

To design saRNAs with improved success rates, the existing rules need to be refined. High-throughput saRNA library screening coupled with promoter context analysis on such information as the presence of overlapping noncoding transcripts, transcription factor binding sites, and DNA accessibility will produce more reliable design rules.

1.5 Targeting Rules for Endogenous RNAa

Computational analysis of potential binding sites on promoters by endogenous miRNAs has revealed that promoters bear as many miRNA targets as does 3′ UTR of mRNA [41, 69, 75]. However, the true number of promoter targets of miRNAs waits to be experimentally validated. These predictions were done using the same prediction rules for miRNA targeting 3′ UTR focusing on the importance of "seed" sequence matches. Using this approach, a number of studies have identified miRNA targets on promoters, which have been validated. For example, by scanning mouse Ccnb1 promoter, our group predicted 22 high-scoring targets for 21 miRNAs. By testing the top five targets, three (60%) can be validated. Due to poor conservation of promoter sequences per se, target conservation is not considered in any of the prediction studies, although some validated targets do have certain degree of evolutionary conservation [16]. It is possible that binding to promoters by miRNAs may use a set of rules which differ from 3′ UTR binding. For example, miRNA has recently found to target DNA by forming a triplex structure [37]. Sequence analysis of DNA pulled down with promoter-targeting miRNAs will allow for the deduction of miRNA-DNA pairing rules.

1.6 RNAa Is Distinct from Nuclear RNAi

RNAa induced by promoter-targeted saRNAs is a process occurring in the nucleus where RNAi can also be executed [46]. Is RNAa a nuclear RNAi in disguise – an effect induced by the RNAi knockdown of noncoding transcripts that overlap promoter targets of saRNA? A number of studies have provided strong evidence against such possibility. Well-annotated long noncoding transcripts (sense or antisense) or cryptic shorter nascent transcripts that exist in low abundance often overlap saRNA targets (Fig. 1.2) and can be associated with the saRNA-Ago complex as revealed by RNA pulldown assay [29, 49]. Several RNAa studies have shown that these noncoding transcripts are not targeted for degradation by saRNA because their abundance remains unchanged and even increases alongside the transcription of their downstream genes [29, 31, 33, 49]. For example, at the promoter of p21, E-cadherin and VEGF genes which are susceptible to RNAa (Fig. 1.2a), very low abundant nascent transcripts were detectable, and after saRNA targeting, were upregulated (p21 and E-cadherin) or remained unchanged (VEGF) (Fig. 1.2b). Therefore, RNAa is not a consequence of derepression of target promoters as has been suggested [34, 64].

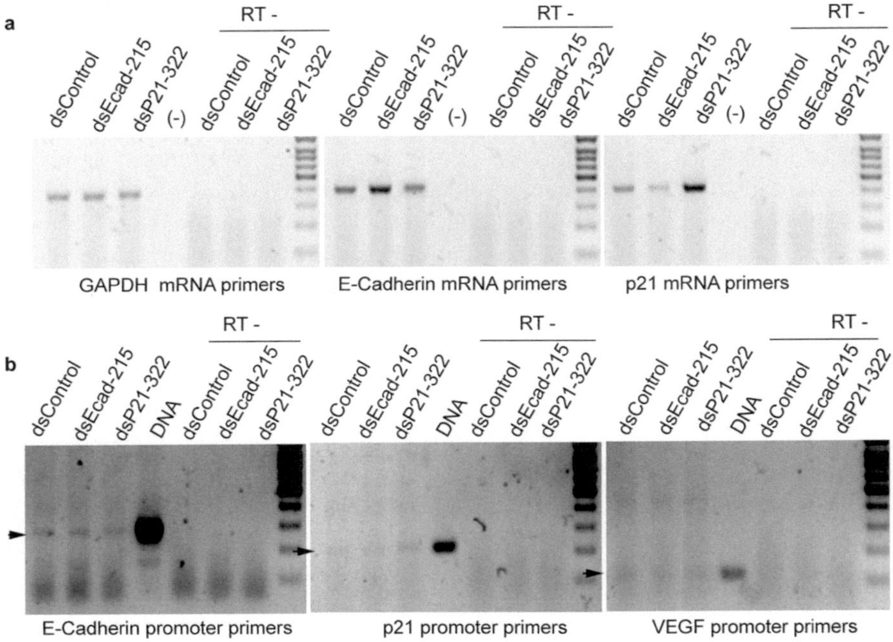

Fig. 1.2 Cryptic promoter transcripts are not degraded by saRNA targeting. Human PC-3 cells were transfected with 50 nM of the indicated saRNAs for 72 h. RNA was isolated using TRI Reagent and 10 μg was reverse transcribed using random hexamer primers. (**a**) cDNA samples were amplified using primer specific for mRNA. The PCRs were cycled 23 times (for GAPDH and p21) and 27 times (for E-cadherin). E-cadherin (*middle panel*) and p21 (*right panel*) were readily activated by dsEcad-215 and dsP21–322, respectively. (**b**) cDNA samples as in (**a**) were amplified using primer specific for promoter sequences surrounding the saRNA target sites. Human genomic DNA (10 ng) was also amplified as positive controls. PCRs were cycled 40 times. Expected size of amplicons for each gene is indicated by an arrowhead. The PCR product sizes for E-cadherin, p21, and VEGF are 289, 220, and 109 bp, respectively. A slight increase in the levels of promoter transcripts was noted for E-cadherin promoter in cells transfected by dsEcad-215 (*left panel*) and p21 promoter in cells transfected by dsP21–322 (*middle panel*). (−), Water; RT–, RT minus

1.7 Is DNA or RNA the Target?

A much debated question for RNAa mechanism is whether DNA or noncoding transcripts that overlap saRNA targets are the cognate sequence which saRNAs or activating miRNAs bind. In this regard, there exist at least three possibilities of promoter targeting by the RNA-Ago complex, including RNA:RNA pairing, RNA: DNA hybrid, and RNA-DNA triplex (Fig. 1.3).

In the RNA:RNA model, if noncoding transcripts overlapping the target exist, the ncRNAs may serve as a molecular scaffold which Ago binds and recruits additional proteins such as histone modifiers to alter local chromatin structure, while keeping the ncRNAs intact [49, 74]. Further, the target ncRNAs can be in the sense or antisense orientation, and can be long (lncRNA) or nascent short

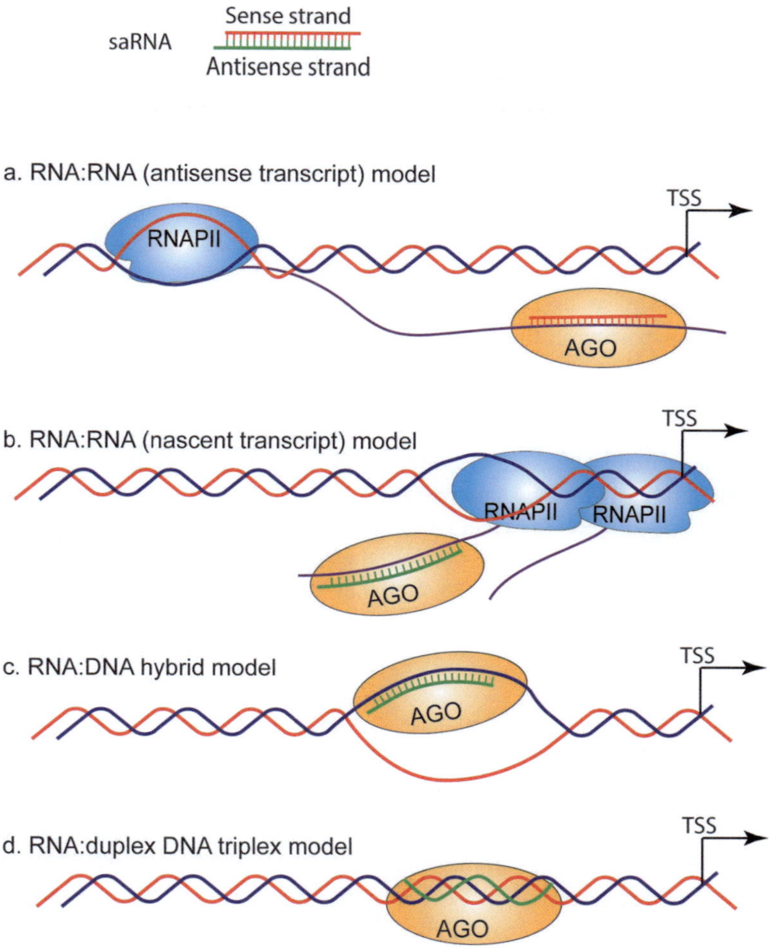

Fig. 1.3 saRNA-promoter binding models. A promoter-targeting saRNA (shown at *top*) is loaded by an Ago protein to form an RNA-Ago complex, which then binds to the saRNA's promoter target in one of the four models. (**a**) The sense RNA strand in the saRNA is the guide and binds to an antisense lncRNA. (**b**) The saRNA guide strand binds to nascent short transcripts (sense or antisense). (**c**) The saRNA guide (sense or antisense) binds to single-stranded antisense or sense DNA. (**d**) The saRNA guide forms a triplex structure with dsDNA

transcripts. Works from the Corey and Rossi Laboratories have provided strong evidence that in the activation of a number of genes, antisense lncRNAs are the saRNA target. A well-studied example is the PR gene activation by saRNA PR-11 in which a known long antisense transcript named AT-2 has been identified as the cognate target of PR-11 [49]. However, this view was challenged by a recent study which showed that AT-2 was dispensable for the activation of PR by PR-11 and also by another upstream saRNA PR-1611 [33]. Since the majority of mammalian genome is transcribed into RNA, for saRNA target promoters without a known overlapping lncRNA, nascent cryptic transcripts in sense and/or antisense orientations could still serve as docking sites for saRNAs.

In the RNA:DNA hybrid model, saRNA guides Ago2 to promoter targets where Ago2 may recruit a DNA helicase such as RHA to unwind duplex DNA, allowing for RNA:DNA pairing.

Lastly, in the RNA:DNA triplex model, purine or pyrimidine-rich miRNAs form triple-helical structures with purine-rich duplex DNA via Hoogsteen or reverse Hoogsteen interaction [37]. In the work by Paugh and colleagues [37], the authors provided multiple lines of physical evidence that miRNAs can directly bind to DNA to form heterotriplex structure with duplex DNA via Hoogsteen and reverse Hoogsteen base-pairing in the major groove of DNA duplex. Mechanistically, the authors speculate that the resulting miRNA:DNA heterotriplexes may loosen local chromatin structure by unwinding DNA or by recruiting triplex-specific binding proteins which alter local topography and allow binding of transcription factors.

Based on available lines of evidence in the literature, targeting mechanism of RNAa cannot be explained by a unified model. It is likely that all models mentioned above exist depending on the local chromatin environment of the target promoters. Before more data is available to support the RNA:DNA models, the RNA:RNA model remains attractive given the fact that RNA is a known substrate of the RNA-Argonaute machinery in RNAi.

1.8 RNAa Occurs at the Transcriptional Level and Acts on Transcription Initiation and Elongation

Gene induction in RNAa directly results from an increase in nascent transcripts as assessed by nuclear run-on assay [42] or by pre-mRNA levels, instead of an effect secondary to changes in mRNA stability [5]. Many studies have provided evidence for the nuclear localization of Agos (Agol and Ago2) [5, 16, 17, 30, 48]. In the nucleus, both saRNA and Ago2 have been shown to associate with the core transcription machinery RNAP II [42]. Evidence from several studies further indicates that promoter-targeting saRNAs/miRNAs act on the early steps of gene transcription. In miRNA-induced RNAa, Zhang et al. [75] found that miRNAs targeting TATA-box motif facilitate binding of TATA-binding protein (TBP) and transcription initiation factor IIA (TFIIA) to core promoters, and the assembly of preinitiation complex (PIC). Our recent work revealed that in cells treated with a p21 saRNA, there is a significant increase in RNAP II binding at the target promoter, suggesting that saRNA-Ago2 targeting may facilitate formation of the transcription initiation complex by associating with the initiating form of RNAP II and possibly by contributing to stabilization of the complex. RNAP II C-terminal domain (CTD) can be phosphorylated at serine 5 (Ser5P) and serine 2 (Ser2P) which correspond to a pausing and elongating polymerase, respectively. Further, by chromatin immunoprecipitation (ChIP) profiling of Ser2P and Ser5P at the p21 promoter and gene body, our study showed that p21 saRNA caused significant accumulation of Ser2P starting from p21 TSS and extending throughout the transcribed region with concurrent decrease in Ser5P, reflecting a transition from RNAP II pausing at the TSS to elongating. These results suggest that promoter targeting by saRNA can stimulate both transcription initiation and productive elongation.

1.9 The RNA-Induced Transcriptional Activation (RITA) Complex

A number of proteins, such as hnRNPs (heterogeneous nuclear ribonucleoproteins) and members of the TNRC6 protein family [13, 49], have been shown to be involved in the RNAa process. By coupling ChIbRP (chromatin isolation by biotinylated RNA pulldown) assay with mass-spectrometry, our recent work further characterized the composition of the RNAa effector complex, which we termed RNA-induced transcriptional activation (RITA) complex. RITA is composed of at least the saRNA guide strand, Ago2, and several proteins involved in RNA/DNA unwinding, transcriptional activation, and histone modification [42], including RHA, CTR9, and PAF1. RHA is a RNA/DNA helicase [52, 71] known to be involved in transcriptional activation by recruiting basal transcription machinery to promoters and modifying chromatin structure [36, 52, 72]. CTR9 and PAF1 are components of the PAF1 complex (RNAP II-associated Paf1 complex, PAF1C) which contains at least four additional proteins (LEO1, CDC73, RTF1, and SKI8) [23, 35]. PAF1C can directly stimulate transcription initiation and elongation [19, 24, 28, 47, 73], and is required for establishing several histone modifications associated with active genes by recruiting histone-modifying factors to the RNAP II complex such as E2/E3 ubiquitin ligase to cause H2B ubiquitination and downstream H3K4 and H3K79 methylation [4, 5, 23, 25, 28, 54, 70]. The observed association of these two components of PAF1C (CTR9 and PAF1) with Ago2 suggests that holo-PAF1C may participate in RNAa.

1.10 A Working Model for RNAa

Figure 1.4 summarizes a working model based on the p21 promoter for saRNA-induced RNAa focusing on pre- and post-saRNA-binding events. In this model, an saRNA exogenous introduced is loaded in the cytoplasm by an Ago protein which incorporates the guide strand in the saRNA to form an active RNA-Ago complex. The RNA-Ago complex then is imported into the nucleus through a yet unknown mechanism or simply gains nuclear access during mitosis when nuclear envelope disappears. In the nucleus, the complex may associate with RHA, which may later facilitate the opening of DNA double helix for base pairing with the saRNA guide

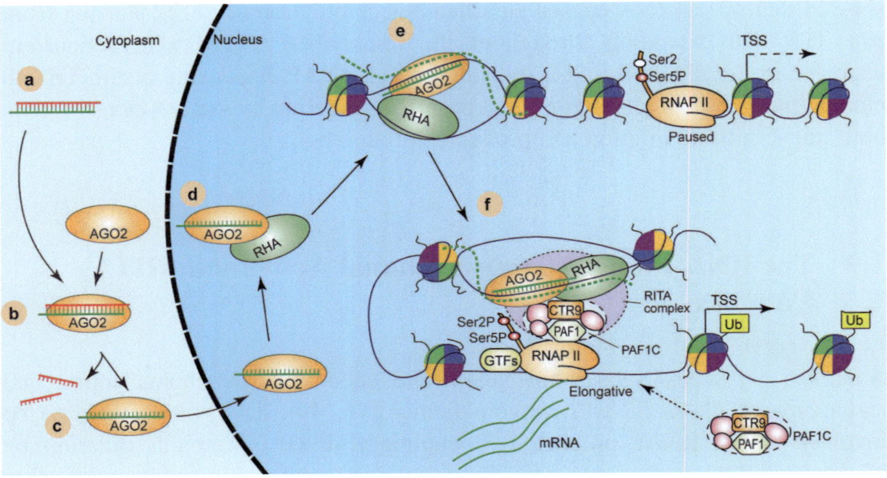

Fig. 1.4 A working model for RNAa. A promoter-targeting duplex RNA (**a**) is loaded onto Ago2, (**b**) which processes the duplex by cleaving the passenger strand and uses the remaining RNA strand as the guide to form an active RNA-Ago2 complex (**c**). After entering the nucleus, the RNA guide strand directs Ago2 to its cognate promoter target. In this step, RHA may be recruited by Ago2 possibly to open up DNA helix by its RNA/DNA helicase activity (**d**) and the RNA guide then base-pairs with target DNA (**e**). Alternatively, noncoding transcripts (*dotted green lines*) overlapping the target, if present, could be bound by the guide RNA. After docking to the promoter target, RNA-programmed and RHA-associated Ago2 serves as a recruitment platform on which components of PAF1C are assembled to stimulate transcription initiation and license RNAP II for productive elongation facilitated by ubiquitination of H2B

(assuming DNA is the target molecule) and/or directly participate in transcription upregulation. Once bound to the target promoter via one of the mechanisms depicted in Fig. 1.3, Ago further recruits RNAP II and components of the PAF1C to the core promoter. PAF1C can directly promote transcription initiation and elongation by recruiting an E2/E3 ubiquitin ligase to ubiquitinate H2B, which further triggers a cascade of permissive histone modifications, leading to transcriptional and epigenetic activation of the target gene.

1.11 Potential Applications of RNAa

Much in a similar way as vector-based gene overexpression, RNAa has applications in such areas as interrogating gene function, cell fate reprogramming, and treating diseases, all of which have already been or are being tested.

1.11.1 Interrogating Gene Function

RNAa has obvious advantages when used as a tool for studying gene function over vector-based ectopic expression of transgenes in that it is easier to implement by omitting the tedious process of constructing the required vector and induces overexpression of endogenous proteins with natural function. However, similar to RNAi, RNAa may suffer from off-target effects which could complicate interpretation of results. In addition, complete silenced genes may pose difficulties for reactivation by current saRNA design. In 2000, our group tested the utility of RNAa in interrogating the function of KLF4 in prostate cancer cells by directly comparing it with vector-based overexpression [58]. We found that the two methods nicely corroborated each other. RNAa may, thus, be routinely used as a surrogate to ectopic overexpression for studying functional genomics [8].

1.11.2 Cell Fate Reprogramming

Since the pioneering work of Yamanaka [51] demonstrating successful reprogramming of somatic cells into induced pluripotent stem (iPS) cells by using a combination of four viral-expressed transcription factors (OCT4, SOX2, MYC, and KLF4), a plethora of subsequent studies achieved cell fate reprogramming including dedifferentiation, differentiation, and transdifferentiation using different combinations of single transcription factor(s). However, the use of foreign genetic materials in these methods has potential safety concerns resulting from insertional mutagenesis if the programmed cells are to be used in human; methods avoiding any foreign genetic materials are thus preferred. In 2012, our group used a saRNA to activate the expression of human NANOG in an embryonic carcinoma cell line [59]. NANOG induction predictably modulated the expression of several known NANOG-regulated genes (e.g., OCT4, FOXH1, REST, REX1) and antagonized retinoic-acid-induced differentiation. In a subsequent study, by screening saRNAs targeting human OCT4 promoter, a potent saRNA was identified which can activate OCT4 expression in several types of human mesenchymal stem cells (MSCs). Introducing the OCT4 saRNA in combination with viruses encoding the remaining three Yamanaka factors (SOX2, MYC, and KLF4) into MSCs was able to convert human mesenchymal stem cells (MSCs) into a partially reprogrammed state. Voutila et al. [57] used saRNAs to induce KLF4 and MYC and observed that their induction resulted in a global expression profile similar to that induced by KLF4 and MYC transgenes. Esseltine et al. [8] used RNAa to induce the expression of GJA1 (Cx43) in oculodentodigital dysplasia (ODDD) patient-derived dermal fibroblasts which express diminished mature Cx43 with increased intracellular content of ECM-associated proteins. RNAa of GJA1 caused reprogramming events in the fibroblasts represented by decreased levels of extracellular matrix (ECM) proteins [8]. In 2013, the Habib group used a saRNA targeting MafA, an islet β-cell transcriptional factor and master regulator of insulin biosynthesis, in adherent CD34+ cells isolated from normal human bone marrow. Activation of MafA profoundly influences the differentiation of human adult CD34+ cells and upregulated several pancreatic endodermal genes including PDX1, Neurogenin 3, NeuroD, and NKX6–1. The differentiated CD34+ cells also expressed several genes required for glucose sensitivity and insulin secretion including glucokinase, glucagon-like peptide 1 receptor (GLP1R), sulfonylurea receptor-1 (SUR1), and phogrin, and processed C-peptide and insulin in response to increasing glucose stimulation [43].

1.11.3 Preclinical Study of RNAa Therapeutics

The first preclinical study of RNAa was published in 2009 in which the pro-angiogenesis activity of Vegfa activation by its promoter-targeted saRNAs expressed as shRNA was tested [54]. By lentiviral transfer of the shRNA into ischemic hindlimbs, vascularity and blood flow were improved [54]. In 2012, our group in collaboration with Alnylam Pharmaceuticals reported a stepwise development of medicinal saRNA targeting the p21 promoter for cancer treatment [40]. By implementing a low-density saRNA screening, a lead saRNA (dsP21–322) which activated p21 by over 14-fold was identified. This lead saRNA was further chemically modified to enhance its serum stability and reduce its immune-stimulatory activity while maintaining its RNAa activity. dsP21–322 was then demonstrated to have potent antitumor activity in an xenograft prostate cancer model by intratumoral injection [40] and in an orthotopic bladder cancer model by intravesical instillation [22]. Table 1.1 shows a list of preclinical developments of RNAa for a number of diseases, including cancer, ischemic conditions, and erectile dysfunction. In these studies, the trigger RNA was in the form of either synthetic saRNA or lentiviral-encoded shRNA. Of particular note is that the first-ever saRNA drug developed by the UK-based MiNA Therapeutics has entered clinical trial in early 2016 to treat liver cancer by upregulating CEBPA gene [6].

1.12 Concluding Remarks

Evidence accumulated in the last 10 years reveals that RNAa is a widespread phenomenon involving different nuclear targeted/localized small RNAs and Ago proteins and conserved in organisms from *C. elegans* to human. For its obvious advantages in reprogramming endogenous gene expression, RNAa has enormous potential in a number of applications, some of which have already been reduced to practice or are under active development. However, before RNAa's full potential can be harnessed, several key questions need to be addressed such as: What are the design rules that govern the sensitivity and specificity of saRNA targeting? What

Table 1.1 List of preclinical studies of RNAa therapeutics

Target gene	Trigger	Species	Disease	Model	Route of administration	Reference
Vegfa	Lentiviral hairpin saRNA	Mouse	Ischemia	Hindlimb ischemia	Local injection	[54]
CDKN1A	Synthetic saRNA	Mouse	Prostate cancer	Xenograft	Intratumoral	[40]
CDKN1A	Synthetic saRNA	Mouse	Bladder cancer	Orthotopic	Intravesical	[22]
NOS2	Adenoviral hairpin saRNA	Rat	Erectile dysfunction	Streptozotocin-induced diabetes	Intracavernous	[60]
NKX3–1	Synthetic saRNA	Mouse	Prostate cancer	Xenograft	Intratumoral	[45]
Vegfa	Lentiviral hairpin saRNA	Mouse	Myocardial Infarction	Myocardial infarction by surgical occlusion of coronary artery	Injection into heart wall	[55]
Cebpa	saRNA in dendrimer	Rat	Liver cirrhosis/HCC	Diethylnitrosamine-induced liver cirrhosis/HCC	Tail vein injection	[44]
CEBPA	saRNA/aptamer conjugates	Mouse	Pancreatic cancer	Xenograft	Tail vein injection	[68]
DPYSL3	saRNA/aptamer conjugates	Mouse	Prostate cancer	Orthotopic xenograft	Intraperitoneal injection	[26]
CEBPA	saRNA-dendrimer	Mouse	HCC	Liver orthotopic xenograft	Intravenous	[14]

are the targeting rules for and how prevalent is nuclear miRNA-mediated RNAa? Fully elucidating the mechanism of RNAa is of great importance in that RNAa not only may provide novel insight into the workings of different biological systems and mechanistic explanation for diseases but can also be used as a technology for integrating gene function and treating disease.

References

1. Alimonti A, Carracedo A, Clohessy JG, Trotman LC, Nardella C, Egia A, Salmena L, Sampieri K, Haveman WJ, Brogi E, Richardson AL, Zhang J, Pandolfi PP (2010) Subtle variations in Pten dose determine cancer susceptibility. Nat Genet 42(5):454–458. doi:10.1038/ng.556ng.556 [pii]
2. Cecere G, Hoersch S, O'Keeffe S, Sachidanandam R, Grishok A (2014) Global effects of the CSR-1 RNA interference pathway on the transcriptional landscape. Nat Struct Mol Biol 21 (4):358–365. doi:10.1038/nsmb.2801nsmb.2801 [pii]
3. Chaluvally-Raghavan P, Jeong KJ, Pradeep S, Silva AM, Yu S, Liu W, Moss T, Rodriguez-Aguayo C, Zhang D, Ram P, Liu J, Lu Y, Lopez-Berestein G, Calin GA, Sood AK, Mills GB (2016) Direct upregulation of STAT3 by MicroRNA-551b-3p deregulates growth and metastasis of ovarian cancer. Cell Rep 15(7):1493–1504. doi:10.1016/j.celrep.2016.04.034S2211-1247(16)30447-8 [pii]
4. Chu Y, Simic R, Warner MH, Arndt KM, Prelich G (2007) Regulation of histone modification and cryptic transcription by the Bur1 and Paf1 complexes. EMBO J 26(22):4646–4656. doi:7601887 [pii]10.1038/sj.emboj.7601887
5. Chu Y, Yue X, Younger ST, Janowski BA, Corey DR (2010) Involvement of argonaute proteins in gene silencing and activation by RNAs complementary to a non-coding transcript at the progesterone receptor promoter. Nucleic Acids Res 38(21):7736–7748. doi:10.1093/nar/gkq648
6. ClinicalTrials.gov (2016) First-in-human safety and tolerability study of MTL-CEBPA in patients with advanced liver cancer. https://ClinicalTrials.gov/show/NCT02716012
7. Conine CC, Moresco JJ, Gu W, Shirayama M, Conte D Jr, Yates JR 3rd, Mello CC (2013) Argonautes promote male fertility and provide a paternal memory of germline gene expression in *C. elegans*. Cell 155(7):1532–1544. doi:10.1016/j.cell.2013.11.032S0092-8674(13)01483-9 [pii]
8. Esseltine JL, Shao Q, Huang T, Kelly JJ, Sampson J, Laird DW (2015) Manipulating Cx43 expression triggers gene reprogramming events in dermal fibroblasts from oculodentodigital dysplasia patients. Biochem J 472(1):55–69. doi:10.1042/BJ20150652BJ20150652 [pii]
9. Fan M, Zhang Y, Huang Z, Liu J, Guo X, Zhang H, Luo H (2014) Optimizations of siRNA design for the activation of gene transcription by targeting the TATA-box motif. PLoS One 9 (9):e108253. doi:10.1371/journal.pone.0108253PONE-D-14-18035 [pii]
10. Gagnon KT, Li L, Chu Y, Janowski BA, Corey DR (2014) RNAi factors are present and active in human cell nuclei. Cell Rep 6(1):211–221. doi:10.1016/j.celrep.2013.12.013S2211-1247 (13)00758-4 [pii]
11. Guenther CA, Tasic B, Luo L, Bedell MA, Kingsley DM (2014) A molecular basis for classic blond hair color in Europeans. Nat Genet 46(7):748–752. doi:10.1038/ng.2991ng.2991 [pii]
12. Gustincich S, Zucchelli S, Mallamaci A (2016) The Yin and Yang of nucleic acid-based therapy in the brain. Prog Neurobiol. doi:S0301-0082(15)30099-X [pii]10.1016/j.pneurobio.2016.11.001
13. Hu J, Chen Z, Xia D, Wu J, Xu H, Ye ZQ (2012) Promoter-associated small double-stranded RNA interacts with heterogeneous nuclear ribonucleoprotein A2/B1 to induce transcriptional activation. Biochem J 447(3):407–416. doi:10.1042/BJ20120256BJ20120256 [pii]

14. Huan H, Wen X, Chen X, Wu L, Liu W, Habib NA, Bie P, Xia F (2016) C/EBPalpha short-activating RNA suppresses metastasis of hepatocellular carcinoma through inhibiting EGFR/beta-catenin signaling mediated EMT. PLoS One 11(4):e0153117. doi:10.1371/journal.pone.0153117PONE-D-16-01908 [pii]
15. Huang V, Qin Y, Wang J, Wang X, Place RF, Lin G, Lue TF, Li LC (2010) RNAa is conserved in mammalian cells. PLoS One 5(1):e8848. doi:10.1371/journal.pone.0008848
16. Huang V, Place RF, Portnoy V, Wang J, Qi Z, Jia Z, Yu A, Shuman M, Yu J, Li LC (2012) Upregulation of cyclin B1 by miRNA and its implications in cancer. Nucleic Acids Res 40 (4):1695–1707. doi:gkr934 [pii]10.1093/nar/gkr934
17. Huang V, Zheng J, Qi Z, Wang J, Place RF, Yu J, Li H, Li LC (2013) Ago1 interacts with RNA polymerase II and binds to the promoters of actively transcribed genes in human cancer cells. PLoS Genet 9(9):e1003821. doi:10.1371/journal.pgen.1003821PGENETICS-D-13-00631 [pii]
18. Huarte M, Guttman M, Feldser D, Garber M, Koziol MJ, Kenzelmann-Broz D, Khalil AM, Zuk O, Amit I, Rabani M, Attardi LD, Regev A, Lander ES, Jacks T, Rinn JL (2010) A large intergenic noncoding RNA induced by p53 mediates global gene repression in the p53 response. Cell 142(3):409–419. doi:10.1016/j.cell.2010.06.040S0092-8674(10)00730-0 [pii]
19. Jaehning JA (2010) The Paf1 complex: platform or player in RNA polymerase II transcription? Biochim Biophys Acta 1799(5–6):379–388. doi:S1874-9399(10)00002-7 [pii]10.1016/j.bbagrm.2010.01.001
20. Janowski BA, Younger ST, Hardy DB, Ram R, Huffman KE, Corey DR (2007) Activating gene expression in mammalian cells with promoter-targeted duplex RNAs. Nat Chem Biol 3 (3):166–173. doi:nchembio860 [pii]10.1038/nchembio860
21. Joshua-Tor L, Hannon GJ (2011) Ancestral roles of small RNAs: an ago-centric perspective. Cold Spring Harb Perspect Biol 3(10):a003772. doi:cshperspect.a003772 [pii]10.1101/cshperspect.a003772
22. Kang MR, Yang G, Place RF, Charisse K, Epstein-Barash H, Manoharan M, Li LC (2012) Intravesical delivery of small activating RNA formulated into lipid nanoparticles inhibits orthotopic bladder tumor growth. Cancer Res 72(19):5069–5079. doi:0008-5472.CAN-12-1871 [pii]10.1158/0008-5472.CAN-12-1871
23. Kim J, Guermah M, Roeder RG (2010) The human PAF1 complex acts in chromatin transcription elongation both independently and cooperatively with SII/TFIIS. Cell 140 (4):491–503. doi:S0092-8674(09)01636-5 [pii]10.1016/j.cell.2009.12.050
24. Krogan NJ, Kim M, Ahn SH, Zhong G, Kobor MS, Cagney G, Emili A, Shilatifard A, Buratowski S, Greenblatt JF (2002) RNA polymerase II elongation factors of *Saccharomyces cerevisiae*: a targeted proteomics approach. Mol Cell Biol 22(20):6979–6992
25. Li LC, Okino ST, Zhao H, Pookot D, Place RF, Urakami S, Enokida H, Dahiya R (2006) Small dsRNAs induce transcriptional activation in human cells. Proc Natl Acad Sci U S A 103 (46):17337–17342
26. Li C, Jiang W, Hu Q, Li LC, Dong L, Chen R, Zhang Y, Tang Y, Thrasher JB, Liu CB, Li B (2016) Enhancing DPYSL3 gene expression via a promoter-targeted small activating RNA approach suppresses cancer cell motility and metastasis. Oncotarget 7(16):22893–22910. doi:10.18632/oncotarget.82908290 [pii]
27. Lopez P, Wagner KD, Hofman P, Van Obberghen E (2016) RNA activation of the Vascular Endothelial Growth Factor gene (VEGF) promoter by double-stranded RNA and hypoxia: role of noncoding VEGF promoter transcripts. Mol Cell Biol 36(10):1480–1493. doi:10.1128/MCB.01096-15MCB.01096-15 [pii]
28. Marton HA, Desiderio S (2008) The Paf1 complex promotes displacement of histones upon rapid induction of transcription by RNA polymerase II. BMC Mol Biol 9:4. doi:1471-2199-9-4 [pii]10.1186/1471-2199-9-4
29. Matsui M, Sakurai F, Elbashir S, Foster DJ, Manoharan M, Corey DR (2010) Activation of LDL receptor expression by small RNAs complementary to a noncoding transcript that

overlaps the LDLR promoter. Chem Biol 17(12):1344–1355. doi:S1074-5521(10)00402-3 [pii]10.1016/j.chembiol.2010.10.009
30. Matsui M, Chu Y, Zhang H, Gagnon KT, Shaikh S, Kuchimanchi S, Manoharan M, Corey DR, Janowski BA (2013a) Promoter RNA links transcriptional regulation of inflammatory pathway genes. Nucleic Acids Res 41(22):10086–10109. doi:10.1093/nar/gkt777
31. Matsui M, Prakash TP, Corey DR (2013b) Transcriptional silencing by single-stranded RNAs targeting a noncoding RNA that overlaps a gene promoter. ACS Chem Biol 8(1):122–126. doi:10.1021/cb300490j
32. Meister G (2013) Argonaute proteins: functional insights and emerging roles. Nat Rev Genet 14(7):447–459. doi:nrg3462 [pii]10.1038/nrg3462
33. Meng X, Jiang Q, Chang N, Wang X, Liu C, Xiong J, Cao H, Liang Z (2016) Small activating RNA binds to the genomic target site in a seed-region-dependent manner. Nucleic Acids Res 44(5):2274–2282. doi:10.1093/nar/gkw076gkw076 [pii]
34. Morris KV, Santoso S, Turner AM, Pastori C, Hawkins PG (2008) Bidirectional transcription directs both transcriptional gene activation and suppression in human cells. PLoS Genet 4(11): e1000258
35. Mueller CL, Jaehning JA (2002) Ctr9, Rtf1, and Leo1 are components of the Paf1/RNA polymerase II complex. Mol Cell Biol 22(7):1971–1980
36. Nakajima T, Uchida C, Anderson SF, Lee CG, Hurwitz J, Parvin JD, Montminy M (1997) RNA helicase a mediates association of CBP with RNA polymerase II. Cell 90(6):1107–1112. doi:S0092-8674(00)80376-1 [pii]
37. Paugh SW, Coss DR, Bao J, Laudermilk LT, Grace CR, Ferreira AM, Waddell MB, Ridout G, Naeve D, Leuze M, LoCascio PF, Panetta JC, Wilkinson MR, Pui CH, Naeve CW, Uberbacher EC, Bonten EJ, Evans WE (2016) MicroRNAs form triplexes with double stranded DNA at sequence-specific binding sites; a eukaryotic mechanism via which microRNAs could directly alter gene expression. PLoS Comput Biol 12(2):e1004744. doi:10.1371/journal.pcbi.1004744PCOMPBIOL-D-15-00361 [pii]
38. Place RF, Li LC, Pookot D, Noonan EJ, Dahiya R (2008) MicroRNA-373 induces expression of genes with complementary promoter sequences. Proc Natl Acad Sci U S A 105 (5):1608–1613. doi:0707594105 [pii]10.1073/pnas.0707594105
39. Place RF, Noonan EJ, Foldes-Papp Z, Li LC (2010) Defining features and exploring chemical modifications to manipulate RNAa activity. Curr Pharm Biotechnol 11(5):518–526. doi:BSP/CPB/E-Pub/0094-11-6aaa [pii]
40. Place RF, Wang J, Noonan EJ, Meyers R, Manoharan M, Charisse K, Duncan R, Huang V, Wang X, Li L-C (2012) Formulation of small activating RNA into lipidoid nanoparticles inhibits xenograft prostate tumor growth by inducing p21 expression. Mol Ther Nucleic Acids 1:e15. doi:10.1038/mtna.2012.5
41. Portnoy V, Huang V, Place RF, Li LC (2011) Small RNA and transcriptional upregulation. Wiley Interdiscip Rev RNA 2(5):748–760. doi:10.1002/wrna.90
42. Portnoy V, Lin SH, Li KH, Burlingame A, Hu ZH, Li H, Li LC (2016) saRNA-guided Ago2 targets the RITA complex to promoters to stimulate transcription. Cell Res 26(3):320–335. doi:10.1038/cr.2016.22cr201622 [pii]
43. Reebye V, Saetrom P, Mintz PJ, Rossi JJ, Kasahara N, Nteliopoulos G, Nicholls J, Haoudi A, Gordon M, Habib NA (2013) A short-activating RNA oligonucleotide targeting the islet beta-cell transcriptional factor MafA in CD34(+) cells. Mol Ther Nucleic Acids 2:e97. doi:10.1038/mtna.2013.23mtna201323 [pii]
44. Reebye V, Saetrom P, Mintz PJ, Huang KW, Swiderski P, Peng L, Liu C, Liu X, Lindkaer-Jensen S, Zacharoulis D, Kostomitsopoulos N, Kasahara N, Nicholls JP, Jiao LR, Pai M, Spalding DR, Mizandari M, Chikovani T, Emara MM, Haoudi A, Tomalia DA, Rossi JJ, Habib NA (2014) Novel RNA oligonucleotide improves liver function and inhibits liver carcinogenesis in vivo. Hepatology 59(1):216–227. doi:10.1002/hep.26669

45. Ren S, Kang MR, Wang J, Huang V, Place RF, Sun Y, Li LC (2013) Targeted induction of endogenous NKX3-1 by small activating RNA inhibits prostate tumor growth. Prostate 73 (14):1591–1601. doi:10.1002/pros.22709
46. Robb GB, Brown KM, Khurana J, Rana TM (2005) Specific and potent RNAi in the nucleus of human cells. Nat Struct Mol Biol 12(2):133–137
47. Rondon AG, Gallardo M, Garcia-Rubio M, Aguilera A (2004) Molecular evidence indicating that the yeast PAF complex is required for transcription elongation. EMBO Rep 5(1):47–53. doi:10.1038/sj.embor.74000457400045 [pii]
48. Rudel S, Flatley A, Weinmann L, Kremmer E, Meister G (2008) A multifunctional human Argonaute2-specific monoclonal antibody. RNA (New York, NY) 14(6):1244–1253
49. Schwartz JC, Younger ST, Nguyen NB, Hardy DB, Monia BP, Corey DR, Janowski BA (2008) Antisense transcripts are targets for activating small RNAs. Nat Struct Mol Biol 15 (8):842–848
50. Seth M, Shirayama M, Gu W, Ishidate T, Conte D Jr, Mello CC (2013) The *C. elegans* CSR-1 argonaute pathway counteracts epigenetic silencing to promote germline gene expression. Dev Cell 27(6):656–663. doi:10.1016/j.devcel.2013.11.014
51. Takahashi K, Tanabe K, Ohnuki M, Narita M, Ichisaka T, Tomoda K, Yamanaka S (2007) Induction of pluripotent stem cells from adult human fibroblasts by defined factors. Cell 131 (5):861–872
52. Tang W, You W, Shi F, Qi T, Wang L, Djouder Z, Liu W, Zeng X (2009) RNA helicase a acts as a bridging factor linking nuclear beta-actin with RNA polymerase II. Biochem J 420 (3):421–428. doi:10.1042/BJ20090402
53. Turner MJ, Jiao AL, Slack FJ (2014) Autoregulation of lin-4 microRNA transcription by RNA activation (RNAa) in *C. elegans*. Cell Cycle 13(5):772–781. doi:10.4161/cc.2767927679 [pii]
54. Turunen MP, Lehtola T, Heinonen SE, Assefa GS, Korpisalo P, Girnary R, Glass CK, Vaisanen S, Yla-Herttuala S (2009) Efficient regulation of VEGF expression by promoter-targeted lentiviral shRNAs based on epigenetic mechanism: a novel example of epigenetherapy. Circ Res 105(6):604–609. doi:CIRCRESAHA.109.200774 [pii]10.1161/CIRCRESAHA.109.200774
55. Turunen MP, Husso T, Musthafa H, Laidinen S, Dragneva G, Laham-Karam N, Honkanen S, Paakinaho A, Laakkonen JP, Gao E, Vihinen-Ranta M, Liimatainen T, Yla-Herttuala S (2014) Epigenetic upregulation of endogenous VEGF-A reduces myocardial infarct size in mice. PLoS One 9(2):e89979. doi:10.1371/journal.pone.0089979PONE-D-13-46801 [pii]
56. Vaishnaw AK, Gollob J, Gamba-Vitalo C, Hutabarat R, Sah D, Meyers R, de Fougerolles T, Maraganore J (2010) A status report on RNAi therapeutics. Silence 1(1):14. doi:1758-907X-1-14 [pii]10.1186/1758-907X-1-14
57. Voutila J, Saetrom P, Mintz P, Sun G, Alluin J, Rossi JJ, Habib NA, Kasahara N (2012) Gene expression profile changes after short-activating RNA-mediated induction of endogenous pluripotency factors in human mesenchymal stem cells. Mol Ther Nucleic Acids 1:e35. doi:10.1038/mtna.2012.20mtna201220 [pii]
58. Wang J, Place RF, Huang V, Wang X, Noonan EJ, Magyar CE, Huang J, Li LC (2010) Prognostic value and function of KLF4 in prostate cancer: RNAa and vector-mediated overexpression identify KLF4 as an inhibitor of tumor cell growth and migration. Cancer Res 70(24):10182–10191. doi:70/24/10182 [pii]10.1158/0008-5472.CAN-10-2414
59. Wang X, Wang J, Huang V, Place RF, Li LC (2012) Induction of NANOG expression by targeting promoter sequence with small activating RNA antagonizes retinoic acid-induced differentiation. Biochem J 443(3):821–828. doi:BJ20111491 [pii]10.1042/BJ20111491
60. Wang T, Li M, Yuan H, Zhan Y, Xu H, Wang S, Yang W, Liu J, Ye Z, Li LC (2013) saRNA guided iNOS up-regulation improves erectile function of diabetic rats. J Urol 190(2):790–798. doi:10.1016/j.juro.2013.03.043S0022-5347(13)03690-2 [pii]
61. Wang J, Huang V, Ye L, Barcena A, Lin G, Lue TF, Li LC (2015a) Identification of small activating RNAs that enhance endogenous OCT4 expression in human mesenchymal stem cells. Stem Cells Dev 24(3):345–353. doi:10.1089/scd.2014.0290

62. Wang J, Place RF, Portnoy V, Huang V, Kang MR, Kosaka M, Ho MK, Li LC (2015b) Inducing gene expression by targeting promoter sequences using small activating RNAs. J Biol Methods 2(1.) doi:e14 [pii]0.14440/jbm.2015.39
63. Wedeles CJ, Wu MZ, Claycomb JM (2013) Protection of germline gene expression by the *C. elegans* argonaute CSR-1. Dev Cell 27(6):664–671. doi:10.1016/j.devcel.2013.11.016
64. Weinberg MS, Morris KV (2013) Long non-coding RNA targeting and transcriptional de-repression. Nucleic Acid Ther 23(1):9–14. doi:10.1089/nat.2012.0412
65. Xiao M, Li J, Li W, Wang Y, Wu F, Xi Y, Zhang L, Ding C, Luo H, Li Y, Peng L, Zhao L, Peng S, Xiao Y, Dong S, Cao J, Yu W (2016) MicroRNAs activate gene transcription epigenetically as an enhancer trigger. RNA Biol 0. doi:10.1080/15476286.2015.1112487
66. Yan H, Dobbie Z, Gruber SB, Markowitz S, Romans K, Giardiello FM, Kinzler KW, Vogelstein B (2002) Small changes in expression affect predisposition to tumorigenesis. Nat Genet 30(1):25–26. doi:10.1038/ng799ng799 [pii]
67. Yi R, Qin Y, Macara IG, Cullen BR (2003) Exportin-5 mediates the nuclear export of pre-microRNAs and short hairpin RNAs. Genes Dev 17(24):3011–3016. doi:10.1101/gad.11588031158803 [pii]
68. Yoon S, Huang KW, Reebye V, Mintz P, Tien YW, Lai HS, Saetrom P, Reccia I, Swiderski P, Armstrong B, Jozwiak A, Spalding D, Jiao L, Habib N, Rossi JJ (2016) Targeted delivery of C/EBPalpha -saRNA by pancreatic ductal adenocarcinoma-specific RNA Aptamers inhibits tumor growth in vivo. Mol Ther. doi:10.1038/mt.2016.60mt201660 [pii]
69. Younger ST, Pertsemlidis A, Corey DR (2009) Predicting potential miRNA target sites within gene promoters. Bioorg Med Chem Lett 19(14):3791–3794. doi:S0960-894X(09)00545-9 [pii] 10.1016/j.bmcl.2009.04.032
70. Yue X, Schwartz JC, Chu Y, Younger ST, Gagnon KT, Elbashir S, Janowski BA, Corey DR (2010) Transcriptional regulation by small RNAs at sequences downstream from 3′ gene termini. Nat Chem Biol 6(8):621–629. doi:nchembio.400 [pii]10.1038/nchembio.400
71. Zhang S, Grosse F (1994) Nuclear DNA helicase II unwinds both DNA and RNA. Biochemistry 33(13):3906–3912
72. Zhang S, Grosse F (2010) Molecular characterization of nuclear DNA helicase II (RNA helicase A). Methods Mol Biol 587:291–302. doi:10.1007/978-1-60327-355-8_21
73. Zhang Y, Sikes ML, Beyer AL, Schneider DA (2009) The Paf1 complex is required for efficient transcription elongation by RNA polymerase I. Proc Natl Acad Sci U S A 106 (7):2153–2158. doi:0812939106 [pii]10.1073/pnas.0812939106
74. Zhang Y, Fan M, Zhang X, Huang F, Wu K, Zhang J, Liu J, Huang Z, Luo H, Tao L, Zhang H (2014a) Cellular microRNAs up-regulate transcription via interaction with promoter TATA-box motifs. RNA 20(12):1878–1889. doi:10.1261/rna.045633.114rna.045633.114 [pii]
75. Zhang X, Li H, Burnett JC, Rossi JJ (2014b) The role of antisense long noncoding RNA in small RNA-triggered gene activation. RNA 20(12):1916–1928. doi:10.1261/rna.043968.113rna.043968.113 [pii]

Part I
Exogenous RNAa

Chapter 2
Enhancing Neuronogenesis and Counteracting Neuropathogenic Gene Haploinsufficiencies by RNA Gene Activation

Antonello Mallamaci

Abstract Small activating RNAs (saRNAs), targeting endogenous genes and stimulating their transcription, are a promising tool for implementing a variety of neurotherapeutic strategies. Among these there is the stimulation of select histogenetic subroutines for purposes of cell-based brain repair, as well as the therapeutic treatment of gene expression deficits underlying severe neurological disorders.

We employed RNA activation (RNAa) to transactivate the *Emx2* transcription factor gene in embryonic cortico-cerebral precursor cells. This led to enhanced self-renewal, delayed differentiation, and reduced death of neuronally committed precursors, resulting in a remarkable expansion of the neuronogenic precursors pool. These results are of paramount interest for purposes of gene-promoted brain repair. As such, RNAa makes therapeutic stimulation of neuronogenesis via *Emx2* overexpression a feasible goal, preventing the drawbacks of exogenous gene copies introduction.

Moreover, we employed RNAa to achieve a gentle transactivation of the *Foxg1* transcription factor gene, specifically in cortico-cerebral cells. This manipulation led to an appreciable biological outcome, while complying with endogenous gene tuning linked to early central nervous system regionalization and late activity of neocortical projection neurons. *Foxg1*-activating miRNAs stimulated RNApolII recruitment, possibly via Ago1. One of them worked promisingly *in vivo*. As such, RNAa can be a valuable approach for therapeutic treatment of the *FOXG1*-haploinsufficiency-linked variant of the Rett syndrome. Remarkably, hemizygosity for specific genes and polygenic chromosomal segments underlies a huge number of neuropathological entities for which no cure is presently available. Based on the results reported above, RNAa might be a simple and scalable approach for fixing this class of problems.

A. Mallamaci (✉)
Lab of Cerebral Cortex Development, Scuola Internazionale Superiore di Studi Avanzati (SISSA), via Bonomea 265, 34136 Trieste, Italy
e-mail: amallama@sissa.it

L.-C. Li (ed.), *RNA Activation*, Advances in Experimental Medicine and Biology 983, DOI 10.1007/978-981-10-4310-9_2

Keywords RNAa • saRNA • *Emx2* • Neuronogenesis • Brain repair • *Foxg1* • Rett syndrome • Gene haploinsufficiency

2.1 Introduction

Small activating RNAs (saRNAs) are micro RNA/small interfering RNA- (miRNA/siRNA)-like molecules able to stimulate gene transcription. Delivered as mature moieties, primary miRNA transcripts (pri-miRNAs) or short hairpin precursors (shRNAs), they target the transcription unit of interest and/or its surroundings, working as effectors of "RNA(-dependent gene) activation" (RNAa) [29, 53, 64]. After the initial RNAa report by Li et al. [35], several genes were transactivated by saRNAs [21, 29, 53, 64]. Despite the complex and heterogeneous epigenetic changes evoked by these effectors [34], their main functional outcome can be fundamentally attributed to two distinct molecular mechanisms. saRNAs can act by destabilizing noncoding RNAs (ncRNAs), which normally dampen mRNA transcription. Alternatively, they can convey the transcriptional machinery to chromatin [47, 58].

Interestingly, small size and moderate power of saRNAs make these molecules promising tools for implementing a variety of neurotherapeutic strategies. These include – for example – the stimulation of select histogenetic subroutines for purposes of cell-based brain repair, as well as the therapeutic treatment of gene expression deficits underlying severe neurological disorders.

In this chapter, we provide an overview of two studies we recently ran in this field [11, 14]. The former was aimed to assess if saRNA-driven overexpression of the *Emx2* transcription factor gene may be employed to enhance the neuronogenic output obtainable from cortico-cerebral precursors [11]. The latter was intended to provide reliable proofs-of-principle that RNAa may compensate for defective expression of the *Foxg1* transcription factor gene leading to a variant of the Rett syndrome, *in vitro* as well as *in vivo*. From a broader perspective, it explored the possibility to employ RNAa as a *scalable tool* for therapy of neuropathogenic gene haploinsufficiencies [14].

2.2 RNAa Stimulation of *Emx2*

Emx2 encodes for a homeobox transcription factor controlling a variety of developmental processes occurring in the mammalian rostral brain, including cerebral cortex specification, arealization, and histogenesis [16]. Its capability to promote self-renewal of neuronogenic precursors [49] and prime neuronal differentiation [3], and prevent astroblasts proliferation [3, 12] makes its overexpression an appealing tool for overcoming two key drawbacks of cell-based brain repair, exaggerated death of transplanted cells and their preferential astrocytic rather than neuronal differentiation [4]. We wondered if *Emx2* overexpression could be achieved by small activating RNAs stimulating the endogenous gene. This was the subject of a dedicated study of ours referred to in this paragraph [11].

To identify potential genomic targets appropriate for *Emx2*-RNAa, we screened the 50kb genomic region encoding for *Emx2*-mRNA and its antisense ncRNA companion for evolutionarily conserved and/or cis-active modules, suitable to miRNA targeting via the *ad hoc* repurposed, pri-miRNA-155-based Block-It platform. We selected 12 high-score candidates and we cloned the cDNAs encoding for the corresponding precursors into a lentiviral constitutive expressor (Fig. 2.1a–c). We acutely delivered the resulting lentiviruses to murine E12.5 neocortical precursors and, 4 days later, we scored the resulting neurospheres for *Emx2*-mRNA levels (Fig. 2.1a). We found that 3 out of 12 miRNAs, 2 antisense-oriented (miR-α Emx2.3 and miR-αEmx2.7) and 1 sense-oriented (miR-αEmx2.12), upregulated *Emx2*. The expression gain ranged from about 1.5-fold (miR-αEmx2.12) to about 2.5-fold (miR-αEmx2.3 and .7) (Fig. 2.1d). As indicated by the concomitant upregulation of *Emx2*-pre-mRNA and hyper-recruitment of RNApolII to the *Emx2* locus (Fig. 2.1e), such gain was likely due to an enhancement of transcription. Moreover, both miR-αEmx2.3 and .7 were directed against bidirectionally transcribed enhancers (Fig. 2.1c). This suggests that they could recognize their targets via nascent RNAs, still attached to chromatin, and could act by destabilizing such RNAs or recruiting transcription effectors to them.

Next, we wondered if the small expression gain elicited by our saRNAs was sufficient to achieve an appreciable biological outcome. Interestingly, both miR-αEmx2.3 and .7 induced a huge absolute oversizing of the engineered culture (about three/fourfold, not shown) as well as a statistically significant, relative enlargement of the neuronogenic lineage, expressing an mCherry reporter under the control of the promoter pTα1 (Fig. 2.1g). This possibly reflected a dramatically increased bias of engineered $Tbr2^{+}Tub\beta3^{-}$ neuronal progenitors to self-renew, so postponing the generation of $Tbr2^{+}Tub\beta3^{+}$ and $Tbr2^{-}Tub\beta3^{+}$ neurons (Fig. 2.1h), as well as a decreased incidence of apoptosis (Fig. 2.1i). All that is of obvious interest for therapeutic enhancement of neuronogenesis.

Finally, we wondered if miR-αEmx2.3 and .7 activities were confined to rostral brain precursors, normally expressing *Emx2*, or was also detectable elsewhere. To address this issue, we alternatively delivered these miRNAs to nonneocortical, rhombo-spinal precursors of a comparable biological stage, which hardly express this gene. Remarkably, no *Emx2* upregulation was detectable at all (Fig. 2.1i). This might reflect a different epigenetic state of the *Emx2* locus in rostral and caudal embryonic brain segments. It suggests that risks of ectopic gene activation upon generalized delivery of saRNAs (as premade molecules or products of constitutively expressed transgenes) can be negligible.

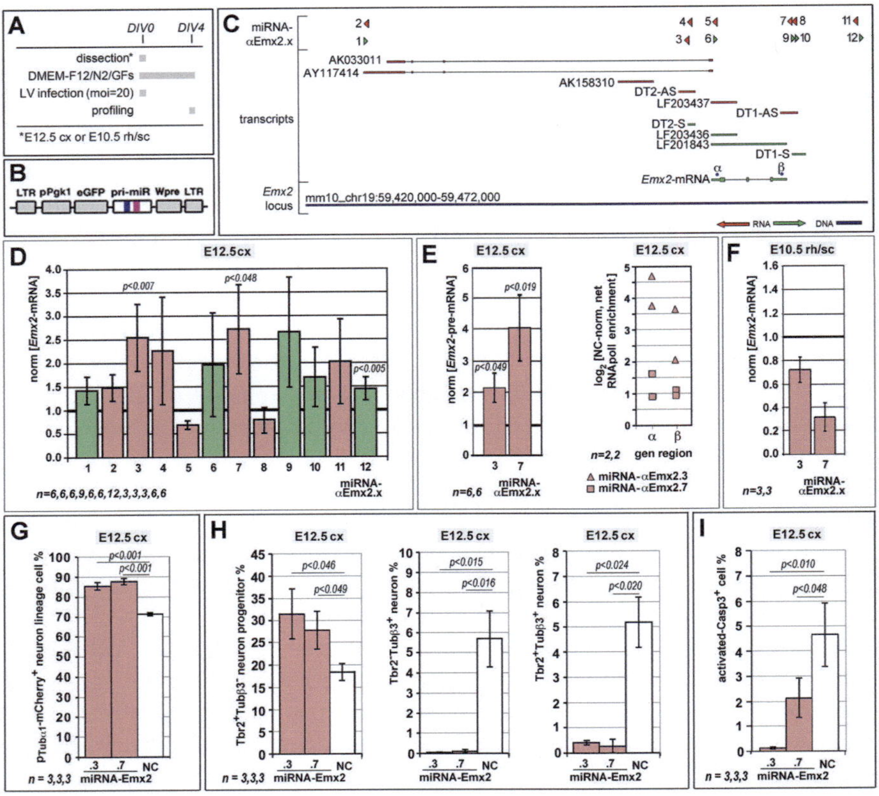

Fig. 2.1 *Emx2*-RNAa and its biological correlate (**a**, **b**) Protocols and lentiviral reagents employed in these assays. (**c**) Schematics of the *Emx2* locus, including endogenous transcripts (According to www.ucsc.org and Diodato et al. [11]), transcription activating miRNAs, and genomic regions, probed in chromatin immunoprecipitation-qRTPCR (ChIP-qRTPCR) assays (α,β). (**d**) *Emx2*-mRNA levels in proliferating neocortical precursors manipulated as in (**a**). (**e**) Molecular mechanisms underlying *Emx2*-RNAa. *Emx2*-pre-mRNA levels and RNA polymerase II enrichment of the *Emx2* locus, in proliferating neocortical precursors, manipulated as in (**a**). Here, the increase of RNApolII binding to chromatin, or "enrichment," induced by bioactive miRNAs was calculated according to the formula $\log2(Ab_X/IgG_X)-\log2(Ab_{NC}/IgG_{NC})$, where Ab and IgG are the number of amplicons immuno-precipitated by α-RNApolII and IgG, respectively, X is the miRNA under examination, and NC is the negative control miRNA. (**f**) Impact of miR-αEmx2.3 and miR-αEmx2..7 on *Emx2*-mRNA levels in proliferating precursors originating from the rh/sc domain. (**g–i**) Fractions of $p_{T\alpha 1}$-mCherry+ neuron lineage cells (**g**), $Tbr2^{+}Tub\beta3^{-}$ neuron progenitors (**h**), $Tbr2^{\pm}Tub\beta3^{+}$ neurons (**h**), and activated-$Caspase3^{+}$ apoptotic elements (**i**) in cultures of neocortical precursors manipulated as described in (**a–c**). Limited to (**d–f**), values double normalized, against *Tbp* and control (NC). cx, neocortex; rh/sc, rhombencephalon/spinal cord; E, embryonic day; DIV, days *in vitro*. Bars represent sem's. n = number of biological replicates. Statistical significance of results evaluated by t-Student assay (one-tail, unpaired)

2.3 saRNAs as General Tools for Therapy of Neuropathogenic Haploinsufficiencies?

Inspired by the moderate transactivating power of saRNAs and their ability to comply with normal regulation of gene expression, we next considered the opportunity to use these molecular devices as tools for precise therapy of neuropathogenic gene haploinsufficiencies.

More than 100 different hemizygous gene deletions underlie a variety of neuropathological conditions, leading to epilepsy, mental retardation, autism, schizophrenia, and neurodegeneration [1, 5, 32, 33, 59]. Their individual prevalence is low; however, their cumulative frequency makes them an issue for social health. A scalable therapeutic approach is needed. How are we achieving this goal? In principle, homologous recombination (HR)-mediated repair of defective genes, evoked by engineered endonucleases (EEN) and driven by a dedicated editor DNA, should be the golden procedure to fix the problem [15, 31, 62]. Actually, the implementation of this approach within the central nervous system (CNS) would be hardly feasible, due to a variety of technical issues [8, 40, 56]. A simpler design, based on therapeutic minigenes, would be problematic as well. In fact, the precise rescue of haploinsufficient gene expression levels is often needed for proper execution of neural gene functions [9, 20, 65]. Moreover, the accurate recapitulation of the gene expression pattern usually requires a number of properly arranged cis-active elements. Fitting all of them into a small transgene, suitable for panneural delivery, would be hardly feasible and/or scalable. Therefore, a different approach is needed.

This might be a gentle stimulation of the spared gene allele, still under the control of the regulatory elements which sculpt its baseline expression profile and drive fine modulation of its levels linked to neuronal physiology. Nowadays, a variety of molecular tools are potentially available for this last approach. They include Zinc finger- (ZF-) [2, 43, 60, 61, 66], TransActivator Like Element- (TALE-) [18, 45, 68], clustered regularly interspaced short palindromic repeat (CRISPR) [6, 19, 27, 30, 37, 51], and NMHV-type [13] transactivators. Unfortunately, despite their capability to stimulate endogenous genes *ad libitum*, the employment of these devices for therapy of neural haploinsufficiencies may be problematic, because of their very large size [17] and ectopic gene activation [26, 27, 30, 37]. Small activating RNAs (saRNAs) might provide a valuable alternative, thanks to their small size, their moderate transactivating power, and their inability to switch on silent genes.

To prove the feasibility of this approach, we employed RNAa for stimulating the haploinsufficient *Foxg1* transcription factor gene [14]. *Foxg1* is a key regulator of cortico-cerebral development and function, implicated in pallial field specification [24], precursors proliferation control [3, 39], and laminar [23] as well as areal [48] neuronal differentiation. In humans, its normal allele dosage is essential to neurological health, as hemideletion and duplication of *Foxg1* lead to Rett and West syndromes, respectively [20]. Briefly, we found that RNAa resulted into a *Foxg1* expression gain suitable for therapeutic purposes and led to an appreciable biological outcome. No ectopic gene activation occurred and endogenous gene tuning was preserved. Finally, a robust *Foxg1* stimulation was also achieved *in vivo* [14].

2.4 Selecting miRNA-Like saRNAs Upregulating *Foxg1*-mRNA

In order to identify potential genomic targets appropriate for *Foxg1*-RNAa, we inspected the 5′ surroundings of *Foxg1*-mRNA transcriptional start sites (TSSs) for sequences specifically amenable to miRNA targeting, via the pri-miRNA-155-based Block-It platform [11]. We selected candidates (Fig. 2.2a) and we cloned the cDNAs encoding for the corresponding precursors into a lentiviral constitutive expressor (Fig. 2.2b, (*a*)). We acutely delivered the resulting lentiviruses to murine E12.5 neocortical precursors, we kept these cells as floating neurospheres in pro-proliferative medium for 4 days, and we eventually scored them for *Foxg1*-mRNA levels by quantitative retrotranscription-polymerase chain reaction (qRTPCR) (Fig. 2.2c). We found that eight out of eight miRNAs, four antisense-oriented (miR-αFoxg1.0650, miR-αFoxg1.1653, miR-αFoxg1.2764, and miR-αFoxg1.3700) and four sense-oriented (miR-αFoxg1.0755, miR-αFoxg1.1694, miR-αFoxg1.2273, and miR-αFoxg1.3795), upregulated *Foxg1,* to different extents. The expression gain ranged from about 1.3-fold (miR-αFoxg1.2764) to about 2.9-fold (miR-αFoxg1.0650) (Fig. 2.2e). Similar results were achieved upon delivery of miR-αFoxg1.0650 and miR-αFoxg1.1694 to NIH3T3 and HEK293T cells, which led to a consistent increase of Foxg1/FOXG1 proteins (not shown).

Next, we wondered if *Foxg1*-RNAa may be also achieved in differentiating derivatives of neocortical precursors. We transferred the pri-miRNA-cDNAs of the four best-performing miRNAs (Fig. 2.2e) into LV_TREt-IRES2EGFP [55], in between the doxycyclin-controlled TREt promoter and an IRES2EGFP reporter gene (Fig. 2.2b, (*b2*)). We employed the resulting lentiviruses paired to a constitutive rtTA^{2S}-M2 transactivator expressor, LV_Pgk1p-rtTA^{2S}-M2 [55] (Fig. 2.1b, (*b1*)), to drive delayed, TetON-controlled miRNA expression. Unexpectedly, we found that only one miRNA (miR-αFoxg1.1694) upregulated *Foxg1*, by about 1.6-fold. The other ones were ineffective (Fig. 2.2f). This may be due to the different epigenetic states of chromatin, generally more accessible in *proliferating* neural precursors than in their postmitotic derivatives [44, 63]. It may specifically reflect a different ncRNA landscape at the *Foxg1* locus.

Finally, to corroborate the significance of these results, we tested if the small expression gain elicited by our saRNAs led to an appreciable biological readout. For this purpose, we stimulated *Foxg1* by miR-αFoxg1.0650 and .1694 in proliferating murine neocortical precursors (Fig. 2.3a, b), and we evaluated the impact of this manipulation on the generation of postmitotic, Tubβ3$^+$ neurons. *Foxg1* – in fact – inhibits the exit of neuronogenic precursors from cell cycle [3, 39] and, as described for a number of other patterning genes [11, 22, 57], even

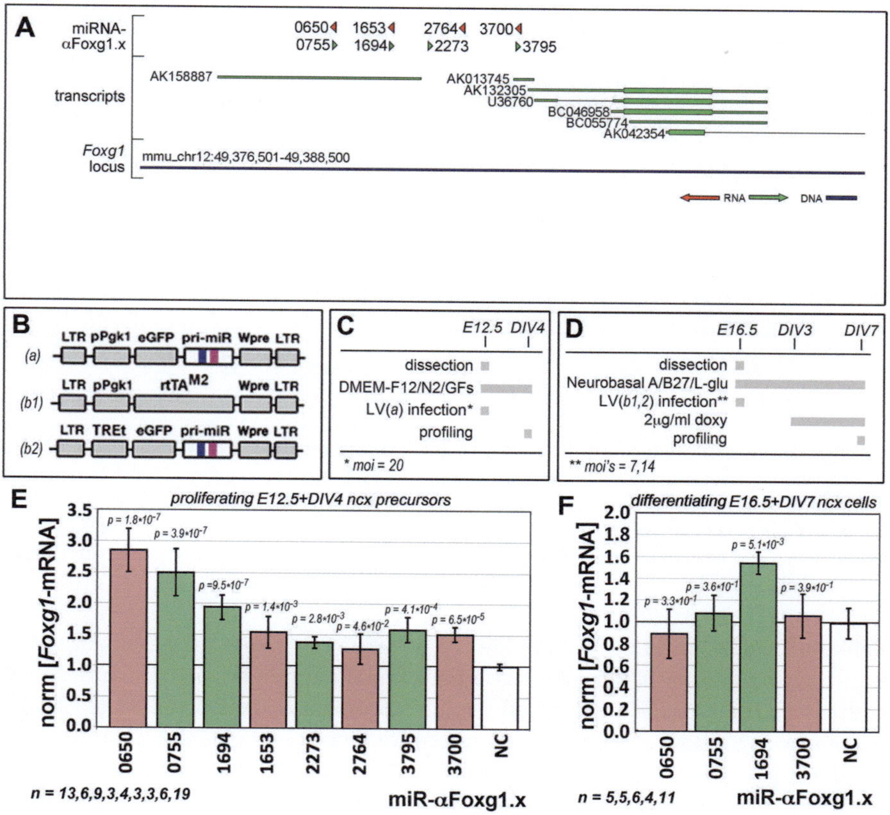

Fig. 2.2 Screening for miRNA-like, small RNAs activating *Foxg1*-mRNA (*Foxg1*-saRNAs) in murine neocortical precursors and derivatives (**a**) Schematics of the *Foxg1* locus including saRNA positions and orientations. (**b–d**) Lentiviral reagents and protocols employed for this screening. (**e, f**) *Foxg1*-mRNA levels in proliferating neocortical precursors and their differentiating derivatives, manipulated as in (**c**) and (**d**), respectively. Values double normalized against *Gapdh* and control (NC). E, embryonic day; DIV, days *in vitro*. Bars represent sem's. $n =$ number of biological replicates. Statistical significance of results evaluated by t-Student assay (one-tail, unpaired) (Adapted from Ref. [14])

a small increase of its expression level may exert a deep impact on neuronogenic differentiation rates [13]. As expected, both *Foxg1*-activating miRNAs halved the neuronal output of the culture, in a highly reproducible fashion (Fig. 2.3c, d).

2.5 Compliance of *Foxg1*-RNAa with Endogenous Tuning of *Foxg1*-mRNA

The therapeutic exploitation of RNAa for the treatment of haploinsufficiencies would be easier if the *activity* of saRNAs would be confined to cells normally expressing the gene in order. To assess the fulfillment of this requirement, we

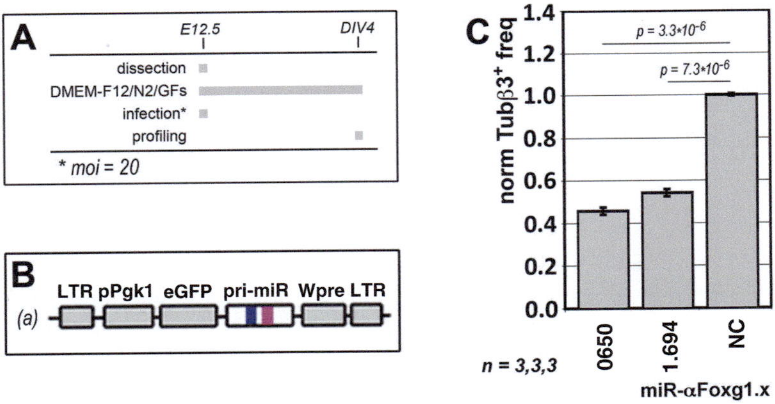

Fig. 2.3 Histogenetic outcome of *Foxg1*-RNAa. (**a**, **b**) Protocols and lentiviral reagents employed for this assay. (**c**) Quantification of cells immunopositive for the neuron-specific Tubβ3 marker, in cultures of neocortical precursors expressing *Foxg1*-saRNAs. E, embryonic day; DIV, days *in vitro*. Bars represent sem's. n = number of biological replicates. Statistical significance of results evaluated by t-Student assay (one-tail, unpaired). Absolute average frequency of Tubβ3$^+$ cells in NC samples was (27.25±0.16)% (Adapted from Ref. [14])

delivered miR-αFoxg1.0650 and .1694 to proliferating neural precursors originating from the murine E10.5 meso-rhombo-cervical neural domain, which does not express *Foxg1* [25]. We employed neural precursors dissected out of the E12.5 neocortex as histogenetically equivalent positive controls (Fig. 2.4a). Interestingly, albeit weakly upregulated by miR-αFoxg1.0650 and .1694, *Foxg1* levels remained about three orders of magnitude lower in meso-rhombo-cervical derivatives, compared to neocortical controls (Fig. 2.4b, c). This suggests that risks of ectopic gene activation upon generalized saRNA delivery can be negligible.

Neuronal genes often undergo fine, electrical activity-related tuning, which may be crucial to proper implementation of their function [50]. An acceptable therapy of neuropathogenic haploinsufficiencies relying on stimulation of the spared gene allele should take into account such physiological gene modulation. Exposure of neocortical neurons to high extracellular [K$^+$] was followed by a prompt arousal of *Foxg1*-mRNA levels (Fig. 2.4d, e), a likely *in vitro* correlate of activity-dependent *Foxg1* stimulation. We reasoned that this phenomenon might provide a valuable opportunity for probing compliance of RNAa with "endogenous" gene tuning. Remarkably, the delivery of miR-αFoxg1.1694 to K$^+$-challenged neocortical neurons elicited a delicate upward shift of the *Foxg1* activation curve under high extracellular [K$^+$]. Interestingly, ANCOVA analysis of data provided no evidences of interaction between K$^+$ stimulation and RNAa (Fig. 2.4e), suggesting that RNAa did not hide activity-driven *Foxg1* tuning.

In summary, we found that the saRNAs achieved a relevant molecular outcome only in primary cultures where the gene of interest was active (Fig. 2.4a–c). This suggests that therapeutic saRNA delivery, driven by a ubiquitous promoter or achieved via straight administration of premade synthetic molecules, should be followed by the activation of the target gene limited to its standard expression

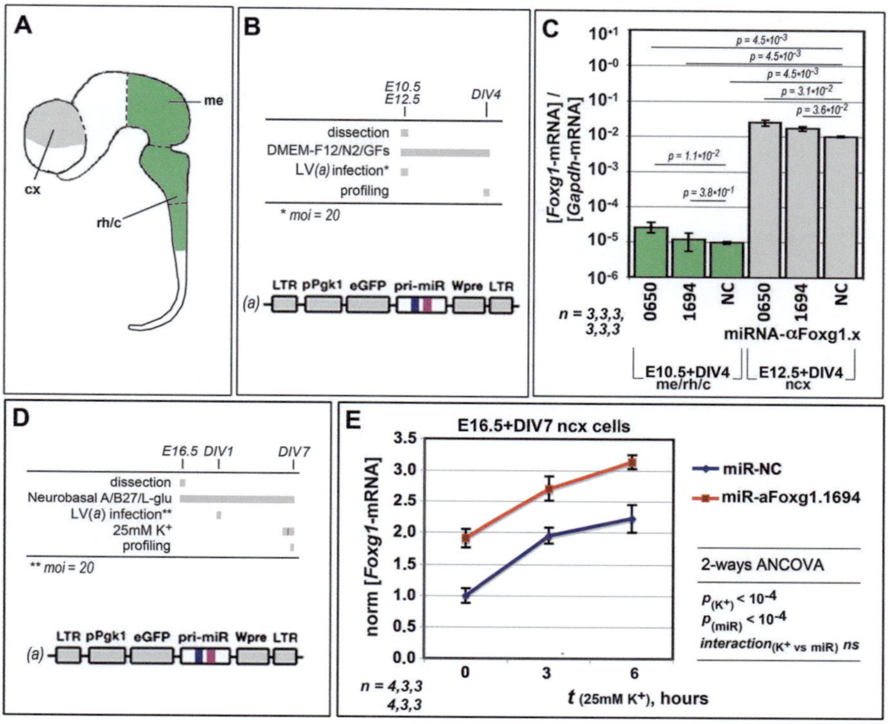

Fig. 2.4 Compliance of *Foxg1*-RNAa with endogenous gene regulation (**a**) Idealized representation of the murine early neural tube, including cortical (cx), mesencephalic (me), and rhombo-cervical (rh/c) domains. (**b**) Protocols and lentiviral reagent employed for the assay referred to in (**c**). (**c**) Impact of miR-αFoxg1.0650 and .1694 on *Foxg1*-mRNA levels in proliferating precursors from the me/rh/c and cx domains. (**d**) Protocols and lentiviral reagent employed for the assay referred to in (**e**). (**e**) *Foxg1*-mRNA modulation by miR-αFoxg1.1694 in differentiating neocortical derivatives upon their timed terminal exposure to 25 mM K^+. E, embryonic day; DIV, days *in vitro*. Bars represent sem's. n = number of biological replicates. Statistical significance of results evaluated by t-Student assay (one-tail, unpaired) (**c**) and ANCOVA (two-ways, unpaired) (**e**) assays. *ns*, not significant (Adapted from ref. [14])

domain. Moreover, within responsive neurons, saRNAs elicited a gentle and reproducible stimulation of the gene in order, which did not hide its fine endogenous tuning (Fig. 2.4d, e). All that strengthens the saRNA suitability for precise and affordable treatment of haploinsufficiencies, with special emphasis on those of neurological interest.

2.6 Molecular Mechanisms Underlying *Foxg1*-RNAa

As reported above, RNAa is a mechanistically heterogeneous process and at least two classes of molecular mechanisms are supposed to underlie it. RNAa may take place via downregulation of ncRNAs, which limit transcription of the associated gene of interest. Alternatively, saRNAs may drive molecular machinery promoting transcription to target chromatin [47, 58]. To cast light on this issue in *Foxg1*-RNAa, we monitored expression levels of the *Foxg1*-associated, sense-oriented AK158887 ncRNA (Fig. 2.5a), following the delivery of antisense-oriented miR-αFoxg1.0650 and miR-αFoxg1.1653. No downregulation of AK158887 was

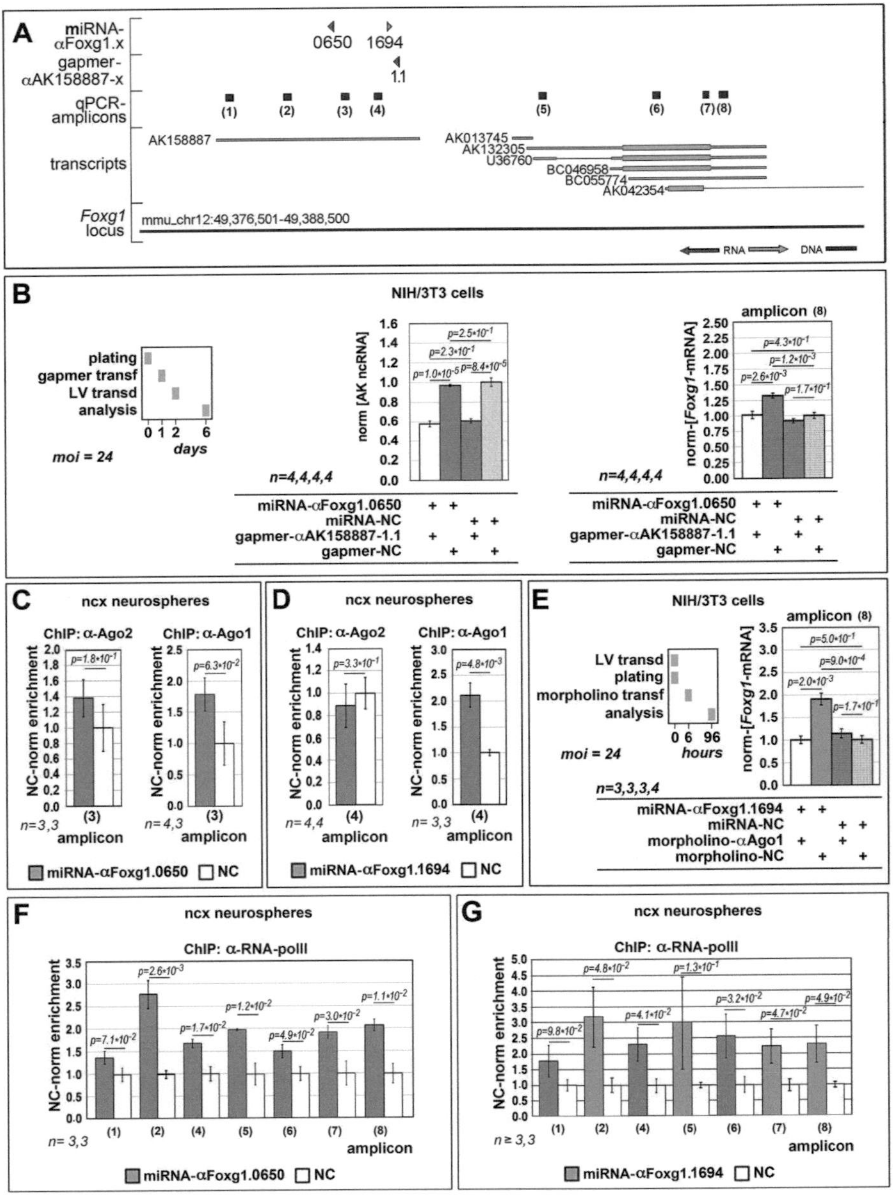

Fig. 2.5 Molecular mechanisms underlying *Foxg1*-RNAa (**a**) Schematics of the *Foxg1* locus including miRNA and gapmer positions and orientations, as well as diagnostic qPCR amplicons. (**b**) AK158887-ncRNA and *Foxg1*-mRNA levels in NIH/3T3 cells upon combined delivery of

found, suggesting that, at least in these cases, the latter mechanism may apply (not shown).

As for recognition of target chromatin, saRNAs might straightly bind to unwound chromosomal DNA. Alternatively, they might pair to nascent RNA molecules stemming from it. To distinguish between these possibilities, we downregulated the putative miR-αFoxg1.0650 target AK158887 RNA by gapmer-αAK158887-1.1 in easily transfectable NIH/3T3 cells (Fig. 2.5a, b). Interestingly, such manipulation fully abolished miR-αFoxg1.0650-dependent *Foxg1* transactivation (Fig. 2.5b), while not affecting *Foxg1* levels in miRNA-NC-treated samples. This suggests that miR-αFoxg1.0650 recognizes its chromatin target via RNA/RNA pairing. Actually, such gapmer-dependent RNAa-suppression assay was already employed in other studies, for unveiling molecular logic of RNAa at different gene loci. According to this assay, if the gapmer targeting the gene-of-interest-associated ncRNA reproduces the saRNA effect, then gene activation should originate from destabilization of such ncRNA target, as described for *Bdnf* [46]. If the gapmer suppresses saRNA activity – as reported for *PR* and *COX2* [42, 58] – then RNAa should rather rely on the recruitment of transactivating effectors to the target locus, via ncRNA docks stemming from it. The latter scenario is what we observed for *Foxg1* upon delivery of miR-αFoxg1.0650 (Fig. 2.5b). Other antisense saRNAs stimulating this gene might work in a similar way. Sense-oriented saRNAs might land on not yet mapped *Foxg1*-associated antisense-ncRNAs or act according to a different molecular logic.

A key role in implementing chromatin-regulatory functions exerted by saRNAs is played by Argonaute (Ago) proteins. Both Ago1 and Ago2 are detectable in the nucleus and can bind miRNAs [28]. Ago2 was also specifically implicated in a number of RNAa cases, possibly acting as a bridge between the saRNA and the supramolecular transactivating complex [42] and resulting to be necessary to RNAa itself [7, 35, 41, 42, 54]. Ago1 binds to TSS surroundings too. Moreover, it interacts with RNApolII and is involved in transcription regulation [28]. However, initial reports implicated it in transcriptional gene silencing (TGS) rather than RNAa [7, 35]. To assess the involvement of Ago2 in *Foxg1*-RNAa, we evaluated its

Fig. 2.5 (continued) miR-αFoxg1.0650 and gapmer-αAK158887-1.1. Values double normalized, against *Gapdh* and control (NC). (**c,d**) qPCR quantification of *Foxg1* chromatin enrichment, upon immunoprecipitation (ChIP) by antibodies against Argonaute 2 (α-Ago2) and Argonaute 1 (α-Ago1). Evaluation performed in neocortical precursors challenged by miR-αFoxg1.0650 (**c**) and miR-aFoxg1.1694 (**d**), according to the protocol shown in Fig. 2.1b,c. Values double normalized against input chromatin and control (NC). (**e**) *Foxg1*-mRNA levels in NIH/3T3 cells upon combined delivery of miR-αFoxg1.1694 and morpholino-αAgo1. Values double normalized against *Gapdh* and control (NC). (**f,g**) qPCR quantification of *Foxg1* chromatin enrichment, upon ChIP by antibodies against RNA polymerase II (α-RNA-polII). Evaluation performed in neocortical precursors challenged by miR-αFoxg1.0650 (**f**) and miR-aFoxg1.1694 (**g**), according to the protocol shown in Fig. 2.1b,c. Values double normalized against input chromatin and control (NC). Bars represent sem's. $n =$ number of biological replicates. Statistical significance of results evaluated by t-Student assay (one-tail, unpaired) (Adapted from Ref. [14])

recruitment to miR-αFoxg1.0650 and .1694 target sequences, upon saRNA delivery to neocortical precursors, by chromatin immunoprecipitation (ChIP). Enrichment for Ago1 was monitored as a specificity control. Unexpectedly, both saRNAs increased the recruitment of Ago1, but not of Ago2 (Fig. 2.5c, d), pointing to a selective involvement of the former in *Foxg1*-RNAa. To corroborate this inference, we antagonized Ago1 translation by a dedicated morpholino in NIH/3T3 cells (Fig. 2.5e). Remarkably, this treatment suppressed miR-αFoxg1.1694-dependent *Foxg1* transactivation (Fig. 2.5e), while not affecting *Foxg1* levels in miRNA-NC-treated samples. This confirms the pivotal role of Ago1 in *Foxg1*-RNAa. It echoes the recent report of Ago1-dependent RNAa at the *IL2* locus [69].

A step further along the RNAa cascade, RNApolII may be recruited to TSS [7, 28, 38, 41, 52, 67] or stimulated to progress downstream of it [36]. To distinguish among these possibilities, we monitored the enrichment of the *Foxg1* locus for RNApolII, upon saRNA delivery to neural precursors. We found that both miR-αFoxg1.0650 and .1694 robustly increased RNApolII recruitment along the entire locus (Fig. 2.5a, f, g), which likely led to augmented transcription rates. Intriguingly, the absolute RNApolII recruitment profile did not display any sudden decrease downstream of *Foxg1*-TSS in control conditions (not shown). Moreover, no abrupt increase of RNApolII recruitment took place in the same position upon saRNA delivery (Fig. 2.5a, f, g). Altogether these data suggest that RNApolII does not normally pause near *Foxg1*-TSS and saRNAs stimulate *Foxg1* transcription by promoting RNApolII recruitment to TSS.

2.7 *In Vivo Foxg1*-RNAa

Finally, to assess the feasibility of *in vivo Foxg1*-RNAa, we administered miR-αFoxg1.1694 to the living brain through AAV9-pseudotyped, self-complementary AAV2-derivative, adeno-associated viral vectors, under the control of a constitutive promoter (Fig. 2.6a). We injected $3*10^{10}$ infecting particles into the right lateral ventricle of P0 mouse pups by free hands. We sacrificed these animals 3 weeks later (P21) and scored their right neocortices for *Foxg1*-mRNA content as well as for the frequency at which $Foxg1^+$ cells were adeno-associated virus (AAV)-transduced (Fig. 2.6b, c). Remarkably, *Foxg1* was upregulated by about 1.66-fold (Fig. 2.6d), in front of a transduction frequency of $Foxg1^+$ cells close to only 0.17 (Fig. 2.6e). In other words, the cumulative *Foxg1* expression gain was about +66%, albeit only one sixth of Foxg1-expressing cells were targeted. This means that the actual expression gain in targeted $Foxg1^+$ cells might be not far from 6*66%, i.e., close to +400%. This suggests that, in a therapeutic scenario, saRNA expression should be dampened to restore physiological *Foxg1*-mRNA expression levels, possibly via a weaker promoter or a tunable transactivating system. Moreover, the employment of more advanced AAV drivers [10] might help targeting the almost totality of telencephalic neural cells.

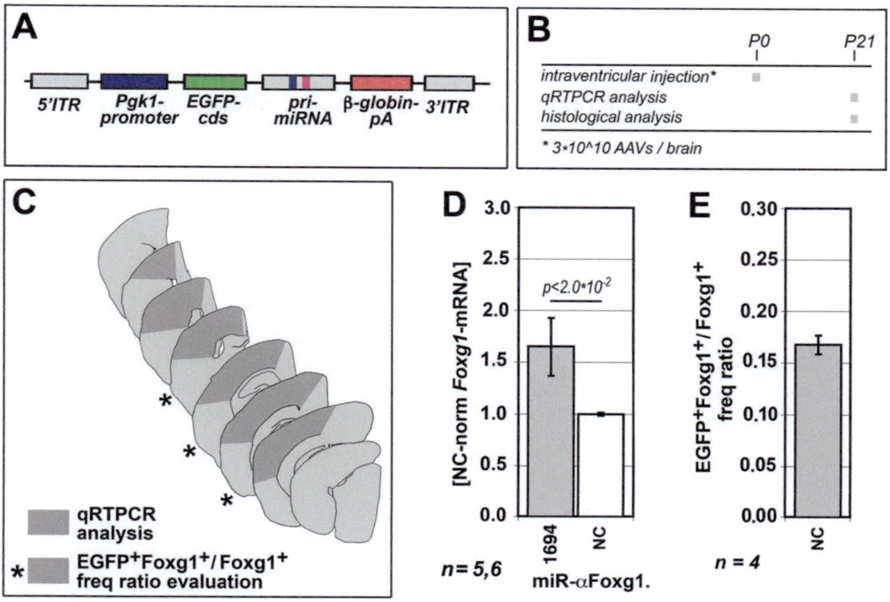

Fig. 2.6 *Foxg1*-RNAa in murine neocortex (**a**) Schematics of AAV9-pseudotyped, self-complementary AAV2-derivative, adeno-associated viral vector, driving constitutive expression of *Foxg1*-activating miRNAs. (**b**) Protocol employed for the assays referred to in (**c–f**). (**c**) Location of neocortical sectors subject of the analyses shown in (**d–f**). (**d**) Quantification of *Foxg1*-mRNA levels in neocortex of juvenile mice previously injected by scAAVs encoding for miR-αFoxg1.1694. (**e**) Evaluation of frequency of $Foxg1^{+}$ cells transduced by EGFP-encoding control virus (NC). P, postnatal day. Bars represent sem's. n = number of biological replicates (i.e., brains). Statistical significance of results evaluated by t-Student assay (one-tail, unpaired) (Adapted from Ref. [14])

2.8 Concluding Remarks

In summary, we have shown that RNAa is a powerful molecular tool potentially suitable for a variety of neurotherapeutic purposes.

We selected a first set of artificial miRNAs, transactivating the *Emx2* transcriptional factor gene, specifically in embryonic cortico-cerebral precursor cells. Their delivery led to enhanced self-renewal, delayed differentiation, and reduced death of neuronally committed precursors, resulting in a remarkable expansion of the neuronogenic precursors pool. These results are of paramount interest for purposes of gene-promoted brain repair. As such, this approach makes therapeutic stimulation of neuronogenesis via *Emx2* overexpression a feasible goal, preventing the drawbacks of exogenous gene copies introduction.

Moreover, we selected a second set of artificial miRNAs, eliciting a gentle transcriptional stimulation of the *Foxg1* transcriptional factor gene, specifically in cortico-cerebral cells. Their delivery led to an appreciable biological outcome, while complying with endogenous gene tuning linked to early CNS regionalization and late activity of neocortical projection neurons. These miRNAs stimulated RNApolII recruitment, possibly via Ago1. One of them worked promisingly *in vivo*. As such, RNAa can be a valuable approach for therapeutic treatment of the *FOXG1*-haploinsufficiency-linked variant of the Rett syndrome. Remarkably, hemizygosity for specific genes and polygenic chromosomal segments underlies a huge number of neuropathological entities [1, 5, 32, 33, 59] for which no cure is presently available. Based on the results reported above, RNAa might be a simple and scalable approach for fixing this class of problems.

Acknowledgments The author thanks all collaborators who contributed to the two primary studies referred to in this chapter [11, 14], as well as the funding bodies providing financial support for their implementation (including Fondation Jérôme Lejeune grant 1176-MA2013A, Fondazione Telethon grant GGP13034, and SISSA intramurary funding to the author).

References

1. Barøy T, Misceo D, Strømme P, Stray-Pedersen A, Holmgren A, Rødningen OK et al (2013) Haploinsufficiency of two histone modifier genes on 6p22.3, ATXN1 and JARID2, is associated with intellectual disability. Orphanet J Rare Dis 8(1):3
2. Beerli RR, Dreier B, Barbas CF 3rd (2000) Positive and negative regulation of endogenous genes by designed transcription factors. Proc Natl Acad Sci U S A 97(4):1495–1500
3. Brancaccio M, Pivetta C, Granzotto M, Filippis C, Mallamaci A (2010) Emx2 and Foxg1 inhibit gliogenesis and promote neuronogenesis. Stem Cells 28(7):1206–1218
4. Burns TC, Verfaillie CM, Low WC (2009) Stem cells for ischemic brain injury: a critical review. J Comp Neurol 515(1):125–144
5. Carvill GL, Mefford HC (2013) Microdeletion syndromes. Curr Opin Genet Dev 23 (3):232–239
6. Cheng AW, Wang H, Yang H, Shi L, Katz Y, Theunissen TW et al (2013) Multiplexed activation of endogenous genes by CRISPR-on, an RNA-guided transcriptional activator system. Cell Res 23(10):1163–1171
7. Chu Y, Yue X, Younger ST, Janowski BA, Corey DR (2010) Involvement of argonaute proteins in gene silencing and activation by RNAs complementary to a non-coding transcript at the progesterone receptor promoter. Nucleic Acids Res 38(21):7736–7748
8. Chu VT, Weber T, Wefers B, Wurst W, Sander S, Rajewsky K et al (2015) Increasing the efficiency of homology-directed repair for CRISPR-Cas9-induced precise gene editing in mammalian cells. Nat Biotechnol 33(5):543–548
9. Decipher database 1q21.1 recurrent microduplication syndrome, accessible at: https://decipher.sanger.ac.uk/syndrome/67#overview
10. Deverman BE, Pravdo PL, Simpson BP, Kumar SR, Chan KY, Banerjee A et al (2016) Cre-dependent selection yields AAV variants for widespread gene transfer to the adult brain. Nat Biotechnol 34(2):204–209
11. Diodato A, Pinzan M, Granzotto M, Mallamaci A (2013) Promotion of cortico-cerebral precursors expansion by artificial pri-miRNAs targeted against the Emx2 locus. Curr Gene Ther 13(2):152–161
12. Falcone C, Filippis C, Granzotto M, Mallamaci A (2015) Emx2 expression levels in NSCs modulate astrogenesis rates by regulating E gf R and F gf9: Emx2 Limits Astrogenesis Repressing EgfR and Fgf9. Glia 63(3):412–422
13. Fimiani C, Goina E, Mallamaci A (2015) Upregulating endogenous genes by an RNA-programmable artificial transactivator. Nucleic Acids Res 43(16):7850–7864

14. Fimiani C, Goina E, Su Q, Gao G, Mallamaci A (2016) RNA activation of haploinsufficient Foxg1 gene in murine neocortex. Sci Rep 6:39311
15. Gaj T, Gersbach CA, Barbas CF (2013) ZFN, TALEN, and CRISPR/Cas-based methods for genome engineering. Trends Biotechnol 31(7):397–405
16. Gangemi RMR, Daga A, Muzio L, Marubbi D, Cocozza S, Perera M et al (2006) Effects of Emx2 inactivation on the gene expression profile of neural precursors. Eur J Neurosci 23 (2):325–334
17. Gasiunas G, Siksnys V (2013) RNA-dependent DNA endonuclease Cas9 of the CRISPR system: holy Grail of genome editing? Trends Microbiol 21(11):562–567
18. Geiβler R, Scholze H, Hahn S, Streubel J, Bonas U, Behrens S-E et al (2011) Transcriptional activators of human genes with programmable DNA-specificity . Shiu S-H, editor. PLoS One 6 (5):e19509
19. Gilbert LA, Larson MH, Morsut L, Liu Z, Brar GA, Torres SE et al (2013) CRISPR-mediated modular RNA-guided regulation of transcription in eukaryotes. Cell 154(2):442–451
20. Guerrini R, Parrini E (2012) Epilepsy in Rett syndrome, and CDKL5- and FOXG1-gene-related encephalopathies. Epilepsia 53(12):2067–2078
21. Gustincich S, Zucchelli S, Mallamaci A (2016) The Yin and Yang of nucleic acid-based therapy in the brain. Prog Neurobiol. epub ahead of print
22. Hamasaki T, Leingärtner A, Ringstedt T, O'Leary DDM (2004) EMX2 regulates sizes and positioning of the primary sensory and motor areas in neocortex by direct specification of cortical progenitors. Neuron 43(3):359–372
23. Hanashima C, Li SC, Shen L, Lai E, Fishell G (2004) Foxg1 suppresses early cortical cell fate. Science 303(5654):56–59
24. Hanashima C, Fernandes M, Hebert JM, Fishell G (2007) The role of Foxg1 and dorsal midline signaling in the generation of Cajal-Retzius subtypes. J Neurosci 27(41):11103–11111
25. Hatini V, Tao W, Lai E (1994) Expression of winged helix genes, BF-1 and BF-2, define adjacent domains within the developing forebrain and retina. J Neurobiol 25(10):1293–1309
26. Hsu PD, Scott DA, Weinstein JA, Ran FA, Konermann S, Agarwala V et al (2013) DNA targeting specificity of RNA-guided Cas9 nucleases. Nat Biotechnol 31(9):827–832
27. Hu J, Lei Y, Wong W-K, Liu S, Lee K-C, He X et al (2014) Direct activation of human and mouse Oct4 genes using engineered TALE and Cas9 transcription factors. Nucleic Acids Res 42(7):4375–4390
28. Huang V, Place RF, Portnoy V, Wang J, Qi Z, Jia Z et al (2012) Upregulation of Cyclin B1 by miRNA and its implications in cancer. Nucleic Acids Res 40(4):1695–1707
29. Jiao AL, Slack FJ (2014) RNA-mediated gene activation. Epigenetics 9(1):27–36
30. Kearns NA, Genga RMJ, Enuameh MS, Garber M, Wolfe SA, Maehr R (2014) Cas9 effector-mediated regulation of transcription and differentiation in human pluripotent stem cells. Development 141(1):219–223
31. Kim H, Kim J-S (2014) A guide to genome engineering with programmable nucleases. Nat Rev Genet 15(5):321–334
32. Lal D, Ruppert A-K, Trucks H, Schulz H, De Kovel CG, Kasteleijn-Nolst Trenité D et al (2015) Burden analysis of rare microdeletions suggests a strong impact of neurodevelopmental genes in genetic generalised epilepsies, Copenhaver GP, editor. PLoS Genet 11(5):e1005226
33. Lee BH, Smith T, Paciorkowski AR (2015) Autism spectrum disorder and epilepsy: disorders with a shared biology. Epilepsy Behav 47:191–201
34. Li L-C (2014) Chromatin remodeling by the small RNA machinery in mammalian cells. Epigenetics 9(1):45–52
35. Li L-C, Okino ST, Zhao H, Pookot D, Place RF, Urakami S et al (2006) Small dsRNAs induce transcriptional activation in human cells. Proc Natl Acad Sci U S A 103(46):17337–17342
36. Liu M, Roth A, Yu M, Morris R, Bersani F, Rivera MN et al (2013) The IGF2 intronic miR-483 selectively enhances transcription from IGF2 fetal promoters and enhances tumorigenesis. Genes Dev 27(23):2543–2548
37. Maeder ML, Linder SJ, Cascio VM, Fu Y, Ho QH, Joung JK (2013) CRISPR RNA–guided activation of endogenous human genes. Nat Methods 10(10):977–979

38. Majid S, Dar AA, Saini S, Yamamura S, Hirata H, Tanaka Y et al (2010) MicroRNA-205-directed transcriptional activation of tumor suppressor genes in prostate cancer. Cancer 116 (24):5637–5649
39. Martynoga B, Morrison H, Price DJ, Mason JO (2005) Foxg1 is required for specification of ventral telencephalon and region-specific regulation of dorsal telencephalic precursor proliferation and apoptosis. Dev Biol 283(1):113–127
40. Maruyama T, Dougan SK, Truttmann MC, Bilate AM, Ingram JR, Ploegh HL (2015) Increasing the efficiency of precise genome editing with CRISPR-Cas9 by inhibition of nonhomologous end joining. Nat Biotechnol 33(5):538–542
41. Matsui M, Sakurai F, Elbashir S, Foster DJ, Manoharan M, Corey DR (2010) Activation of LDL receptor expression by small RNAs complementary to a noncoding transcript that overlaps the LDLR promoter. Chem Biol 17(12):1344–1355
42. Matsui M, Chu Y, Zhang H, Gagnon KT, Shaikh S, Kuchimanchi S et al (2013) Promoter RNA links transcriptional regulation of inflammatory pathway genes. Nucleic Acids Res 41 (22):10086–10109
43. Mattei E, Corbi N, Di Certo MG, Strimpakos G, Severini C, Onori A et al (2007) Utrophin up-regulation by an artificial transcription factor in transgenic mice. PLoS One 2(8):e774
44. Meshorer E, Misteli T (2006) Chromatin in pluripotent embryonic stem cells and differentiation. Nat Rev Mol Cell Biol 7(7):540–546
45. Miller JC, Tan S, Qiao G, Barlow KA, Wang J, Xia DF et al (2010) A TALE nuclease architecture for efficient genome editing. Nat Biotechnol 29(2):143–148
46. Modarresi F, Faghihi MA, Lopez-Toledano MA, Fatemi RP, Magistri M, Brothers SP et al (2012) Inhibition of natural antisense transcripts in vivo results in gene-specific transcriptional upregulation. Nat Biotechnol 30(5):453–459
47. Morris KV, Santoso S, Turner A-M, Pastori C, Hawkins PG (2008) Bidirectional transcription directs both transcriptional gene activation and suppression in human cells. PLoS Genet 4(11): e1000258
48. Muzio L, Mallamaci A (2005) Foxg1 confines Cajal-Retzius neuronogenesis and hippocampal morphogenesis to the dorsomedial pallium. J Neurosci 25(17):4435–4441
49. Muzio L, Soria JM, Pannese M, Piccolo S, Mallamaci A (2005) A mutually stimulating loop involving emx2 and canonical wnt signalling specifically promotes expansion of occipital cortex and hippocampus. Cereb Cortex 15(12):2021–2028
50. Okuno H (2011) Regulation and function of immediate-early genes in the brain: beyond neuronal activity markers. Neurosci Res 69(3):175–186
51. Perez-Pinera P, Kocak DD, Vockley CM, Adler AF, Kabadi AM, Polstein LR et al (2013) RNA-guided gene activation by CRISPR-Cas9–based transcription factors. Nat Methods 10 (10):973–976
52. Place RF, Li L-C, Pookot D, Noonan EJ, Dahiya R (2008) MicroRNA-373 induces expression of genes with complementary promoter sequences. Proc Natl Acad Sci 105(5):1608–1613
53. Portnoy V, Huang V, Place RF, Li L-C (2011) Small RNA and transcriptional upregulation. Wiley Interdiscip Rev RNA 2(5):748–760
54. Portnoy V, Lin SHS, Li KH, Burlingame A, Hu Z-H, Li H et al (2016) saRNA-guided Ago2 targets the RITA complex to promoters to stimulate transcription. Cell Res 26(3):320–335
55. Raciti M, Granzotto M, Duc MD, Fimiani C, Cellot G, Cherubini E et al (2013) Reprogramming fibroblasts to neural-precursor-like cells by structured overexpression of pallial patterning genes. Mol Cell Neurosci 57:42–53
56. Robert F, Barbeau M, Éthier S, Dostie J, Pelletier J (2015) Pharmacological inhibition of DNA-PK stimulates Cas9-mediated genome editing. Genitourin Med 7(1):93
57. Sansom SN, Griffiths DS, Faedo A, Kleinjan D-J, Ruan Y, Smith J et al (2009) The level of the transcription factor Pax6 is essential for controlling the balance between neural stem cell self-renewal and neurogenesis. PLoS Genet 5(6):e1000511
58. Schwartz JC, Younger ST, Nguyen N-B, Hardy DB, Monia BP, Corey DR et al (2008) Antisense transcripts are targets for activating small RNAs. Nat Struct Mol Biol 15 (8):842–848

59. Silva AM, Brown JM, Buckle VJ, Wade-Martins R, Lufino MMP (2015) Expanded GAA repeats impair FXN gene expression and reposition the FXN locus to the nuclear lamina in single cells. Hum Mol Genet 24(12):3457–3471
60. Snowden AW, Zhang L, Urnov F, Dent C, Jouvenot Y, Zhong X et al (2003) Repression of vascular endothelial growth factor A in glioblastoma cells using engineered zinc finger transcription factors. Cancer Res 63(24):8968–8976
61. Sohn JH, Yeh B-I, Choi J-W, Yoon J, Namkung J, Park K-K et al (2010) Repression of human telomerase reverse transcriptase using artificial zinc finger transcription factors. Mol Cancer Res 8(2):246–253
62. Song M, Kim Y-H, Kim J-S, Kim H (2014) Genome engineering in human cells. Methods Enzymol 546:93–118
63. Takizawa T, Meshorer E (2008) Chromatin and nuclear architecture in the nervous system. Trends Neurosci 31(7):343–352
64. Turner A-M, Morris K (2010) Controlling transcription with noncoding RNAs in mammalian cells. BioTechniques 48(6):ix–xvi
65. Verrotti A, Carelli A, Di Genova L, Striano P (2015) Epilepsy and chromosome 18 abnormalities: a review. Seizure 32:78–83
66. Wilson KA, Chateau ML, Porteus MH (2013) Design and development of artificial zinc finger transcription factors and zinc finger nucleases to the hTERT locus. Mol Ther Nucleic Acids 2: e87
67. Yue X, Schwartz JC, Chu Y, Younger ST, Gagnon KT, Elbashir S et al (2010) Transcriptional regulation by small RNAs at sequences downstream from 3′ gene termini. Nat Chem Biol 6 (8):621–629
68. Zhang F, Cong L, Lodato S, Kosuri S, Church GM, Arlotta P (2011) Efficient construction of sequence-specific TAL effectors for modulating mammalian transcription. Nat Biotechnol 29 (2):149–153
69. Zhang Y, Fan M, Zhang X, Huang F, Wu K, Zhang J et al (2014) Cellular microRNAs up-regulate transcription via interaction with promoter TATA-box motifs. RNA 20 (12):1878–1889

Chapter 3
Target-Recognition Mechanism and Specificity of RNA Activation

Huiqing Cao, Xing Meng, Xiaoxia Wang, and Zicai Liang

Abstract Small activating RNA (saRNA)-mediated gene activation has opened a new avenue for upregulating the expression of target genes by promoting endogenous transcription, a phenomenon known as RNA activation (RNAa). RNAa is distinct from the established RNAi mechanistic framework, although AGO2 is required by both. The precise mechanism of RNAa is currently disputable and has become a bottleneck in the development of this new technology. saRNA may achieve activation of target genes by directly binding to DNA targets in promoter, or interacting with antisense transcripts transcribed from overlapping promoter sequences, or by silencing other genes. In this chapter, we focused on recent development in our understanding of the target-recognition mechanism in RNAa. Conflicting results on saRNA targets are also discussed. Despite that the target mechanism of RNAa is more complex than expected and not completely understood so far, independent lines of evidence have suggested that saRNAs work by an "on-site" mechanism by binding to target genomic DNA in a "seed-region"--dependent manner. Finally, "off-target" effects of saRNA are observed and should be carefully controlled in designing experiments for and interpreting results from RNAa-related studies.

Keywords RNAa • saRNA • CRISPR/Cas • Progesterone receptor • Off-target effects

3.1 Introduction

The role of RNA molecules in gene regulation has been suggested for decades. Increasing importance has been attested for the role of RNA in gene expression when multiple small regulatory RNA pathways were identified. Small RNAs are approximately 18–30 nt noncoding molecules which are classified into microRNAs (miRNAs), small interfering RNAs (siRNAs), and PIWI-interacting RNAs (piRNAs). miRNAs and siRNAs have been identified as posttranscriptional gene

H. Cao • X. Meng • X. Wang • Z. Liang (✉)
Institute of Molecular Medicine, Peking University, Beijing, China
e-mail: liangzc@ribolia.com

L.-C. Li (ed.), *RNA Activation*, Advances in Experimental Medicine and Biology 983, DOI 10.1007/978-981-10-4310-9_3

silencers since the groundbreaking discovery of RNA interference (RNAi) by Andrew Fire and Craig Mello in 1998 [4]. Numerous studies have focused on the mechanistic dissection of RNAi, and the RNAi pathways in the cytoplasm have now been well established.

Recent findings in small RNA research showed that small RNAs could regulate target gene expression in a binary manner, i.e., upregulate gene expression in a process known as RNA activation (RNAa) [14]. Such small RNAs are termed as small activating RNAs (saRNAs). RNAa possesses distinct functional features, kinetic properties, and target mechanisms compared with those of RNAi. saRNA-mediated regulation occurred at the transcriptional level in the nucleus of cells by targeting the promoter region of genes. Moreover, naturally occurring small RNAs such as miRNAs were also found to mediate RNAa [7, 15, 21]. Given the capacity of promoting endogenous transcription, RNAa thus opens a new avenue for targeted activation of genes at will. While serving as potential regulators in cellular processes, saRNAs are also a promising new class of therapeutics for the treatment of diverse diseases.

However, the features of promoter-targeted saRNA fit less into the established RNAi mechanistic framework. The precise mechanism of RNAa is disputable and remains to be elucidated. In this chapter, we reviewed the current understanding on target-recognition mechanism of saRNA with comparison to that of RNAi. The discussion is limited to studies carried out in mammalian cells.

3.2 Target-Recognition Mechanism of RNAi

RNAi is a powerful approach to almost all of the gene expression by using small RNA molecules, including siRNA and miRNA. Posttranscriptional gene silencing (PTGS), including mRNA degradation and translational inhibition, was induced by small RNAs through a sequence-specific target mechanism. RNAi has been a game-changer for modern biomedical research in the last 15 years. Its effecter molecules, siRNA and miRNA, have been widely used as a powerful tool of loss-of-function experiments in a majority of biomedical research laboratories. RNAi-based drugs exhibited promising clinical application for a variety of diseases.

The mechanism that siRNAs recognize and cleave target mRNAs has been well established [8]. The antisense strand of siRNA, also known as the guide strand, serves as a key factor for gene silencing by targeting cellular mRNA sequence through base-pairing. The guide strand RNA is first loaded into proteins of the Argonaute family [8] to form the RNA-induced silencing complex (RISC), binds to its target mRNA with perfect match and forms an A-form helix structure, and then triggers enzymatic cleavage of target mRNA via the slicer activity of Argonaute 2 (AGO2), the catalytic component of the RISC. The mRNA target is cleaved at a position corresponding to 10–11 nt from the 5′ end of the guide strand.

It was found later that the guide strand can also be supplied by endogenously expressed small RNAs such as miRNA. Synthetic duplex RNAs are normally configured to share perfect sequence complementarity with their target mRNAs, while miRNAs recognize target sequences through imperfect complementarity. The 3′-untranslated regions (3′-UTRs) of many genes are found to be preferred sites for miRNA targeting. Bulge structures were usually formed due to mismatches between miRNA guide strand and its target sequence. The specificity of miRNA targeting is still very much beyond the comprehension of the RNAi field, although it was known that a main determinant is the region spanning nucleotides 2–8 of its 5′ region, which was known as the "seed" region [1]. This short-seed-region-based target-recognition manner theoretically would enable a single miRNA to recognize multiple mRNA via 3′-UTR sequences, but it is now clear that not all mRNAs with the matching sequence of the seed region will be regulated.

Since RNAi could work through short sequence recognition instead of the full sequence, the target specificity of RNAi, especially when mediated by siRNA, should be taken into consideration. The guide strand of siRNA may recognize unintended mRNA sequences with partial complementarity to the 5′-end region and silence their expression via a mechanism closely mirroring miRNA, which is referred to "off-target" effect of siRNA [9, 10].

While most RNAi studies in mammalian cells in the past years were focused on gene silencing through posttranscriptional gene silencing by small RNAs, there have been studies showing that small RNAs can also induce transcriptional gene silencing (TGS) by targeting transcription start sites (TSS) or promoter region of target genes in the nucleus [12, 19, 28]. The mechanism of TGS is reported to be associated with histone modification that facilitates the repressive heterochromatin formation at the target promoter.

3.3 Discovery of RNAa

The "Yin Yang" theory in ancient Chinese culture suggests that all phenomena consist of two opposite but complementary aspects; therefore, it is not surprising that in late 2006, a report by Li et al. showed that small RNAs could also induce gene expression at transcriptional level [14]. These small RNAs target the promoter regions of genes and exert an effect opposite to that of RNAi. The discovery of RNAa/RNAi represents bidirectional roles of small RNAs in the regulation of gene expression. To date, RNAa of dozens of genes has been reported including several prototype genes (e.g., p21, E-cadherin, progesterone receptor [PR]) used for mechanistic studies of RNAa. Similar to RNAi, RNAa was perceived both as a powerful molecular tool for gene expression manipulation and as a potential therapeutic technology for the treatment of human diseases. However, further studies to elucidate the mechanism of RNAa could be critical for accelerating the development and application of this new technology.

3.4 Target-Recognition Mechanism of RNAa

The finding of the promoter-targeted saRNAs is of significant interest for the field of the regulatory small RNAs, but the target molecule of RNAa has not been revealed for quite several years. It is well known that small RNAs targeted to coding mRNAs in the cytoplasm by the RISC complex could mediate gene silencing, whereas small RNAs targeted to upstream promoter regions could cause TGS in the nucleus via an epigenetic mechanism. Corroborating the complexity of the interaction of small RNAs with their targets, it was further reported that synthetic RNAs complementary to gene promoters, which were also called antigene RNAs (agRNAs), have indeed the function of either activating or suppressing gene transcription [13, 27]. Gene regulation induced by all these small RNAs requires AGO2, suggesting that they might share a general mechanism. The bottom line question is what the target molecule of saRNA in RNAa is: DNA or the cognate RNA it encodes.

3.4.1 Distribution of Target Sequence in Gene Promoter

Activity of saRNAs is sensitive to their position of targets. saRNAs with a single base shift on its target could lead to totally different activity for gene activation [13]. It has been proposed that saRNAs targeting sequences around the TSS or the TATA-box-centered region in a promoter could have higher gene activation efficiency [2, 6, 35].

To identify a potent saRNA, usually at least six saRNAs within 200 bp upstream of the TSS need to be designed and tested [11]. Other studies further reported that saRNAs targeting 200–700 bp upstream of the TSS are still effective [14]. Actually, most of the known active saRNA target sites are located upstream of no more than 1 kb from the TSS. saRNAs targeting downstream of 3′-UTR also showed successful RNAa, rather than the silencing of the target genes [34]. For gene activation of the PR gene, Janowski et al. tested a series of small dsRNAs targeting the −56–+17 region of PR promoter and found a cluster of effective saRNAs, among which PR-11 showed the strongest activity.

Our lab has attempted to expand the target region of effective saRNAs for PR gene [18]. The promoter and TSS of PR are well characterized. Twenty saRNAs targeting different sites in a ~2 kb region upstream of the PR TSS were designed and individually transfected into MCF-7 cells, and PR expression was assessed after 72 h. The previously reported effective saRNA, PR-11, was used as a positive control. Five of the 20 saRNAs were able to elevate the PR expression by at least three- to fivefold at the mRNA level. Among these five saRNAs, one saRNA, PR-1611, designed to target a distal target site 1611 bp upstream of the TSS, exhibited the strongest activating effect on PR expression. Consequently, 31 tiling saRNAs targeting a 30-bp region around the PR-1611 target (from −1596 site to

−1626 site) were designed to investigate whether such a remote target site could indeed mediate RNAa. It was found that six saRNAs clustered within a 10-bp region around the PR-1611 target led to apparent up-regulation of PR. These remote sites around −1611 seemed to constitute a distal hotspot for RNAa. These data suggest that saRNA-mediated gene activation could be achieved by targeting sites beyond the regions close to the TSS or TATA boxes. This finding is different from earlier reports that most saRNA target sites are within the proximal 1-kb region of promoters [13, 14, 24–26, 29–32]. Considering that the exact promoter regions for most genes were not well characterized, it raised an interesting possibility that RNAa might occur within a wider area beyond the core promoter region.

We considered the PR-1611 saRNA cluster to be an ideal model for the mechanistic studies of RNAa and speculated that this remote RNAa hotspot, compared to proximal targets, might be a cleaner site because it might be free from complex transcriptional events around the TSS.

3.4.2 A "Seed Region" in Antisense Strand of the saRNA Is Pivotal for Target Recognition

In the literature, it has not been well established which strand of the duplex saRNA guides the targeting and triggers RNAa. Answering this question could provide critical clues for understanding the mechanism of RNAa. There have been controversial reports about the relative contributions of the two strands of saRNAs from some of the pioneering labs in the field [14, 16, 22, 27]. In the first study of saRNA, Li et al. showed that mismatches between the 5′-end of the antisense strand of an saRNA and the target DNA almost completely abolished the activation of p21 and E-cadherin. In contrast, 3′-termini disruption retained RNAa activity [14]. It was suggested that 5′-portion of the antisense strand of saRNAs is essential for RNAa. Moreover, blocking the 5′-end of the antisense strand by chemical modification almost completely inhibited RNAa activity, but modification in the 5′-end of the sense strand did not [22]. These results suggested the importance of the antisense strand in RNAa. This was further corroborated by our experiments by mutation analysis. In our PR activation model, two groups of mutated saRNAs with single point mutations across the length of PR-1611 and PR-11 were synthesized and assessed for RNAa activity. Interestingly, mutations at sites 4, 6, or 8 bp counted from the 5′-end of the antisense strand seriously disrupted their ability to mediate PR activation, whereas mutations in the rest part of the saRNAs did not. In addition, when single-stranded mutations in the 4, 6, or 8 sites of the antisense strand (5′-end) or the sense strand (3′-end) of PR-1611 were created, saRNAs with mutations in the antisense strand had stronger inhibitory effects on RNAa than those with mutations in the sense strand. Moreover, effective saRNA had their antisense strand efficiently integrated into AGO2, whereas such a phenomenon was not found for the sense strand. These studies suggested that the "seed region" in the 5′-end of the antisense strand of saRNA is critically important for its RNAa activity. This feature of saRNAs indicates that the mechanism of target recognition by saRNAs might be similar to that of miRNA [3, 20].

3.4.3 Interaction Between saRNA and Promoter DNA

RNAa is a nuclear process. Conflicting results have been reported about the interaction between saRNA and its potential targets. The following working models of RNAa have been suggested: (1) saRNAs could recognize and bind to DNA targets in the promoter region [5, 21]; (2) saRNAs might bind to noncoding transcripts transcribed from the promoter sequence [16, 27] or induce the cleavage of antisense transcripts complementary to mRNA or the mRNA of upstream genes [33]; and (3) RNAa might be resulted from silencing other genes by saRNA.

3.4.3.1 Antisense Transcripts Mediated RNAa

More than 30% of genes were predicted to have antisense transcripts overlapping with their promoters. Early studies from the Corey group have shown that some saRNAs targeting PR promoter activated while others silenced PR expression. It was speculated that the activating saRNAs may induce cleavage of antisense transcripts and thus remove their repressive effect on PR transcription. The Corey group studied the mechanism using a biotin-labeled saRNA, PR-11; they showed that saRNAs bound to the antisense transcript acted as a functional regulator or a scaffold of regulators for the PR gene [27]. Similarly, Morris et al. demonstrated that a gene-specific antisense transcript mediated p21 activation; suppression of the antisense transcript resulted in a loss of epigenetic inhibition for the p21 mRNA transcription and thereby increased its expression. The balance between p21 sense and antisense transcription thus bidirectionally controlled gene expression [17]. In these studies, the antisense noncoding RNAs serve as endogenous transcriptional regulators to mediate the interaction between the saRNA and the target promoter.

3.4.3.2 RNA Activation Is an "On-Site" Process in Promoter DNA

Another possible target mechanism is that saRNAs directly recognize and bind genomic DNA. Solid evidence to clarify the molecular interaction of saRNAs with their targets was largely lacking before we tackled this question by taking advantage of the CRISPR/Cas technology. We created a subline of cells (Clone-5) which harbor an 8-bp deletion in the PR-1611 target site in one allele and a 1-bp insertion in the other. In the mutated cells in which the -1611 site was disrupted, the activation of PR by PR-1611 was significantly reduced, indicating that RNAa is

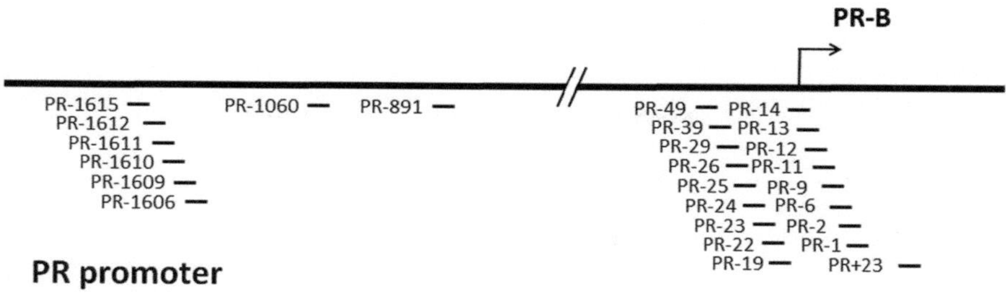

Fig. 3.1 Distribution of saRNA target sequence for RNAa in PR promoter

most likely an "on-site" process that relies on the target site in the promoter. Similar results were reported at almost the same time by Li et al. that deletion of the saRNA target on the p21 promoter by CRISPR technique significantly reduced gene activation effects by the saRNA [23]. Together, these data provided direct evidence that in RNAa mediated by saRNAs, a sequence-specific process occurring "on-site" in the promoter region, no matter whether this process occurred at the DNA or RNA level, would serve as a major mechanism of RNA activation (Fig. 3.1).

3.4.3.3 saRNA Binds to Promoter DNA

The antisense strand of an saRNA can mediate RNAa by an "on-site" process through binding to either any sense transcripts (RNA) originating from the promoter region or the promoter DNA itself. To discriminate between the two possibilities, an saRNA pull-down assay was carried out in both wild-type and the target DNA-mutated cell line, by using PR-1611 labeled with biotin at the 3′-end of the antisense strand. The results showed a fourfold enrichment of promoter DNA by biotin-labeled PR-1611 in wide-type cells, indicating an interaction between saRNA and promoter DNA. Importantly, the amount of target promoter DNA pulled down by Bio-PR-1611 was significantly reduced in mutated cells. Together, these results suggested that saRNA PR-1611 bound directly to genomic DNA of the PR promoter target site in the process of RNAa. The target mechanism of PR activation was summarized in Fig. 3.2 [18].

3.4.4 Complexity of Target-Recognition Mechanism for RNA Activation

Earlier reports suggested that PR activation induced by saRNA, PR-11, was mediated by AT-2, an antisense transcript of PR [27]. Considering the possibility that RNA transcript and DNA may both act as the target of saRNA, possible transcripts covering the PR-1611 target site were surveyed. No other mature transcript was

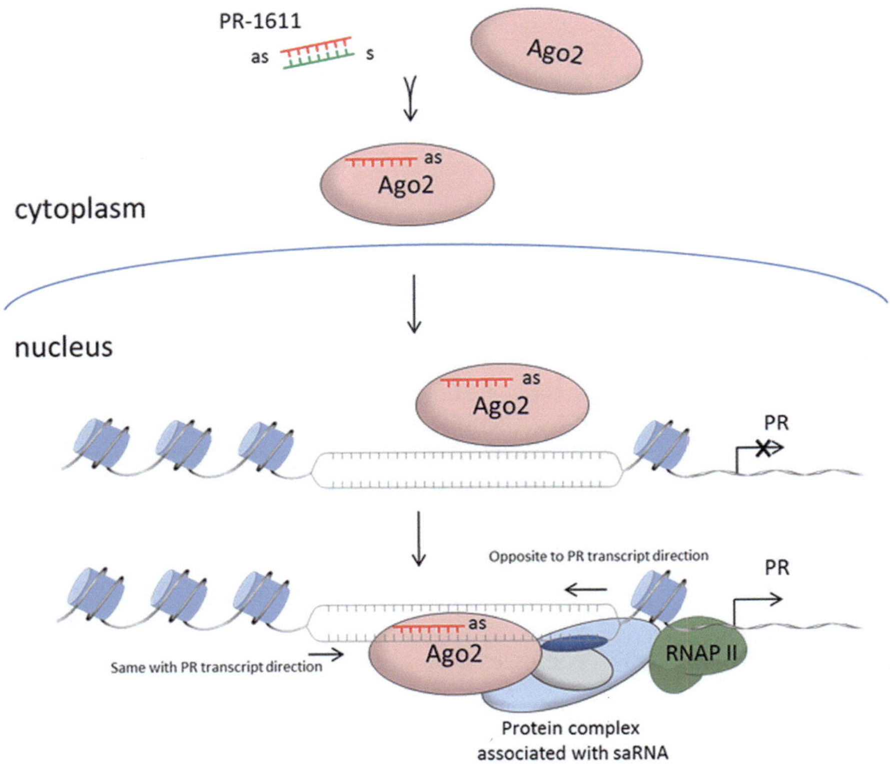

Fig. 3.2 A working model of target mechanism of saRNA

found in the area, except for AT-2. Since the PR-1611 site is located in the intron region of AT-2, the mature sequence of AT-2 does not contain the PR-1611 target. The data suggest that PR-1611 would not have a direct interaction with the mature AT-2. Experimental evidence also demonstrated that treatment of cells by PR-1611 could achieve gene activation without causing any reduction in the level of AT-2. To further rule out a role of AT-2 in PR-1611-mediated RNAa, two effective siRNAs were used to knock down the level of AT-2. The results showed that although more than 80% of AT-2 transcript was knocked down, the RNAa activity of PR-1611 was not affected, indicating that this antisense transcript does not play a critical role in PR-1611-induced gene activation. PR-1611 works directly on the DNA target site without the need of an RNA transcript as a scaffold [18].

3.5 Specificity of saRNA

It was well known that siRNAs have off-target effects if not properly modified. It is not surprising to notice that saRNA could illicit the upregulation of hundreds of genes (Meng, Cao and Liang, data not shown). The off-target effects of saRNA might be ascribed as the consequence of the saRNA-induced downregulation of an inhibitor [33]. However, this argument is not fully convincing for there are obviously differences between RNAi "off-target" and RNAa. The kinetics and duration of RNAi and RNAa are very different. siRNA-induced gene expression change is observed at 6–12 h and reached the peak during 24–48 h, whereas saRNAs increase gene expression after 24 h and reached the peak at 72 h. The effect of RNAa lasts much longer than RNAi [14, 22, 31, 32]. In our experiments, we have observed "seed region" matched sequences on the promoters of a number of genes that were "nonspecifically" upregulated by PR-1611.

3.6 Conclusions

RNAa is an attractive approach to specifically upregulate the expression of an endogenous gene by saRNAs. Although the target mechanism of RNAa is not completely understood so far, several lines of evidence suggest that saRNAs work with an "on-site" mechanism by binding to target genomic DNA in a "seed-region"-dependent manner. An "off-target" effect of saRNA has been observed, and this should be carefully controlled in designing experiments for and interpreting results from RNAa-related studies.

References

1. Doench JG, Sharp PA (2004) Specificity of microRNA target selection in translational repression. Genes Dev 18:504–511
2. Fan M, Zhang Y, Huang Z, Liu J, Guo X, Zhang H, Luo H (2014) Optimizations of siRNA design for the activation of gene transcription by targeting the TATA-box motif. PLoS One 9: e108253
3. Farh KK, Grimson A, Jan C, Lewis BP, Johnston WK, Lim LP, Burge CB, Bartel DP (2005) The widespread impact of mammalian MicroRNAs on mRNA repression and evolution. Science 310:1817–1821
4. Fire A, Xu SQ, Montgomery MK, Kostas SA, Driver SE, Mello CC (1998) Potent and specific genetic interference by double-stranded RNA in *Caenorhabditis elegans*. Nature 391:806–811
5. Hu J, Chen Z, Xia D, Wu J, Xu H, Ye ZQ (2012) Promoter-associated small double-stranded RNA interacts with heterogeneous nuclear ribonucleoprotein A2/B1 to induce transcriptional activation. Biochem J 447:407–416
6. Huang V, Qin Y, Wang J, Wang X, Place RF, Lin G, Lue TF, Li LC (2010) RNAa is conserved in mammalian cells. PLoS One 5:e8848

7. Huang V, Place RF, Portnoy V, Wang J, Qi Z, Jia Z, Yu A, Shuman M, Yu J, Li LC (2011) Upregulation of Cyclin B1 by miRNA and its implications in cancer. Nucleic Acids Res 40:1695–1707
8. Hutvagner G, Zamore PD (2002) A microRNA in a multiple-turnover RNAi enzyme complex. Science 297:2056–2060
9. Jackson AL, Bartz SR, Schelter J, Kobayashi SV, Burchard J, Mao M, Li B, Cavet G, Linsley PS (2003) Expression profiling reveals off-target gene regulation by RNAi. Nat Biotechnol 21:635–637
10. Jackson AL, Burchard J, Schelter J, Chau BN, Cleary M, Lim L, Linsley PS (2006) Widespread siRNA "off-target" transcript silencing mediated by seed region sequence complementarity. Rna-a Publ Rna Soc 12:1179–1187
11. Janowski BA, Corey DR (2010) Minireview: switching on progesterone receptor expression with duplex RNA. Mol Endocrinol 24:2243–2252
12. Janowski BA, Hu J, Corey DR (2006) Silencing gene expression by targeting chromosomal DNA with antigene peptide nucleic acids and duplex RNAs. Nat Protoc 1:436–443
13. Janowski BA, Younger ST, Hardy DB, Ram R, Huffman KE, Corey DR (2007) Activating gene expression in mammalian cells with promoter-targeted duplex RNAs. Nat Chem Biol 3:166–173
14. Li LC, Okino ST, Zhao H, Pookot D, Place RF, Urakami S, Enokida H, Dahiya R (2006) Small dsRNAs induce transcriptional activation in human cells. Proc Natl Acad Sci U S A 103:17337–17342
15. Majid S, Dar AA, Saini S, Yamamura S, Hirata H, Tanaka Y, Deng G, Dahiya R (2010) MicroRNA-205-directed transcriptional activation of tumor suppressor genes in prostate cancer. Cancer 116:5637–5649
16. Matsui M, Sakurai F, Elbashir S, Foster DJ, Manoharan M, Corey DR (2010) Activation of LDL receptor expression by small RNAs complementary to a noncoding transcript that overlaps the LDLR promoter. Chem Biol 17:1344–1355
17. McManus MT, Morris KV, Santoso S, Turner A-M, Pastori C, Hawkins PG (2008) Bidirectional transcription directs both transcriptional gene activation and suppression in human cells. PLoS Genet 4:e1000258
18. Meng X, Jiang Q, Chang N, Wang X, Liu C, Xiong J, Cao H, Liang Z (2016) Small activating RNA binds to the genomic target site in a seed-region-dependent manner. Nucleic Acids Res 44:2274–2282
19. Morris KV, Chan SW, Jacobsen SE, Looney DJ (2004) Small interfering RNA-induced transcriptional gene silencing in human cells. Science 305:1289–1292
20. Nielsen CB, Shomron N, Sandberg R, Hornstein E, Kitzman J, Burge CB (2007) Determinants of targeting by endogenous and exogenous microRNAs and siRNAs. Rna-a Publ Rna Soc 13:1894–1910
21. Place RF, Li LC, Pookot D, Noonan EJ, Dahiya R (2008) MicroRNA-373 induces expression of genes with complementary promoter sequences. Proc Natl Acad Sci U S A 105:1608–1613
22. Place RF, Noonan EJ, Foldes-Papp Z, Li LC (2010) Defining features and exploring chemical modifications to manipulate RNAa activity. Curr Pharm Biotechnol 11:518–526
23. Portnoy V, Lin SH, Li KH, Burlingame A, Hu ZH, Li H, Li LC (2016) saRNA-guided Ago2 targets the RITA complex to promoters to stimulate transcription. Cell Res 26:320–335
24. Qin Q, Lin YW, Zheng XY, Chen H, Mao QQ, Yang K, Huang SJ, Zhao ZY (2012) RNAa-mediated overexpression of WT1 induces apoptosis in HepG2 cells. World J Surg Oncol 10:11
25. Ren S, Kang MR, Wang J, Huang V, Place RF, Sun Y, Li LC (2013) Targeted induction of endogenous NKX3-1 by small activating RNA inhibits prostate tumor growth. Prostate 73:1591–1601
26. Sakurai F, Nanjo Y, Okamoto S, Tachibana M, Mizuguchi H (2014) Upregulation of RECK gene expression by small double-stranded RNA targeting the promoter region. Cancer Gene Ther 21:164–170

27. Schwartz JC, Younger ST, Nguyen NB, Hardy DB, Monia BP, Corey DR, Janowski BA (2008) Antisense transcripts are targets for activating small RNAs. Nat Struct Mol Biol 15:842–848
28. Ting AH, Schuebel KE, Herman JG, Baylin SB (2005) Short double-stranded RNA induces transcriptional gene silencing in human cancer cells in the absence of DNA methylation. Nat Genet 37:906–910
29. Wang J, Place RF, Huang V, Wang X, Noonan EJ, Magyar CE, Huang J, Li LC (2010) Prognostic value and function of KLF4 in prostate cancer: RNAa and vector-mediated overexpression identify KLF4 as an inhibitor of tumor cell growth and migration. Cancer Res 70:10182–10191
30. Wang T, Li M, Yuan H, Zhan Y, Xu H, Wang S, Yang W, Liu J, Ye Z, Li LC (2013) saRNA guided iNOS up-regulation improves erectile function of diabetic rats. J Urol 190:790–798
31. Wang J, Place RF, Portnoy V, Huang V, Kang MR, Kosaka M, Ho MK, Li LC (2015a) Inducing gene expression by targeting promoter sequences using small activating RNAs. J Biol Methods 2(1):e14
32. Wang J, Huang V, Ye L, Barcena A, Lin G, Lue TF, Li LC (2015b) Identification of small activating RNAs that enhance endogenous OCT4 expression in human mesenchymal stem cells. Stem Cells Dev 24:345–353
33. Weinberg MS, Barichievy S, Schaffer L, Han J, Morris KV (2007) An RNA targeted to the HIV-1 LTR promoter modulates indiscriminate off-target gene activation. Nucleic Acids Res 35:7303–7312
34. Yue X, Schwartz JC, Chu Y, Younger ST, Gagnon KT, Elbashir S, Janowski BA, Corey DR (2010) Transcriptional regulation by small RNAs at sequences downstream from 3′ gene termini. Nat Chem Biol 6:621–629
35. Zhang Y, Fan M, Zhang X, Huang F, Wu K, Zhang J, Liu J, Huang Z, Luo H, Tao L, Zhang H (2014) Cellular microRNAs up-regulate transcription via interaction with promoter TATA-box motifs. Rna-a Publ Rna Soc 20:1878–1889

Chapter 4
Promoter-Targeted Small Activating RNAs Alter Nucleosome Positioning

Bin Wang and Yunzhang Hu

Abstract Epigenetic modification of target promoters has been identified as a mechanism underlying RNA activation (RNAa) induced by promoter-targeting small activating RNAs (saRNAs), but it is unclear how the chromosomal environment influences gene expression. In a study of the activation of the OCT4, SOX2, and NANOG genes by saRNAs, we found that saRNA targeting induced nucleosome-depleted region (NDRs) and the accumulation of RNA polymerase II (RNAPII) near or at the saRNA target sites. Additionally, promoters containing certain cis-regulatory elements such as the TATA box and CpG islands (CGIs) appeared to be more susceptible to RNAa. These results provide novel insight into the mechanism underlying RNAa in that saRNAs induce NDRs in the target promoter to remove nucleosome barriers between RNAPII-binding sites and the transcription start site (TSS), resulting in rapid assembly of transcription preinitiation complex (PIC) and subsequent activation of transcription.

Keywords RNAa • Nucleosome • saRNA • Nucleosome barrier • Open chromatin structure

4.1 Introduction

RNAs transcribed from noncoding regions in the genome have been proposed to act as an "activator RNA" in the nucleus [1]. In 2004, it was reported that endogenously occurring noncoding, double-stranded (ds)RNAs could activate the expression of target genes by binding to the neuron-restrictive silencer element in their promoters and thereby induce stem cell differentiation [11]. A study published in 2006 used 21-bp synthetic dsRNAs to target the promoters of several genes and activate their expression [12]. This phenomenon has been termed RNA activation (RNAa) and

B. Wang
Department of Biology, Kunming University, Kunming, China

Y. Hu (✉)
Institute of Medical Biology, Chinese Academy of Medical Sciences and Peking Union Medical College, Kunming, China
e-mail: huyunzhangym@126.com

L.-C. Li (ed.), *RNA Activation*, Advances in Experimental Medicine and Biology 983, DOI 10.1007/978-981-10-4310-9_4

the involved dsRNAs as small activating RNA (saRNA) [12]. Soon after, dsRNAs called antigene (ag)RNAs targeting the promoter of the progesterone receptor (PR) gene were found to activate PR expression in human T47D and MCF7 cells [9]. In 2008, a microRNA (miRNA), miR-373, was also shown to activate E-cadherin and cold shock domain-containing C2 expression by targeting their promoters [16], indicating that RNAa is a widespread phenomenon.

However, the cell-type specificity of RNAa [9] makes it difficult to generate a universal algorithm for designing saRNAs [23]. It has been suggested that distinct proteins in each cell type and epigenetic modification (e.g., DNA methylation status) of the promoter in different cell types influence RNAa. In addition, the basal transcription level of target genes is also thought to impact RNAa efficiency [9].

In this chapter, we discuss current efforts in understanding the RNAa mechanism by examining saRNA-induced changes in chromatin structure, especially nucleosome positioning. This study designed three dsRNAs targeting the SOX2 promoter: S134 and S227 were located upstream of the furthest TATA box, whereas S278 was located between the TATA box and a CpG island (CGI). Five dsRNAs for OCT4 (O302, O619, O134, O38, and O111) and four for NANOG (N35, N116, N185, and N80) were also designed. O134 was located in the proximal enhancer of OCT4; O38 and N185 were located in the proximal promoters of OCT4 and NANOG, respectively. After screening, S278 for SOX2; O134, O38, O111 for OCT4; and N185, N80 for NANOG were identified to have RNAa activity by inducing their respective target gene expression at both the mRNA and protein levels (Fig. 4.1).

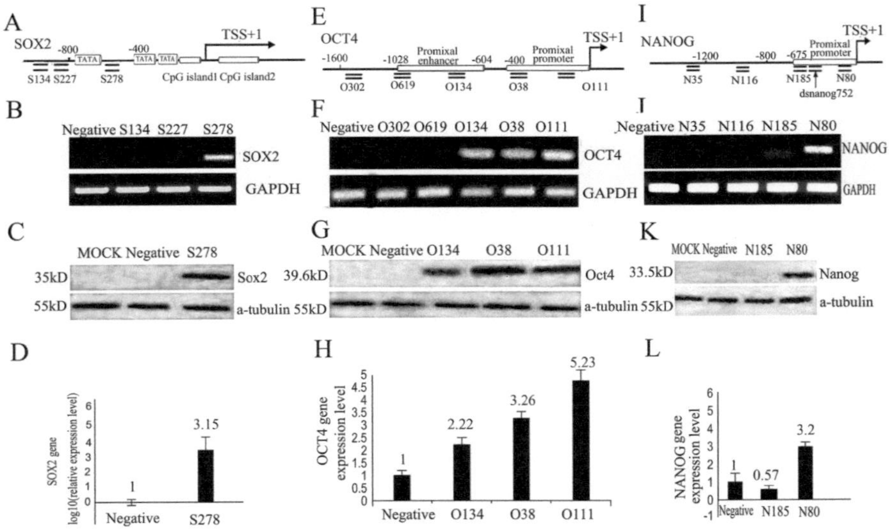

Fig. 4.1 SaRNA target sites and potential role of cis-acting elements in the RNAa of OCT4, SOX2, and NANOG. (**a–d**) The SOX2 promoter containing the TATA box and CGI showed a 1000-fold activation by saRNA relative to the negative control dsRNA; (**e–l**) OCT4 and NANOG, which contain a proximal promoter and enhancer, were activated by less than tenfold by their respective saRNAs

4.2 Nucleosome Repositioning Induces an Open Chromatin Structure in RNAa

Epigenetic modifications in promoter regions are associated with RNA activation (RNAa) [9, 12, 18]; however, it is unclear how the modifications contribute to RNAa [8]. It has also been suggested that guide RNAs facilitate the localization of AGO proteins to target DNA where AGO serves as a recruiting platform for histone modification, chromatin remodeling, and transcriptional activation [17]. In theory, an open chromatin structure is achieved by saRNA-induced epigenetic modification, which provides a basis for the RNAa mechanism. However, the nature of open chromatin as it relates to RNAa has not been clearly defined.

To investigate whether the binding of an saRNA to its target DNA can induce chromatin remodeling and ultimately affect nucleosome positioning, nucleosome binding to the promoter regions was investigated in human primary fibroblasts transfected with saRNAs for OCT4, SOX2, and NANOG as mentioned above by chromatin immunoprecipitation (ChIP), using an anti-histone H3 antibody, followed by quantitative (q)PCR [21]. Since H3 is one of the core histones in the nucleosome, its absence in the promoter can reflect a nucleosome-depleted region (NDR), which is involved in nucleosome positioning [10]. Most loci have regular nucleosomal arrays with 195-bp periodicity (Fig. 4.2); therefore, nucleosomes were analyzed accordingly, starting from the TSS, since the proximal promoter region is generally an NDR [10].

The ChIP-PCR results indicated that nucleosomes were generally depleted in two hotspots. The first was the region spanning the saRNA target site and TSS. In this region, nucleosomes in the SOX2 promoter were depleted and formed an NDR close to the TSS where the TATA box and CGI were located. Upon activation of OCT4 and

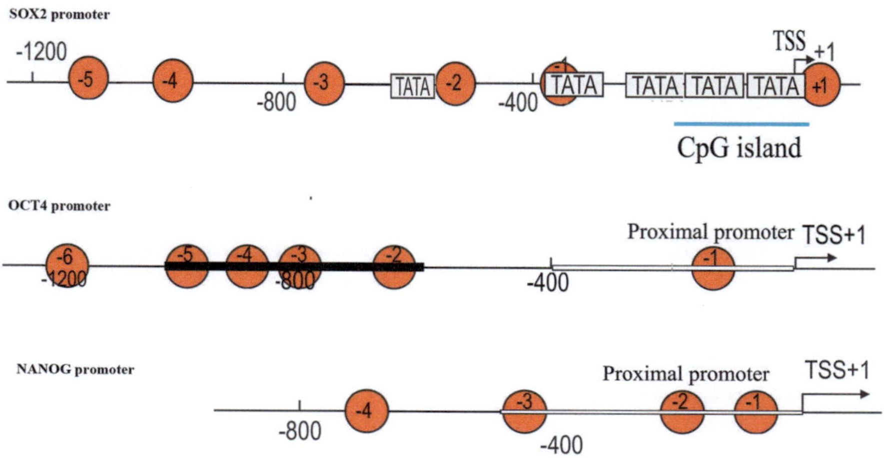

Fig. 4.2 Nucleosomal arrays with 195-bp periodicity assayed in the promoters of the SOX2, OCT4, and NANOG genes before RNAa as detected by ChIP-PCR

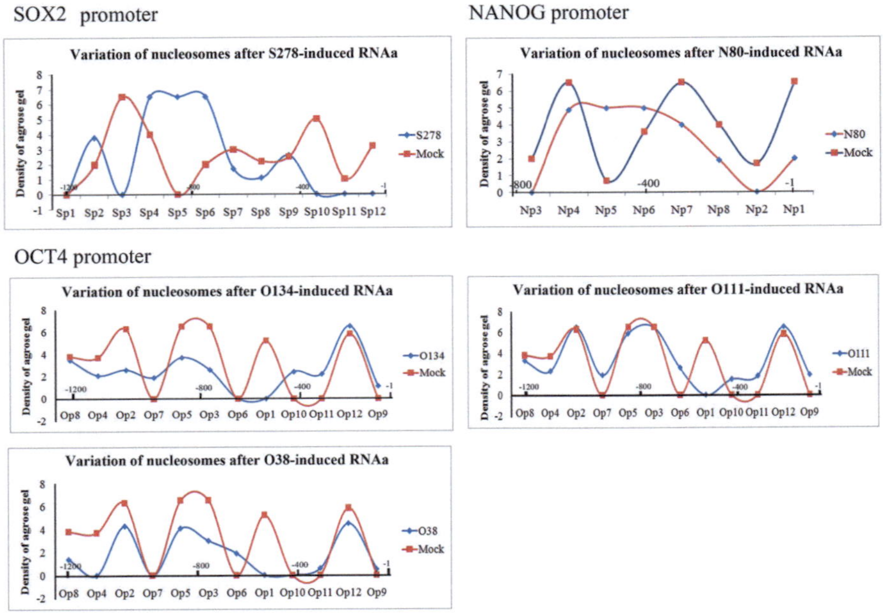

Fig. 4.3 Nucleosome position changes in the OCT4, SOX2, and NANOG promoters after RNAa. In the densitograms, the horizontal and vertical axes indicate promoter regions upstream of the TSS of activated genes and the density of the bands in agarose gels, as determined by ChIP-PCR, respectively. From Wang et al. [21]

NANOG promoters, quantitative, rather than all-or-nothing, changes in nucleosomes were observed. These also occurred in the first hotspot of the OCT4 promoter, where RNAa resulted in the loss of nucleosomes in a region spanning the proximal promoter and enhancer. An saRNA targeting the natural NDR near the TSS of OCT4 triggered nucleosome depletion at nucleotide position −300 to −500. In the NANOG promoter, an saRNA caused a reduction of nucleosomes (position −1 to −400) in the first hotspot, forming an NDR in the proximal promoter region (Fig. 4.3). The second hotspot was in the proximal enhancer in the promoter of OCT4; the resultant activation also caused decreased nucleosome binding in this region.

4.3 RNAPII Is Recruited to the Open Chromatin Structure in RNAa

Given the possible involvement of the open chromatin region in RNAPII recruitment, we investigated whether RNAa induced the recruitment of RNAPII to saRNA target sites by ChIP, using an anti-RNAPII antibody. The results showed that RNAPII binding was higher in the S278 target region. ChIP-PCR analysis of the 1-kb upstream region of the TSS revealed that RNAPII was widely distributed throughout the promoter before RNAa, but was bound to the proximal promoter region after RNAa [21]. After O38-induced OCT4 activation, RNAPII binding decreased at sites upstream of the O38 target site, but increased at the region between the O38 target site and the TSS in the proximal promoter region of OCT4. Similarly, RNAPII binding was also increased in the proximal promoter of NANOG. In general, proximal promoter regions facilitate the formation of transcription initiation complexes; thus, the recruitment of RNAPII to this region of the OCT4 promoter implies transcriptional activation. Distribution mapping of RNAPII recruitment and NDRs after RNAa revealed RNAPII enrichment at these sites in the OCT4, SOX2, and NANOG promoters, which showed nucleosome depletion (Fig. 4.4). These results indicate that RNAPII was recruited to NDRs targeted by saRNA, as also indicated by previous studies [7, 15].

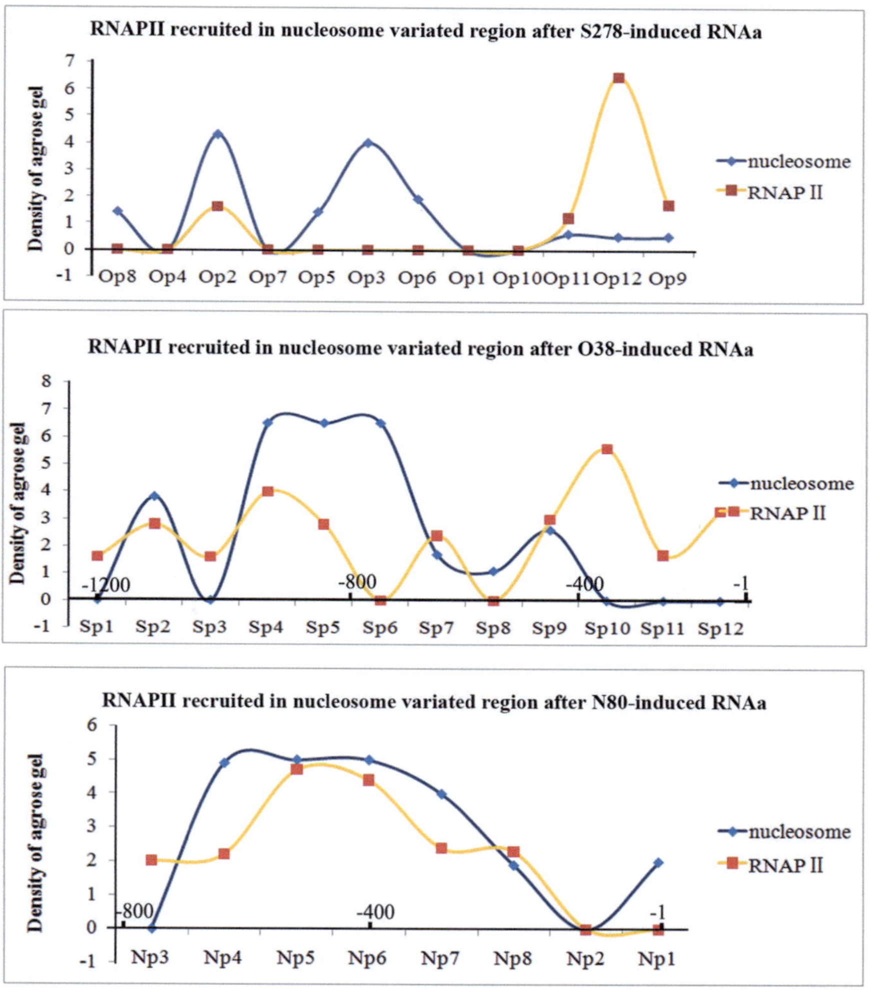

Fig. 4.4 Distribution mapping of RNAPII recruitment and NDRs in or near the saRNA target sites in the SOX2, OCT4, and NANOG promoters after RNAa

4.4 Potential Role of Cis-Acting Elements in RNAa

The NDR formed in the promoter of SOX2 contains a TATA box and a CGI in which the TSS is located. The promoters of OCT4 and NANOG contain a core promoter in the proximal region near the TSS; moreover, in the OCT4 promoter, an enhancer is distributed in the −604 to −1028-bp region relative to TSS where the NDR is induced by saRNA O134 [21].

The SOX2 promoter showed a 1000-fold activation by saRNA relative to the negative control dsRNA (Fig. 4.1a–d). In contrast, OCT4 and NANOG were activated by less than tenfold by their respective saRNAs; these genes contain a proximal promoter and enhancer. O38 targeting the proximal promoter of OCT4 induced a 3.26-fold activation, whereas O134 targeting the enhancer induced a 2.22-fold activation. Similarly, N80 induced a 3.2-fold activation of NANOG (Fig. 4.1e–l). These results suggest that TATA-box- and CGI-containing promoters are likely more susceptible to RNAa.

Although the TATA box is the main sequence recognized by RNAPII, CGIs can also recruit the PIC to promoters. They are also involved in nucleosome positioning, migration, and transcription factor binding [6]. In promoters containing a TATA box, the nucleosome at the +1 location was always located in the region crossed by PIC [20]; moreover, the TATA box in the mammalian β-globin gene promoter was found to be involved in nucleosome migration [4]. Similarly, CGIs play an important role in chromatin remodeling and modification. CGI- or GC-rich DNA preferentially recruits active chromatin-modifying CXXC domain proteins such as the Cfp1 subunit of the H3K4me3 methylase complex and KDM2A, a H3K36me2 demethylase [20].

Only 35% of genes in the human genome, most of which have a housekeeping function, contain a TATA box [22]. In contrast, 72% of genes contain CGIs [19]. CGIs selectively recruit chromatin, which is accompanied by loss of cysteine methylation, low levels of H1 histone, high levels of histone acetylation, and DNase I hypersensitivity [13]. In CGI-containing promoters, the region upstream of the TSS is always a nucleosome-depleted region (NDR). Nucleosome identification, positioning, transformation, and histone protein modification during gene activation differ between promoters with and without CGIs.

The proximal promoter is in the region upstream of the TSS through which the PIC passes; it is also the region of PIC anchoring and is effective as an saRNA target [9]. In eukaryotic cells, many cis-regulatory elements are distributed in the proximal promoters; in RNAPII transcriptional coding and noncoding genes, these elements determine the position and speed of the PIC [2]. Transcription factors bind to cis-regulatory elements in the NDR to activate nearby genes. For example, an NDR is present in the proximal promoter of the NANOG gene in the NCCIT human embryonic cell line, and transcription factor binding to this site induces NANOG gene expression.

In summary, the cis-regulatory elements present in the promoter not only facilitate formation of the NDR, but also determine the position and speed of RNAPII binding.

4.5 Mechanism of Small RNA Activation

When an saRNA binds to its complementary target site in the promoter, chromatin-modifying enzymes or the chromatin-remodeling complex is recruited, leading to nucleosome repositioning and NDR formation near the saRNA targeting region. However, it is unclear how NDRs influence gene activation. Here, we proposed a model in which saRNAs interplay with some cis-regulatory elements such as the TATA box, proximal promoter, and CGIs, and induce the formation of NDRs near their target sites. Normally, the nucleosomes are distributed every 195 bp upstream of TSS; however, sites with some cis-regulatory elements can be induced to form NDR, which is named as short-locus open chromatin [5]. When an saRNA targets the appropriate site in the promoter, the short-locus open chromatin can be extended to large open chromatin (Fig. 4.5). The formation of an extended open chromatin structure involves removal of the nucleosome barrier between the RNAPII-binding site and TSS, which enables rapid assembly and function of the PIC. Rapid activation of the PIC can lead to transcriptional activation and upregulation of target gene expression [3, 14]. If the regulatory element is an enhancer, binding of saRNAs could lead to the recruitment of trans-acting factors and consequently to increased gene expression.

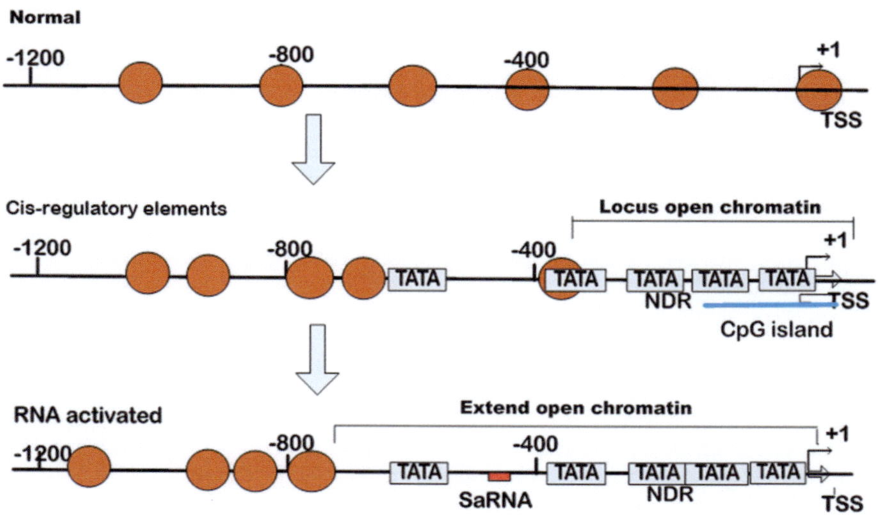

Fig. 4.5 Proposed mechanism model for RNAa. saRNAs interplay with certain cis-regulatory elements such as the TATA box, proximal promoter, and CGIs, and induce the formation of NDRs near their target sites. Circles represent nucleosomes and horizontal axes indicate promoter regions upstream of the TSS

References

1. Britten RJ, Davidson EH (1969) Gene regulation for higher cells: a theory. Science 165 (3891):349–357
2. Dikstein R (2012) The unexpected traits associated with core promoter elements. Transcription 2(5):201–206
3. Floer M, Wang X, Prabhu V, Berrozpe G, Narayan S, Dan S, Alvarez D, Kendall J, Krasnitz A, Stepansky A (2010) A RSC/nucleosome complex determines chromatin architecture and facilitates activator binding. Cell 141(3):407–418
4. Gui CY, Dean A (2003) A major role for the TATA box in recruitment of chromatin modifying complexes to a globin gene promoter. Proc Natl Acad Sci U S A 100(12):7009–7014
5. Higgs DR (1998) Do lcrs open chromatin domains? Cell 95(3):299–302
6. Hornung G, Barziv R, Rosin D, Tokuriki N, Tawfik DS, Oren M, Barkai N (2012) Noise-mean relationship in mutated promoters. Genome Res 22(12):2409–2417
7. Hu J, Chen Z, Xia D, Wu J, Xu H, Ye ZQ (2012) Promoter-associated small double-stranded RNA interacts with heterogeneous nuclear ribonucleoprotein A2/B1 to induce transcriptional activation. Biochem J 447(3):407–416
8. Huang V, Qin Y, Wang J, Wang X, Place RF, Lin G, Lue TF, Li LC (2010) RNAa is conserved in mammalian cells. PLoS One 5(1):e8848
9. Janowski BA, Younger ST, Hardy DB, Ram R, Huffman KE, Corey DR (2007) Activating gene expression in mammalian cells with promoter-targeted duplex RNAs. Nat Chem Biol 3 (3):166–173
10. Kharchenko PV, Woo CJ, Tolstorukov MY et al (2008) Nucleosome positioning in human HOX gene clusters. Genome Res 18(10):1554–1561
11. Kuwabara T, Hsieh J, Nakashima K, Taira K, Gage FH (2004) A small modulatory dsRNA specifies the fate of adult neural stem cells. Cell 116(6):779–793
12. Li LC, Okino ST, Zhao H, Pookot D, Place RF, Urakami S, Enokida H, Dahiya R (2006) Small dsRNAs induce transcriptional activation in human cells. Proc Natl Acad Sci U S A 103 (46):17337–17342. doi:0607015103 [pii]10.1073/pnas.0607015103
13. Lin JC, Jeong S, Liang G, Takai D, Fatemi M, Tsai YC, Egger G, Gal-Yam EN, Jones PA (2007) Role of nucleosomal occupancy in the epigenetic silencing of the MLH1 CpG island. Cancer Cell 12(5):432–444
14. Lu B, Morozov AV (2011) Gene regulation by nucleosome positioning. Trends Genet 26 (11):476–483
15. Mahajan MC, Weissman SM (2005) Heterogeneous nuclear ribonucleoprotein C1/C2, MeCP1, and SWI/SNF form a chromatin remodeling complex at the -globin locus control region. Proc Natl Acad Sci U S A 102(42):15012–15017
16. Place RF, Li LC, Pookot D, Noonan EJ, Dahiya R (2008) MicroRNA-373 induces expression of genes with complementary promoter sequences. Proc Natl Acad Sci U S A 105 (5):1608–1613
17. Portnoy V, Huang V, Place RF, Li LC (2011) Small RNA and transcriptional upregulation. Wiley Interdisciplinary Rev-RNA 2(5):748–760
18. Saito Y, Saito H, Liang G, Friedman JM (2014) Epigenetic alterations and microRNA misexpression in cancer and autoimmune diseases: a critical review. Clin Rev Allergy Immunol 47(2):128–135
19. Sandelin A, Carninci P, Lenhard B, Ponjavic J, Hayashizaki Y, Hume DA (2007) Mammalian RNA polymerase II core promoters: insights from genome-wide studies. Nat Rev Genet 8 (6):424–436
20. Vavouri T, Lehner B (2012) Human genes with CpG island promoters have a distinct transcription-associated chromatin organization. Genome Biol 13(11):212–220
21. Wang B, Sun J, Shi J, Guo Q, Tong X, Zhang J, Hu N, Hu Y (2016) Small-activating RNA can change nucleosome positioning in human fibroblasts. J Biomol Screen 21(6):634–642

22. Yang C, Bolotin E, Jiang T, Sladek FM, Martinez E (2007) Prevalence of the initiator over the TATA box in human and yeast genes and identification of DNA motifs enriched in human TATA-less core promoters. Gene 389(1):52–65. doi:S0378-1119(06)00623-8 [pii]10.1016/j.gene.2006.09.029
23. Zhang LT, Wang YG, Liu XY (2008) RNAa: a new contra-standard model for ncRNA regulating gene expression. Chinese J Cell Biol 30(3):281–286

Part II
Endogenous RNAa

Chapter 5
Endogenous miRNAa: miRNA-Mediated Gene Upregulation

Vera Huang

Abstract The phenomenon of RNA activation (RNAa) was initially discovered by Li and colleagues about a decade ago. Subsequently, gene activation by exogenously expressed small activating RNA has been demonstrated in different cellular contexts by a number of laboratories. Conceivably, endogenously expressed microRNAs may also utilize RNA activation as a cellular mechanism for gene regulation, which may be dysregulated in disease states such as cancer. RNA activation can be applied to gain-of-function studies and holds great promise for disease intervention. This chapter will discuss examples of promoter-targeting microRNAs discovered in recent years and their pathophysiological relevance. I will also briefly touch upon other novel classes of microRNAs with positive gene regulatory roles, including TATA-box-activating microRNAs and enhancer-associated microRNAs.

Keywords Argonautes • miRNAa • Noncoding RNA • Promoter-targeting miRNAs • RNAa • Transcriptional activation

5.1 Introduction

High-throughput sequencing technologies and computational methods have revealed that the vast majority of the mammalian genome is pervasively transcribed, but a large percentage of transcripts lack protein-encoding capability [12, 17]. It is becoming evident that these noncoding RNAs (ncRNAs) have a huge impact on gene regulation at different levels. Many types of ncRNAs have been identified, including microRNAs (miRNAs), long noncoding RNAs (lncRNAs), Piwi-interacting RNAs (piRNAs), enhancer RNAs (eRNAs), and extracellular RNAs (exRNAs). miRNAs, which are typically 20–24 nucleotides long, belong to a major functional class of short noncoding RNAs which control gene expression at the posttranscriptional level via translational repression and/or messenger RNA (mRNA) degradation. It is estimated that miRNAs constitute nearly 1% of all

V. Huang (✉)
Molecular Stethoscope, Inc., 10835 Road to the Cure, Suite 100, San Diego, CA 92121, USA
e-mail: vhuang2013@gmail.com

L.-C. Li (ed.), *RNA Activation*, Advances in Experimental Medicine and Biology 983, DOI 10.1007/978-981-10-4310-9_5

predicted genes in mammalian genomes, and approximately >60% of all mammalian protein-coding genes can be regulated by miRNAs [1]. miRNAs are important regulators of developmental and differentiation processes and play key roles in many diseases [38]. To date, more than 2000 human miRNAs have been identified and cataloged in miRbase [6]. It is commonly accepted that miRNAs exert negative regulatory effects on gene expression by binding to the 3′ untranslated regions (UTR) in mRNA via partial complementarity with their seed sequence, which results in posttranscriptional gene silencing. In addition to 3′UTR targeting, miRNAs can also enhance protein translation during amino acid starvation by targeting 5′UTR of mRNAs encoding ribosomal proteins [31]. As another nonclassical example, miR-483-5p, embedded within the IGF2 gene, induces the transcription of its host gene by binding to the 5′UTR region [23].

miRNA biogenesis consists of multiple steps of processing involving both nuclear and cytoplasmic compartments. miRNAs encoded in different regions of the genome are typically transcribed by RNA polymerase II (RNAPII) to form primary miRNAs, or pri-miRNAs. Pri-miRNAs are processed by the Drosha–DGCR8 complex to approximately 70 nt pre-miRNA hairpin structures in the nucleus and subsequently exported to the cytoplasm by an Exportin-5-mediated mechanism where they are further processed by the Dicer–TRBP complex to form the mature 22 nt miRNA duplexes. In the cytoplasm, the miRNA duplex is unwound and the passenger strand is released and discarded. The mature single-stranded miRNA is then incorporated into the miRNA-containing RNA-induced silencing complex (miRISC), including Argonaute proteins (Ago) as the core component. In the classical RNA interference (RNAi) pathway, the mature miRNA in the complex guides the miRISC to the target mRNAs to downregulate gene expression at the posttranscriptional level. Although miRNAs are predominantly regarded as suppressors of gene expression within the cytoplasm, emerging evidence has shown that RNAi factors are present in the mammalian cell nuclei [5, 11] and a subset of miRNAs are enriched in the nucleus [22]. Ago proteins have also been shown to interact with the RNAPII core transcription machinery, implying that miRNAs may have a transcriptional role and beyond [2, 11]. Promoter-targeted double-stranded duplex RNA (dsRNAs) or miRNAs have both positive and negative regulation on gene expression at the transcriptional/epigenetic level [7], through a process called RNA activation (RNAa) [13, 20] or transcriptional gene silencing [18, 29], respectively.

The discovery of RNAa was serendipitous when Li et al. attempted to knock down E-cadherin in human cells using synthetic dsRNAs and intended to test whether they could be used to induce DNA methylation. To their surprise, they noticed the transcript level of E-cadherin was increased. Although they did not observe the expected change in DNA methylation at the target sites, loss of H3K9me2 and H3K9me3 was noted at the E-cadherin promoter [20]. Shortly after, Janowski and colleagues independently found that the expression of the progesterone receptor (PR) gene could be activated by specific dsRNAs targeting the promoter and downstream of its 3′UTR [13]. To date, more than a handful of genes have been discovered to be susceptible to RNAa, which include Ccnb1,

CDH1, CDKN1A, CSDC2, DPYSL3, KLF4, LDL4, NANOG, NKX3.1, OCT4, p53, PR, and VEGF [9, 13, 20, 21, 24, 26, 37, 44, 45, 47]. The phenomenon of RNAa was proven to be a sequence-dependent gene regulation mechanism evolutionarily conserved from *C. elegans* to mammals, including mice, rats, and nonhuman primates [9]. Synthetically designed dsRNAs capable of turning on gene expression are termed small/short activation RNAs (saRNAs), or antigene RNAs (agRNAs). Over the past few years, saRNAs have greatly facilitated gain-of-function studies by selectively activating genes of interest. As a common theme, RNAa-mediated gene activation seems to depend on Ago protein and correlates with activating chromatin modification at the saRNA-targeted site. In terms of its targeting mechanism(s), saRNA can either act on promoter DNA or nascent transcripts as there is literature to support both cases [28, 35, 39, 52].

Following the discovery of saRNA-mediated gene activation, endogenous miRNAs have also been shown to elicit a similar phenomenon called miRNA-induced RNAa (miRNAa) (reviewed in [7]). In fact, 2 years prior to the discovery of RNAa, another group had identified an endogenous RNA duplex that was shown to transcriptionally upregulate a set of neuronal genes to induce neural differentiation [19]. Early studies related to RNAa had been largely overlooked due to lack of understanding of its mechanism. However, accumulating experimental evidence suggests that RNAa has a mechanism distinct from RNAi that is not entirely due to secondary effects from gene-silencing mechanisms (e.g., transcriptional derepression) [28, 35]. In addition, a number of studies have shown that activation by saRNA occurs without the cleavage of promoter transcripts [39], suggesting that the RNA serves as a scaffold to recruit other protein components to the promoter. However, the physiological relevance of miRNAa had been questionable and underappreciated due to the lack of in vivo evidence. In 2014, Turner et al. reported the first in vivo evidence of RNAa by an endogenous miRNA, lin4, which activates its own transcription through an autoregulatory mechanism by directly binding to its own promoter [42]. In this chapter, I will specifically discuss representative examples of mammalian promoter-targeting miRNAs, their potential targeting mechanism, and physiological relevance in diseases (summarized in Table 5.1).

5.2 Promoter-Targeting miRNAs

5.2.1 miR-373

miR-373 was initially identified as an embryonic stem cell-specific miRNA and was subsequently validated as an oncomiR by having pleiotropic functions in tumorigenesis [40]. In 2008, miR-373 was the first endogenous miRNA identified to exert positive regulation *in trans* by activating the expression of E-cadherin and cold shock domain-containing protein c2 (CSDC2) by binding to complementary sequences in the promoter region [33]. Thus, miR-373 served as the first proof-

Table 5.1 Examples of promoter-targeting miRNAs

miRNA	Target region (relative to TSS[a])	Physiological consequence	Reference
Hsa-miR-373	CSDC2 (−787/−763)	n/a	[33]
	CDH1(−645/−622)		
Mmu-miR-744	Ccnb1 (−192/−171)	Inhibited cell proliferation in vitro and induced chromosomal instability in a mouse xenograft tumor model	[10]
Mmu-miR-1186	Ccnb1 (−699/−678)		
Hsa-miR-324-3p	RelA (−66/−45)	Induced apoptosis in neuronal cell lines	[4]
Hsa-miR-589	COX-2 (−53/−37)	n/a	[27]
	COX-2 (−34/−13)		
	PLA2G4F[b]		
Hsa-miR-370	CDKN1A (−552/−537)	Inhibited cell proliferation and induced apoptosis and senescence in bladder and renal cancer cells	[46]
Hsa-miR-1180	CDKN1A (−397/−379)		[49]
Hsa-miR-1236	CDKN1A (−243/−226)		
Hsa-miR-6734	CDKN1A (−322/−304)	Inhibited cell proliferation and induced cell cycle arrest in colon cancer cell lines	[16]
Hsa-miR-551-3p	STAT3 (−530/−503)	Reduced apoptosis and increased proliferation in 2D and 3D cell cultures and increased ascites formation in animal models	[3]

[a]TSS: transcription start site
[b]Coactivated by miR-589

of-principle study to demonstrate that miRNAs complementary to promoter sequences are capable of activating transcription in human cells. Place et al. showed that transfection of a mature miR-373 mimic or its precursor (pre-miR-373) into prostate cancer cell lines induced E-cadherin and CSDC2 expression, and enrichment of RNAPII was detected at both gene promoters. This study spurred further interest in the physiological context of miRNAa and raised several key questions: (i) Is miRNAa naturally exploited by cells for physiological functions? (ii) What are the phenotypic consequences and functional significance of promoter-targeting miRNAs in diseases?

5.2.2 *miR-744*

Similar to protein-coding genes, miRNAs have been known to be involved in cancer by acting either as tumor suppressors or oncogenes. Conceivably, endogenous miRNAs may be involved in tumorigenesis via nuclear functions that impact gene transcription and epigenetic states. In 2012, Huang et al. identified two oncomiRs (miR-744 and miR-1186) in mouse cell lines that are able to promote cell cycle progression and cause chromosomal instability by inducing the expression of Cyclin B1 (Ccnb1) [10]. Initially, the authors set out to identify genes that are potentially activated by miRNAs in a physiological context. Interestingly, they found that the expression of mouse Ccnb1 is partially dependent on the miRNA biogenesis and processing machinery (i.e., Drosha, Dicer, Ago1, and Ago2). The expression of Ccnb1 was reduced when Drosha, Dicer, or Ago1 was depleted by RNAi; conversely, the expression of Ccnb1 was increased upon ectopic overexpression of Ago1. These results indicated that Ccnb1 transcription may be positively regulated by an miRNA-dependent gene regulation. By performing in silico analysis of Ccnb1 promoter, miR-744 and miR-1186 were identified as the top candidates containing seed sequences highly complementary to the Ccnb1 promoter. These miRNAs possess significant sequence homology with the Ccnb1 promoter both in their seed and extended regions, but lack putative target sites in either the 3′ or 5′UTR of the Ccnb1 transcript.

To understand the phenotypic consequences and functional significance of promoter-targeting miRNAs, subsequent studies were focused on miR-744 using mouse cell lines and xenograft models. miR-744, located in chromosome 17, targets position-192 in the Ccnb1 promoter relative to the transcription start site (TSS) with an overall sequence identity of 93%. Transient overexpression of either mature or precursor miR-744 induced Ccnb1 expression, while knockdown of endogenous pre-miR-744 resulted in a decreased level of Ccnb1 transcript. As commonly seen in other genes, Ccnb1 activation was accompanied by increased enrichment of Ago1, RNAPII, as well as trimethylated H3K4 (H3K4me3) at the Ccnb1 promoter, a classic histone marker of active transcription. miR-744 plays a positive role in promoting cell growth in mouse cell lines since overexpression of miR-744 enhanced 2D cell proliferation. Given that overexpression of Ccnb1 can result in chromosome instability, the authors asked whether prolonged expression of miR-744 affected cytogenetic profiles of prostate carcinoma cell lines derived from the TRAMP (transgenic adenocarcinoma of mouse prostate) mouse model. As expected, prolonged expression of miR-744 induced cytogenetic changes in TRAMP C1 cells in a manner similar to Ccnb1 overexpression.

Ccnb1 is a key cell cycle regulator known to promote tumor growth in various cancer types [43, 55]. To test whether Ccnb1 activation by miR-744 and miR-1186 has any physiological consequences in modulating tumorigenicity, TRAMP C1 cells infected with lentiviral constructs encoding miR-744 or miR-1186 were inoculated subcutaneously into immunodeficient mice. Consistent with the proliferation data, short-term expression of miR-744 and miR-1186 promoted in vivo tumor growth through Ccnb1 activation. Unexpectedly, prolonged activation of Ccnb1 via activation of miR-744 and miR-1186 seemed to inhibit tumor growth in TRAMP C1 cells, likely due to chromosomal instability. The authors speculated that the cytogenetic aberration resulting from Ccnb1 activation may lead to loss of tumorigenic feature and inhibit tumor growth, since high levels of genome instability have been reported to have a tumor-suppressive effect [50]. Karyotypic instability is one of the features of human prostate cancer cell lines. Similarly, chromosomal abnormalities including polyploidy, aneuploidy, and other structural rearrangements were also found in mouse prostate cancer cells. Huang et al. showed that Ccnb1 overexpression appeared to enhance chromosomal instability in TRAMP C1 cells by promoting chromosomal alteration. Overexpression of miR-744 and miR-1186 also led to similar chromosomal irregularities, suggesting that they might impair chromosome segregation via activation of Ccnb1 and render cancer cells more susceptible to cytogenetic alterations over time.

Since the seminal discovery of miRNAa, this is the first study where a physiologically relevant role for miRNA-mediated gene activation was revealed. We speculate that endogenous miRNAs may facilitate RNAa in cancer to fine-tune gene expression and manipulate cancer cell phenotype. Further study is needed to identify additional examples and elucidate mechanism(s) by which miRNAs directly activate gene expression and oncogenic pathways. In a follow-up study [8, 11], genome-wide analysis of potential miRNA binding in human prostate cancer cells further revealed that Ago1 interacts with the transcriptional machinery to sustain the expression of hundreds of genes which are important for cell proliferation, evading apoptosis, angiogenesis, and DNA damage repair. These findings support that miRNAs could be implicated in tumorigenesis through a noncanonical miRNA-targeting mechanism.

5.2.3 miR-589

miR-589 is another endogenous miRNA shown to induce cyclooxygenase-2 (COX2), a critical regulator of inflammation, at the transcriptional level in human cancer cells. Matsui et al. identified RNA transcripts that overlap the COX2 promoter and contain two adjacent binding sites for miR-589 [27]. Mechanistically, miR-589 activates COX2 by binding to the promoter RNAs derived from the COX2 promoter. Transfection of anti-miR-589-5p resulted in reduced basal expression of COX2. Conversely, transfection of miR-589 mimic resulted in increased levels of COX2. These results suggested that COX2 expression is partly regulated by miR-589-5p. Enrichment of RNAPII at the COX2 promoter was also observed, implying that the gene regulation occurs at the transcriptional level. The authors also showed biochemically that RNAi factors Ago2 and GW182 associate in the nucleus and are required for transcriptional activation of the COX2 gene by miR-589. The authors proposed that the RNA transcripts at the COX2 promoter act as a scaffold for small RNA-mediated recruitment of Ago2 and GW182 near the COX2 promoter *in cis*. Intriguingly, the authors also found that miR-589 can also activate another gene in the same pathway, PLA2G4A, which is 149 kb away from COX2. By performing chromatin conformation analysis (3C), the authors showed that there is a direct physical interaction between the PLA2G4A and COX2 promoters through a looping mechanism. In conclusion, miRNAa provides a mechanism that allows coordinated expression of two key enzymes within the same inflammatory pathway. It would be interesting to explore the physiological relevance of miR-589 in inflammation.

5.2.4 *miR-324-3p*

Dharap et al. demonstrated the relevance of RelA gene regulation by a promoter-targeting RNA following cerebral ischemia [4]. By performing genome-wide search of miRNA binding sites within the putative promoters of the whole rat genome, miR-324-3p and RelA (p65, a subunit of NF-κB) came up as one of the strongest miRNA: promoter interactions. Transfection of pre- or mature miR-324-3p into PC12 cells derived from a pheochromocytoma of the rat adrenal medulla induces the expression RelA at both the mRNA and protein level in an Ago2-dependent manner. Phenotypically, induction of RelA by miR-324-3p induced apoptosis in PC12 cells as shown by increased levels of cleaved caspase 3, suggesting that miR-324-3p might be responsible in part for the induction of RelA after ischemia. By chromatin immunoprecipitation experiments, the authors showed that miR-324-3p targets the RelA promoter within 100b of the TSS. It is plausible to speculate that Ago2–miR-324-3p complex might recruit chromatin modifiers (e.g., EZH1) to establish activating chromatin signatures to activate transcription at the TSS. This study indicates that miRNAa-based therapeutics can be applied to control inflammation-induced injury and cell death in neurons.

5.2.5 *miR-551b-3p*

In a recent report, Chaluvally-Raghvan et al. [3] identified miR-551b-3p as an oncomiR, which confers growth advantage to ovarian cancer cells through transcriptional activation of STAT3 via RNAa. miR-551b-3p is located in the 3q26.2 locus, a region frequently amplified in high-grade serous epithelial ovarian cancer (HGSEOC), which is the most common and aggressive form of ovarian cancer. miR-551b-3p is the key mature form derived from pre-miR-551b and is highly expressed in both ovarian tumor samples and cancer cell lines, which also correlates with poor patient outcome. 3q26.2 amplicon leads to increased miR-551b-3p expression among other oncogenes and contributes to cell proliferation and survival.

Overexpression of miR-551-3p in ovarian cancer cell lines resulted in reduced apoptosis and increased the number and size of spheroids in 3D culture conditions. miR-551b-3p also increased proliferation of a number of ovarian and breast cancer cell lines in 2D adherent culture conditions. Consistent with the in vitro cell proliferation data, intraperitoneal injection of miR-551b-3p into nude mice showed a significant increase in the number of tumor nodules and tumor weight, as well as ascite formation, which is one of the important features of HGSEOC.

STAT3 was one of the computationally predicted targets of miR-551b-3p. Upon overexpression of miR-551b-3p, the authors showed that both STAT3 mRNA and protein levels were upregulated. Conversely, overexpression of anti-miR-551b decreased STAT3 levels in cell lines with high endogenous level of miR-551b-3p. Mechanistically, the authors showed that miR-551b-3p binds to a complementary sequence in the loop of the STAT3 promoter, which in turn recruits RNAPII and the TWIST1 transcription factor to activate STAT3 transcription. The loop structures in the STAT3 promoter are critical as deletion of the loop structure or sequences nearby abolished the effects of miR-551b-3p on STAT3 promoter-induced transcriptional activity. Alternatively, miR-551b-3p can also regulate mRNA stability or translation through canonical mechanisms involving binding to 3′UTR of mRNAs. These effects of miR-551b-3p could, in turn, indirectly regulate the action of miR-551b-3p on the STAT3 promoter. Further studies are required to determine whether miR-551b-3p alters transcriptional activation of other genes that along with STAT3 combine to mediate the effects of miR-551b-3p on ovarian cancer pathophysiology. As one of the most important highlights of the paper, the authors explored the therapeutic utility of miR-551b-3p as a therapeutic target. They demonstrated that liposomal delivery of anti-miR-551b reduced tumor burden and downregulated STAT3 in nude mice bearing tumors in the peritoneal cavity. The anti-miR approach against miR-551b-3p may be a good strategy for treating human ovarian cancers harboring the 3q26.2 amplification.

5.2.6 miR-1236 and miR-6734

p21 (CDKN1A) is a well-characterized model gene for RNAa studies [15, 20, 35]. p21, a transcriptional target of p53, is a cyclin-dependent kinase inhibitor and functions to inhibit cell cycle progression at G1 and S phase. In addition to growth arrest, p21 also mediates cellular senescence. A number of promoter-targeting miRNAs have been identified for p21 [46, 49]. In a 2014 study by Wang and colleagues, the authors identified three promoter-targeting miRNAs (miR-370, miR-1180, and miR-1236) with complementary binding sites in the p21 promoter region. Overexpression of these three miRNAs in bladder cancer cells readily induced p21 gene expression by targeting the predicted sites in the p21 promoter and led to potent inhibition of cell migration. In a subsequent study by Wang and colleagues [49], they confirmed that the overexpression of miR-1236 induced the expression of p21, which resulted in decreased expression of CDK4/6 and Cyclin D1 and inhibited cell proliferation in renal carcinoma cells. They also found that in tumor samples, low expression of p21 and miR-1236 was associated with poor survival in renal carcinoma patients. Thus, overexpression of miR-1236 via RNAa may offer a potential therapeutic opportunity for renal cancer. In the most recent study, Kang et al. identified another p21-targeting miRNA [16]. They found that overexpression of miR-6734 upregulated p21 expression at mRNA and protein levels, and chromatin immunoprecipitation analysis confirmed the association of miR-6734 with the p21 promoter. Transient overexpression of miR-6734 resulted in cell cycle arrest and apoptosis in colon cancer cell lines. Collectively, these findings support the notion that miRNAa may be an endogenous mechanism exploited by cancer cells to gain growth advantage during tumorigenesis.

5.2.7 TATA-Box-Binding miRNAs

There are also a few examples of cellular miRNAs upregulating transcription via interaction with TATA-box motifs within 50 bp upstream of the TSS. In 2014, Zhang et al. reported that let-7i binds to the TATA-box motif of the interleukin-2 (IL-2) gene and upregulates IL-2 mRNA and protein production in both human and mouse $CD4^+$ T-lymphocytes [54]. Unlike the classical RNAa, histone methylation changes seemed to be dispensable for the IL-2 activation. Nascent transcripts derived from the IL-2 promoter region were not responsible for the upregulation of IL-2 promoter activity induced by let-7i. The authors proposed that as TATA-binding protein (TBP) binds to the TATA-box motif and unwinds the DNA, small RNA base pairs with its single-stranded DNA target, which stabilizes the preinitiation complex (PIC) to improve the processivity of RNAPII. Alternatively, small RNAs might interact with the promoter DNA, which further recruit TBP and other general transcription factors to the core promoter to facilitate the assembly of PIC to accelerate transcription initiation. A similar mechanism has been previously reported in viruses. HIV-1-encoded miRNA miR-H3 enhances viral production by targeting the TATA box in HIV-1 5′ LTR to enhance the promoter activity [53]. Taken together, these findings reveal a novel role for miRNAs in regulating viral replication through transcriptional activation.

5.3 Enhancer-Activating miRNAs

Recently, a subset of miRNAs termed nuclear activating miRNAs or namiRNAs have been reported. namiRNA loci are usually associated with enhancers in genomic regions that are often enriched with H3K27ac modification. Interestingly, H3K27ac-enriched miRNAs (e.g., miR-24-1, miR-26a-1) tend to be present in the nucleus and upon their overexpression result in activation of their neighboring genes. By using the TALEN (transcription activator-like effector nucleases) genome-editing approach, the authors showed that transcriptional activation of miR-24-1 neighboring genes is dependent on the presence of the intact enhancer sequence as deletion of the enhancer sequence abolished its activity. Apart from regulating local gene expression, namiRNAs can also activate global transcription by upregulating eRNA expression, increasing the enrichment of p300, RNAPII, as well as Ago2 at the enhancer locus. The biological functions of namiRNAs remain to be further explored.

5.4 Concluding Remarks and Perspectives

It turns out that miRNAs are more than just suppressors of gene expression. Since the discovery of RNAa in 2006, there have been numerous reports on miRNAa in different cellular contexts via a myriad of mechanisms which are not completely understood. Based on the work from the Li and Corey laboratories, miRNAa requires either Ago1 or Ago2 and recruitment of RNAPII, and it is often associated with chromatin modifications at the target site [10, 11, 14, 39]. However, it remains uncertain whether such changes are the cause or consequence of transcriptional activation. Some of the features of RNAa (e.g., distinct kinetics and prolonged duration of activation) can be attributed to epigenetic changes. There are several lines of evidence that suggest that RNAa-induced transcriptional activation is driven by chromatin remodeling. A recent study by Wang et al. proposed that saRNA induces nucleosome positioning through the formation of open chromatin, allowing RNAPII and other transcription factors to bind and initiate transcription [48]. Moreover, namiRNAs activate gene transcription through chromatin remodeling at the enhancer region [51]. Apart from the involvment of epigenetic changes in RNAa, how the miRNAs interact with the gene promoter remains an open question. Several miRNA binding models have been proposed. As one of the possible mechanisms, miRNAs can bind directly to promoter DNA through Watson–Crick base pairing [30] or triple helix formation [41]. Early work from fission yeast showed that a heterochromatic ncRNA, transcribed from both strands of centromeric *dh* repeats, was associated with chromatin via the formation of a DNA–RNA hybrid to provide a platform for the RNAi-directed heterochromatin assembly [30]. Recently, computational analysis showed that mammalian and nonmammalian genomes are enriched for strong miRNA triplex binding sites. In primary leukemia cells, there is a positive correlation between gene upregulation and the overrepresentation of miRNA triplex binding sites [32]. In other words, genes containing miRNA triplex binding sites are more frequently associated with increased gene expression. By using biochemical and biophysical approaches, they also provided multiple lines of physical evidence of heterotriplex formation involving miRNA and duplex DNA, suggesting that this could be one of the mechanisms by which miRNA may promote gene transcription. Thus, miRNAa may also involve formation of RNA–DNA hybrid to form a transcriptional activation platform. As another alternative mechanism, miRNAs may also regulate transcription by binding to nascent transcripts in proximity to gene promoters. Several promoter transcripts have been implicated in RNAa, including PR [39], COX2 [27], and LDLR [26]. More work is needed to further dissect the precise molecular mechanisms by which miRNAs activate transcription.

There has been a tremendous effort in utilizing RNAi-based approaches to treat diseases in recent years [25]. miRNA-based approaches are appealing given that miRNAs can modulate multiple genes in the same disease pathway. From a therapeutic standpoint, the ability to turn on genes would be clinically beneficial to activate targets that are otherwise undruggable by other means. MTL-CEBPA lipid nanoparticle is the first saRNA clinical candidate developed against CEBPA by Mina Therapeutics in 2016 and is currently being evaluated in a Phase I study in patients with liver cancer [36]. miRNAs discussed in this chapter may represent promising targets for therapeutic intervention in various disease indications, many of which may also have diagnostic and prognostic values. For example, nanoliposome-based anti-miRNA therapies against miR-551-3p warrant evaluation as an approach to improve outcomes for ovarian cancer patients. Previous work has shown that lipidoid-encapsulated nanoparticle delivery of p21 saRNA is efficacious at activating p21 in mouse xenografts [34] and inhibiting bladder tumor growth in an orthotropic model [15]. Alternatively, selective modulation of p21 expression by overexpressing miR-6734 might be another potential option for the treatment for colon cancer. miRNAa-based therapeutics can also be used to control pathological consequences of inflammation-induced injury in the brain by overexpression of inflammation-associated miRNAs. Many lessons can be learned from *C. elegans*. Turner et al. [42] showed that miRNAa is not limited to protein-coding genes, and may apply to the thousands of noncoding RNAs throughout the genome. Therefore, RNAa-based technology can also be used to activate naturally occurring miRNAs or ncRNAs for treating certain types of diseases, including cancer. For example, RNAa can facilitate miRNA replacement therapy to restore tumor suppressor miRNA levels to inhibit tumor growth. Like other nucleic acid therapeutics, miRNAa-based therapeutics will face some common obstacles, including stability, delivery, and undesired off-target effect and adverse immune responses. Regardless, it would be a truly exciting time in the small RNA history as more RNAa-based research is making its way into clinics in the imminent future.

References

1. Bartel DP (2009) MicroRNAs: target recognition and regulatory functions. Cell 136 (2):215–233. doi:S0092-8674(09)00008-7 [pii] 10.1016/j.cell.2009.01.002
2. Cernilogar FM, Burroughs AM, Lanzuolo C, Breiling A, Imhof A, Orlando V (2013) RNA-interference components are dispensable for transcriptional silencing of the drosophila bithorax-complex. PLoS One 8(6):e65740. doi:10.1371/journal.pone.0065740 PONE-D-12-39913 [pii]
3. Chaluvally-Raghavan P, Jeong KJ, Pradeep S, Silva AM, Yu S, Liu W, Moss T, Rodriguez-Aguayo C, Zhang D, Ram P, Liu J, Lu Y, Lopez-Berestein G, Calin GA, Sood AK, Mills GB (2016) Direct upregulation of STAT3 by MicroRNA-551b-3p deregulates growth and metastasis of ovarian cancer. Cell Rep 15(7):1493–1504. doi:S2211-1247(16)30447-8 [pii] 10.1016/j.celrep.2016.04.034
4. Dharap A, Pokrzywa C, Murali S, Pandi G, Vemuganti R (2013) MicroRNA miR-324-3p induces promoter-mediated expression of RelA gene. PLoS One 8(11):e79467. doi:10.1371/journal.pone.0079467 PONE-D-13-28574 [pii]

5. Gagnon KT, Li L, Chu Y, Janowski BA, Corey DR (2014) RNAi factors are present and active in human cell nuclei. Cell Rep 6(1):211–221. doi:S2211-1247(13)00758-4 [pii] 10.1016/j.celrep.2013.12.013
6. Griffiths-Jones S, Grocock RJ, van Dongen S, Bateman A, Enright AJ (2006) miRBase: microRNA sequences, targets and gene nomenclature. Nucleic Acids Res 34 (Database issue):D140–144. doi:34/suppl_1/D140 [pii] 10.1093/nar/gkj112
7. Huang V, Li LC (2012) miRNA goes nuclear. RNA Biol 9(3):269–273. doi:19354 [pii] 10.4161/rna.19354
8. Huang V, Li LC (2014) Demystifying the nuclear function of Argonaute proteins. RNA Biol 11(1):18–24. doi:27604 [pii] 10.4161/rna.27604
9. Huang V, Qin Y, Wang J, Wang X, Place RF, Lin G, Lue TF, Li LC (2010) RNAa is conserved in mammalian cells. PLoS One 5(1):e8848. doi:10.1371/journal.pone.0008848
10. Huang V, Place RF, Portnoy V, Wang J, Qi Z, Jia Z, Yu A, Shuman M, Yu J, Li LC (2012) Upregulation of cyclin B1 by miRNA and its implications in cancer. Nucleic Acids Res 40 (4):1695–1707. doi:gkr934 [pii] 10.1093/nar/gkr934
11. Huang V, Zheng J, Qi Z, Wang J, Place RF, Yu J, Li H, Li LC (2013) Ago1 interacts with RNA polymerase II and binds to the promoters of actively transcribed genes in human cancer cells. PLoS Genet 9(9):e1003821. doi:10.1371/journal.pgen.1003821 PGENETICS-D-13-00631 [pii]
12. Jacquier A (2009) The complex eukaryotic transcriptome: unexpected pervasive transcription and novel small RNAs. Nat Rev Genet 10(12):833–844. doi:nrg2683 [pii] 10.1038/nrg2683
13. Janowski BA, Younger ST, Hardy DB, Ram R, Huffman KE, Corey DR (2007) Activating gene expression in mammalian cells with promoter-targeted duplex RNAs. Nat Chem Biol 3 (3):166–173. doi:nchembio860 [pii] 10.1038/nchembio860
14. Kalantari R, Chiang CM, Corey DR (2016) Regulation of mammalian transcription and splicing by nuclear RNAi. Nucleic Acids Res 44(2):524–537. doi:gkv1305 [pii] 10.1093/nar/gkv1305
15. Kang MR, Yang G, Place RF, Charisse K, Epstein-Barash H, Manoharan M, Li LC (2012) Intravesical delivery of small activating RNA formulated into lipid nanoparticles inhibits orthotopic bladder tumor growth. Cancer Res 72(19):5069–5079. doi:0008-5472.CAN-12-1871 [pii] 10.1158/0008-5472.CAN-12-1871
16. Kang MR, Park KH, Yang JO, Lee CW, Oh SJ, Yun J, Lee MY, Han SB, Kang JS (2016) miR-6734 Up-regulates p21 gene expression and induces cell cycle arrest and apoptosis in colon cancer cells. PLoS One 11(8):e0160961. doi:10.1371/journal.pone.0160961 PONE-D-16-17979 [pii]
17. Kapranov P, Cheng J, Dike S, Nix DA, Duttagupta R, Willingham AT, Stadler PF, Hertel J, Hackermuller J, Hofacker IL, Bell I, Cheung E, Drenkow J, Dumais E, Patel S, Helt G, Ganesh M, Ghosh S, Piccolboni A, Sementchenko V, Tammana H, Gingeras TR (2007) RNA maps reveal new RNA classes and a possible function for pervasive transcription. Science 316(5830):1484–1488. doi:1138341 [pii] 10.1126/science.1138341
18. Kawasaki H, Taira K (2004) Induction of DNA methylation and gene silencing by short interfering RNAs in human cells. Nature 431(7005):211–217. doi:10.1038/nature02889 nature02889 [pii] (retracted)
19. Kuwabara T, Hsieh J, Nakashima K, Taira K, Gage FH (2004) A small modulatory dsRNA specifies the fate of adult neural stem cells. Cell 116(6):779–793. doi:S009286740400248X [pii]
20. Li LC, Okino ST, Zhao H, Pookot D, Place RF, Urakami S, Enokida H, Dahiya R (2006) Small dsRNAs induce transcriptional activation in human cells. Proc Natl Acad Sci U S A 103 (46):17337–17342. doi:0607015103 [pii] 10.1073/pnas.0607015103
21. Li C, Jiang W, Hu Q, Li LC, Dong L, Chen R, Zhang Y, Tang Y, Thrasher JB, Liu CB, Li B (2016) Enhancing DPYSL3 gene expression via a promoter-targeted small activating RNA approach suppresses cancer cell motility and metastasis. Oncotarget 7(16):22893–22910. doi:8290 [pii] 10.18632/oncotarget.8290

22. Liao JY, Ma LM, Guo YH, Zhang YC, Zhou H, Shao P, Chen YQ, Qu LH (2010) Deep sequencing of human nuclear and cytoplasmic small RNAs reveals an unexpectedly complex subcellular distribution of miRNAs and tRNA 3′ trailers. PLoS One 5(5):e10563. doi:10.1371/journal.pone.0010563
23. Liu M, Roth A, Yu M, Morris R, Bersani F, Rivera MN, Lu J, Shioda T, Vasudevan S, Ramaswamy S, Maheswaran S, Diederichs S, Haber DA (2013) The IGF2 intronic miR-483 selectively enhances transcription from IGF2 fetal promoters and enhances tumorigenesis. Genes Dev 27(23):2543–2548. doi:27/23/2543 [pii] 10.1101/gad.224170.113
24. Lopez P, Wagner KD, Hofman P, Van Obberghen E (2016) RNA activation of the vascular endothelial growth factor gene (VEGF) promoter by double-stranded RNA and hypoxia: role of noncoding VEGF promoter transcripts. Mol Cell Biol 36(10):1480–1493. doi:MCB.01096-15 [pii] 10.1128/MCB.01096-15
25. Matsui M, Corey DR (2016) Non-coding RNAs as drug targets. Nat Rev Drug Discov 16 (3):167–179. doi:nrd.2016.117 [pii] 10.1038/nrd.2016.117
26. Matsui M, Sakurai F, Elbashir S, Foster DJ, Manoharan M, Corey DR (2010) Activation of LDL receptor expression by small RNAs complementary to a noncoding transcript that overlaps the LDLR promoter. Chem Biol 17(12):1344–1355. doi:S1074-5521(10)00402-3 [pii] 10.1016/j.chembiol.2010.10.009
27. Matsui M, Chu Y, Zhang H, Gagnon KT, Shaikh S, Kuchimanchi S, Manoharan M, Corey DR, Janowski BA (2013) Promoter RNA links transcriptional regulation of inflammatory pathway genes. Nucleic Acids Res 41(22):10086–10109. doi:gkt777 [pii] 10.1093/nar/gkt777
28. Meng X, Jiang Q, Chang N, Wang X, Liu C, Xiong J, Cao H, Liang Z (2016) Small activating RNA binds to the genomic target site in a seed-region-dependent manner. Nucleic Acids Res 44(5):2274–2282. doi:gkw076 [pii] 10.1093/nar/gkw076
29. Morris KV, Chan SW, Jacobsen SE, Looney DJ (2004) Small interfering RNA-induced transcriptional gene silencing in human cells. Science 305(5688):1289–1292. doi:10.1126/science.1101372 1101372 [pii]
30. Nakama M, Kawakami K, Kajitani T, Urano T, Murakami Y (2012) DNA-RNA hybrid formation mediates RNAi-directed heterochromatin formation. Genes Cells 17(3):218–233. doi:10.1111/j.1365-2443.2012.01583.x
31. Orom UA, Nielsen FC, Lund AH (2008) MicroRNA-10a binds the 5′UTR of ribosomal protein mRNAs and enhances their translation. Mol Cell 30(4):460–471. doi:S1097-2765(08)00328-6 [pii] 10.1016/j.molcel.2008.05.001
32. Paugh SW, Coss DR, Bao J, Laudermilk LT, Grace CR, Ferreira AM, Waddell MB, Ridout G, Naeve D, Leuze M, LoCascio PF, Panetta JC, Wilkinson MR, Pui CH, Naeve CW, Uberbacher EC, Bonten EJ, Evans WE (2016) MicroRNAs form triplexes with double stranded DNA at sequence-specific binding sites; a eukaryotic mechanism via which microRNAs could directly alter gene expression. PLoS Comput Biol 12(2):e1004744. doi:10.1371/journal.pcbi.1004744 PCOMPBIOL-D-15-00361 [pii]
33. Place RF, Li LC, Pookot D, Noonan EJ, Dahiya R (2008) MicroRNA-373 induces expression of genes with complementary promoter sequences. Proc Natl Acad Sci U S A 105 (5):1608–1613. doi:0707594105 [pii] 10.1073/pnas.0707594105
34. Place RF, Wang J, Noonan EJ, Meyers R, Manoharan M, Charisse K, Duncan R, Huang V, Wang X, Li LC (2012) Formulation of small activating RNA into lipidoid nanoparticles inhibits xenograft prostate tumor growth by inducing p21 expression. Mol Ther Nucleic Acids 1:e15. doi:mtna20125 [pii] 10.1038/mtna.2012.5
35. Portnoy V, Lin SH, Li KH, Burlingame A, Hu ZH, Li H, Li LC (2016) saRNA-guided Ago2 targets the RITA complex to promoters to stimulate transcription. Cell Res 26(3):320–335. doi:cr201622 [pii] 10.1038/cr.2016.22
36. Reebye V, Saetrom P, Mintz PJ, Huang KW, Swiderski P, Peng L, Liu C, Liu X, Lindkaer-Jensen S, Zacharoulis D, Kostomitsopoulos N, Kasahara N, Nicholls JP, Jiao LR, Pai M, Spalding DR, Mizandari M, Chikovani T, Emara MM, Haoudi A, Tomalia DA, Rossi JJ, Habib

NA (2014) Novel RNA oligonucleotide improves liver function and inhibits liver carcinogenesis in vivo. Hepatology 59(1):216–227. doi:10.1002/hep.26669
37. Ren S, Kang MR, Wang J, Huang V, Place RF, Sun Y, Li LC (2013) Targeted induction of endogenous NKX3-1 by small activating RNA inhibits prostate tumor growth. Prostate 73 (14):1591–1601. doi:10.1002/pros.22709
38. Sayed D, Abdellatif M (2011) MicroRNAs in development and disease. Physiol Rev 91 (3):827–887. doi:91/3/827 [pii] 10.1152/physrev.00006.2010
39. Schwartz JC, Younger ST, Nguyen NB, Hardy DB, Monia BP, Corey DR, Janowski BA (2008) Antisense transcripts are targets for activating small RNAs. Nat Struct Mol Biol 15 (8):842–848. doi:nsmb.1444 [pii] 10.1038/nsmb.1444
40. Suh MR, Lee Y, Kim JY, Kim SK, Moon SH, Lee JY, Cha KY, Chung HM, Yoon HS, Moon SY, Kim VN, Kim KS (2004) Human embryonic stem cells express a unique set of microRNAs. Dev Biol 270(2):488–498. doi:10.1016/j.ydbio.2004.02.019 S0012160604001381 [pii]
41. Toscano-Garibay JD, Aquino-Jarquin G (2014) Transcriptional regulation mechanism mediated by miRNA-DNA*DNA triplex structure stabilized by Argonaute. Biochim Biophys Acta 1839(11):1079–1083. doi:S1874-9399(14)00204-1 [pii] 10.1016/j.bbagrm.2014.07.016
42. Turner MJ, Jiao AL, Slack FJ (2014) Autoregulation of lin-4 microRNA transcription by RNA activation (RNAa) in C. elegans. Cell Cycle 13(5):772–781. doi:27679 [pii] 10.4161/cc.27679
43. Wang A, Yoshimi N, Ino N, Tanaka T, Mori H (1997) Overexpression of cyclin B1 in human colorectal cancers. J Cancer Res Clin Oncol 123(2):124–127
44. Wang J, Place RF, Huang V, Wang X, Noonan EJ, Magyar CE, Huang J, Li LC (2010) Prognostic value and function of KLF4 in prostate cancer: RNAa and vector-mediated overexpression identify KLF4 as an inhibitor of tumor cell growth and migration. Cancer Res 70 (24):10182–10191. doi:70/24/10182 [pii] 10.1158/0008-5472.CAN-10-2414
45. Wang X, Wang J, Huang V, Place RF, Li LC (2012) Induction of NANOG expression by targeting promoter sequence with small activating RNA antagonizes retinoic acid-induced differentiation. Biochem J 443(3):821–828. doi:BJ20111491 [pii] 10.1042/BJ20111491
46. Wang C, Chen Z, Ge Q, Hu J, Li F, Xu H, Ye Z, Li LC (2014) Up-regulation of p21(WAF1/CIP1) by miRNAs and its implications in bladder cancer cells. FEBS Lett 588(24):4654–4664. doi:S0014-5793(14)00786-8 [pii] 10.1016/j.febslet.2014.10.037
47. Wang J, Huang V, Ye L, Barcena A, Lin G, Lue TF, Li LC (2015) Identification of small activating RNAs that enhance endogenous OCT4 expression in human mesenchymal stem cells. Stem Cells Dev 24(3):345–353. doi:10.1089/scd.2014.0290
48. Wang B, Sun J, Shi J, Guo Q, Tong X, Zhang J, Hu N, Hu Y (2016a) Small-activating RNA can change nucleosome positioning in human fibroblasts. J Biomol Screen 21(6):634–642. doi:1087057116637562 [pii] 10.1177/1087057116637562
49. Wang C, Tang K, Li Z, Chen Z, Xu H, Ye Z (2016b) Targeted p21(WAF1/CIP1) activation by miR-1236 inhibits cell proliferation and correlates with favorable survival in renal cell carcinoma. Urol Oncol 34(2):59 e23–34. doi:S1078-1439(15)00413-5 [pii] 10.1016/j.urolonc.2015.08.014
50. Weaver BA, Cleveland DW (2008) The aneuploidy paradox in cell growth and tumorigenesis. Cancer Cell 14(6):431–433. doi:S1535-6108(08)00377-2 [pii] 10.1016/j.ccr.2008.11.011
51. Xiao M, Li J, Li W, Wang Y, Wu F, Xi Y, Zhang L, Ding C, Luo H, Li Y, Peng L, Zhao L, Peng S, Xiao Y, Dong S, Cao J, Yu W (2016) MicroRNAs activate gene transcription epigenetically as an enhancer trigger. RNA Biol 0. doi:10.1080/15476286.2015.1112487
52. Younger ST, Corey DR (2011) Transcriptional gene silencing in mammalian cells by miRNA mimics that target gene promoters. Nucleic Acids Res 39(13):5682–5691. doi:gkr155 [pii] 10.1093/nar/gkr155
53. Zhang Y, Fan M, Geng G, Liu B, Huang Z, Luo H, Zhou J, Guo X, Cai W, Zhang H (2014a) A novel HIV-1-encoded microRNA enhances its viral replication by targeting the TATA box region. Retrovirology 11:23. doi: 10.1186/1742-4690-11-23 [pii]

54. Zhang Y, Fan M, Zhang X, Huang F, Wu K, Zhang J, Liu J, Huang Z, Luo H, Tao L, Zhang H (2014b) Cellular microRNAs up-regulate transcription via interaction with promoter TATA-box motifs. RNA 20(12):1878–1889. doi:rna.045633.114 [pii] 10.1261/rna.045633.114
55. Zhao M, Kim YT, Yoon BS, Kim SW, Kang MH, Kim SH, Kim JH, Kim JW, Park YW (2006) Expression profiling of cyclin B1 and D1 in cervical carcinoma. Exp Oncol 28(1):44–48. doi:25/494 [pii]

Chapter 6
miRNA-Mediated RNA Activation in Mammalian Cells

Ramani Ramchandran and Pradeep Chaluvally-Raghavan

Abstract MicroRNA (miRNA or miR) is a small noncoding RNA molecule ~22 nucleotides in size, which is found in plants, animals, and some viruses. miRNAs are thought to primarily down regulate gene expression by binding to 3′ untranslated regions of target transcripts, thereby triggering mRNA cleavage or repression of translation. Recently, evidence has emerged that miRNAs can interact with the promoter and activate gene expression. This mechanism, called RNA activation (RNAa), is a process of transcriptional activation where the direct interaction of miRNA on the promoter triggers the recruitment of transcription factors and RNA-Polymerase-II on the promoter to activate gene transcription. To date, very little is known about the mechanism by which miRNA regulates RNA activation (RNAa) and their role in tumor progression. This is an emerging field in RNA biology. In this chapter, we describe the mechanisms utilized by miRNAs to activate transcription.

Keywords RNA activation (RNAa) • microRNA • Duplex RNA • Promoter and RNA interference (RNAi)

6.1 Introduction

Current research in the microRNA (miRNA) field largely focuses on the actions of miRNAs on the 3′ untranslated regions of genes that result in the suppression of target expression either by degradation of mRNA or inhibition of translation [1, 5]. Recently, several groups including ours have demonstrated that small

R. Ramchandran
Department of Obstetrics and Gynecology, Medical College of Wisconsin, Milwaukee, WI 53226, USA

Department of Pediatrics, Medical College of Wisconsin, Milwaukee, WI 53226, USA

P. Chaluvally-Raghavan (✉)
Department of Obstetrics and Gynecology, Medical College of Wisconsin, Milwaukee, WI 53226, USA

Department of Physiology, Medical College of Wisconsin, Milwaukee, WI 53226, USA
e-mail: pchaluvally@mcw.edu

L.-C. Li (ed.), *RNA Activation*, Advances in Experimental Medicine and Biology 983, DOI 10.1007/978-981-10-4310-9_6

noncoding RNAs can influence gene transcription by direct interaction of small RNAs with the promoter [8, 16, 19]. This mechanism, also called RNA activation (RNAa), is a process of transcriptional activation where microRNAs through their direct interaction with the promoter sequence activate gene transcription. Most of the studies reported on RNAa are on cancer models [17, 18, 25]; however, recent studies have extended these findings to other conditions such as ischemia [28, 31] and erectile dysfunction in diabetes [34]. Together, the current research provides evidence that RNA can precisely target selective genes to activate signaling pathways.

RNA interference caused by siRNA and/or miRNA leads to suppression of transcription or degradation of mRNA and/or inhibition of translation within hours after the transfection of small RNAs [9]. In contrast, transcriptional activation induced by miRNAs requires several days to turn on, but can persist for several days once it is activated [26]. This delay could be due to the epigenetic changes mediated by miRNA as the promoter of target genes, which then facilitates long-term effects on gene expression. Thus RNAa is potentially an important unexplored mechanism employed by cells to activate gene expression for a persistent period of time. This chapter will address the current state of RNAa research and the underlying mechanisms that are responsible for this process.

6.2 History

In 1998, Craig Mello and Andrew Fire reported the gene-silencing effects of double-stranded small RNAs (dsRNAs) in *C. elegans*, where they demonstrated that dsRNAs successfully silenced the targeted gene and coined the term RNA interference (RNAi) [10]. This striking discovery in RNA biology was rewarded with a Nobel Prize in Physiology or Medicine in 2006. In contrast to the term RNAi, RNA activation (RNAa) refers to the process of transcriptional activation mediated by small RNAs. The new term RNAa was coined first by Long-Cheng Li and Rajvir Dahiya [17] who designed and synthesized 21-nt dsRNAs targeting selected promoter regions of human genes *E-CADHERIN*, $p21^{WAF1/CIP1}$ *(p21)*, and vascular endothelial growth factor (VEGF). Intriguingly, transfection of those dsRNAs into human prostate cancer cell lines caused long-lasting and sequence-specific induction of all three genes [17]. Shortly thereafter, Janowski et al. identified the induction of progesterone receptor (PR) and major vault protein by promoter-targeting dsRNA [16]. Since then, similar observations have been made by several other groups in different mammalian species including human [17], rat [11], and mouse [12], suggesting that RNAa is a general mechanism of gene regulation conserved across mammalian species. Although RNAa was proposed by Li et al., in 2006, the theory of gene activation by the action of RNAs was initially proposed by Britten and Davidson in their article entitled “Gene regulation for higher cells: This theory postulates the existence of gene regulation by small non-coding RNA” published in *Science* [6]. In this theory, they proposed that RNA molecules form a sequence-specific complex with the nontranscribed sequences that reside upstream of the sequences transcribed into RNA molecules [6]. To date, very little is known about the precise mechanism of how RNAa occurs in normal cells and the importance of this process in physiology and disease. Because small RNAs can interact with gene promoters, which influence various transcriptional processes, the current hypothesis is that the RNAa mechanism includes changes in the occupancy of transcription factors (TFs) or RNA polymerases on the promoter, and the epigenetic changes associated with this occupancy process. We recently found some evidence in support of this hypothesis. We identified that miR551b, an miRNA located in the 3q26.2 amplicon, a region that is frequently amplified in ovarian cancer patients, interacts with complementary sequences on the promoter of STAT3 transcription factor. This miR551b interaction with STAT3 promoter affects STAT3 transcription by the recruitment of RNA Polymerase II (RNA Pol II) and the Twist1 transcription factor to the STAT3 promoter. To help understand the RNAa mechanism, we will outline next the various steps in the process of miRNA synthesis and biology that are directly relevant to RNAa.

6.3 Maturation of miRNAs

In northern blot analyses for most miRNAs, two species of nucleotides (nt) are often noticed: a larger (~70 nt) and a smaller (~22 nt) RNA. The smaller RNA (~22 nt) is the mature form of miRNAs [1, 24]. It is known that more than two thirds of all human miRNAs are encoded in the intervening regions (introns) of protein-coding genes as well as in long noncoding transcripts. MiRNAs can also be encoded in exons or introns, or can be located in intergenic region, and within the chromosomal regions encompassing two genes [2, 29]. During the biogenesis, the primary precursor (pri-miRNA) is processed into an approximately 70-nt-long stem-loop structure by nuclear RNase III Drosha present in the microprocessor complex, which is the complex of proteins involved in the processing and maturation of miRNAs. The two RNase domains of Drosha help cleave the 5′ and 3′ ends of the pri-miRNA, which is then exported to the cytoplasm by Exportin-5 and Ran-GTP complexes. Later the final maturation of miRNA occurs with Dicer, another RNase III nuclease that processes the pre-miRNA into a 22-bp dsRNA in the cytoplasm. Subsequently, Dicer-cleaved miRNAs are loaded with RNA-induced silencing complex (RISC) that includes Argonaute (AGO) proteins, transactivating response RNA-binding protein (TRBP), protein kinase R-activating protein (PACT), and Dicer.

6.3.1 Transport of Mature miRNA to the Nucleus for RNAa

Since the major process of miRNA maturation occurs in the cytoplasm, it is expected that miRNAs are needed to translocate into the nucleus for RNAa. However, the exact mechanism underlying the transport of miRNA from the cytoplasm to the nucleus is not well understood. Argonaute (AGO) proteins that facilitate the transport of miRNAs from the cytoplasm to the nucleus are widely considered to promote this process. AGO proteins are highly conserved proteins and ubiquitously expressed in all higher eukaryotes. In humans, eight AGO genes are known to exist, of which family members AGO1, 2, 3, and 4 have been extensively studied [14]. AGO proteins encode four functional domains, namely, N-terminal, PAZ, Mid, and C-terminal PIWI domains [14]. Using all of its functional domains, AGO proteins bind to different classes of small RNAs including miRNAs, siRNAs, and PIWI-interacting RNAs (piRNAs), which further bind with their specific targets through sequence complementarity base pairing [3, 15]. Among the domains in AGO protein, the PAZ domain binds to the 3′ end of both siRNA and miRNA, which facilitates the binding of small RNAs on target mRNA by base pair interaction [22]. The PIWI domain of AGO proteins contains the catalytic residues which are essential for cleavage [22], and also mediates protein-protein interaction with the RNase III domain of Dicer protein, suggesting that PIWI is a critical domain that is responsible for miRNA maturation function. Within the mid domain, AGO protein encodes for MC motif and is involved in binding cap structures of RNA, thereby suggesting a role in controlling translation of capped mRNA. Altogether, the AGO protein family with multiple domains and multiple protein interactive motifs allows miRNA-guided AGO proteins to interact with several proteins directly or indirectly. For example, AGOs recruit TFs and RNA Pol II to the promoter of the target genes, and this complex recruitment leads to conformational changes that enhance the occupancy of TFs and RNA Pol II to STAT3 oncogene promoter to turn on transcription (Fig. 6.1) [8]. In addition to AGO proteins, a second class of proteins, namely, importin8 (IPO8), a member of the karyopherin family [35], has been identified for miRNA transport from the cytoplasm to the nucleus. Therefore, both IPO8 and AGO proteins have an important role in the transport of mature miRNA from the cytoplasm to the nucleus [35]. AGO proteins have been more extensively studied in the context of miRNA transport. When AGO1 or AGO2 proteins are depleted individually, the respective member of its family (AGO2 for AGO1 depleted and AGO1 for AGO2 depleted) is preferentially retained in the nucleus [20]. Furthermore, Matsui et al. demonstrated that silencing AGO1 reduces nuclear Dicer levels by 90% and nuclear TRBP levels by >50% [20], suggesting that in addition to nuclear import of miRNAs, the AGO protein family may also be involved in the maturation of miRNAs in the nucleus.

6.4 Target Specificity and Selection for miRNA-Mediated RNAa

Once in the nucleus, the miRNA needs to interact with its target. The principles governing the genomic target site recognition of small RNAs are thought to be sequence dependent in the sense of antisense strand [27]. One of the first studies reported on RNAa, were used duplex RNAs target promoter sequences designed based on the rules of functional siRNAs, where the rules included low GC content, lack of

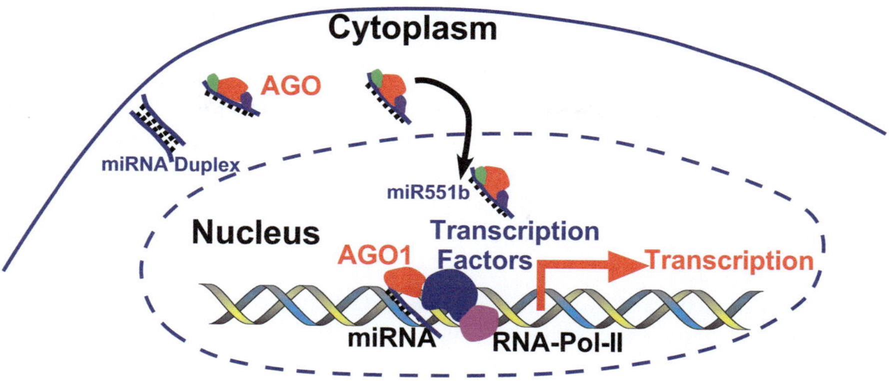

Fig. 6.1 Proposed model of microRNA-mediated RNA activation

repeated or inverted sequences, and avoidance of targeting CpG-rich regions [17]. In consequence, authors have identified that, four dsRNAs that met these rules targeting E-cadherin, p21, and VEGF promoters activated the expression of E-cadherin, p21, and VEGF at the mRNA and protein levels [17]. The same group later identified miR-373 complementary target sites in the promoters of E-cadherin and cold shock domain containing protein C2 (CSDC2), thereby inducing the expression of both E-Cadherin and CSDC2 [25]. Of note, the site-specific mutations on the miR-373 sequences abrogated any increase in the expression of both E-Cadherin and CSDC [25]. In a recent study, Vera Huang and her team tested the effects of miR-774 and miR-1186 which have sequence complementarity to the sequences located in the CyclinB1 promoter [12]. Similarly, miRNA-589 has complementarity to adjacent sites in the COX2 gene promoter where it binds to and activates COX2 transcription. Our recent studies concur with these reports, and indeed, miR551b-3p interacts with the promoter sequences on the STAT3 promoter and activates STAT3 transcription [8].

Although most studies report proximal targeting of the miRNA on the promoter site, for example, overlapping the transcriptional start site (TSS) [11, 16, 33], there are reports of miRNA targeting 200–1200 bps upstream of the TSS. This upstream targeting region is considered optimal for RNAa. Altogether, the studies in the past 10 years on the transcriptional activation by small RNAs suggest that sequence specificity is a critical factor and miRNA-mediated transcriptional activation largely depends on the sequence complementarity between the miRNAs and the promoter so as to engage the interaction between miRNAs and the promoter for transcriptional activation.

6.4.1 Mechanism of Transcriptional Activation

The majority of studies report miRNAs' role in gene inhibition, but only a handful of reports have described their role in the RNAa mechanism. Exactly how RNAa is facilitated is not clear, but studies are beginning to show that the mechanism includes binding of miRNA on the promoter and improve occupancy of transcription factor and RNA Polymerase II (RNA Pol II). We hypothesize that AGO proteins that are responsible for transporting miRNAs to the nucleus are also critical for facilitating RNAa. This hypothesis is partly based on published studies where AGO1 has been shown to interact with RNA Pol II [13, 20]. Similarly, we have also reported that miRNA-551b-AGO1 complex interacts with RNA Pol II, which in turn facilitates the recruitment of TWIST1 transcription factor to the STAT3 promoter to activate STAT3 transcription [8]. Interestingly, the miR551b-3p-mediated STAT3 activation occurs within 24 h but lasts for more than 10 days. Others have reported similar effects with RNAa occurring within 24 h, and further elevated to the maximum levels by 72 h, which persists for at least *one and a half* weeks in cells [26]. The persistent nature of this activation is intriguing given that gene-silencing effects by siRNA last for 24 h, and are turned off within 3–4 days [9]. One potential explanation is the alteration of chromatin structures or epigenetic machineries by RNAa. The transcriptional state of genes is dependent on the state of chromatin occupancy, which is modified via histone protein and DNA modifications. Recent CHIP data for AGO1 protein identified AGO1 peaks that mapped within 65 kb of transcription start sites (TSS), and overlapped with H3K4me3 marks [13]. Trimethylation of lysine 4 on histone H3 protein subunit (H3K4me3) is an important histone marker in epigenetic studies, and is indicative of the active gene promoters [4]. Similar increases in trimethylated H3K4 were also noticed in PR and CyclinB1 promoters that were facilitated by duplex short RNAs [11, 16]. In addition to the H3K4 methylation, there is a second class of methylation, H3K27 group, that is responsible for transcriptional silencing. In contrast to the methylation of H3K4, trimethyation of H3K27 at the promoter is associated with inactive gene promoters. H3K27 trimethylation is catalyzed by a class of proteins that belongs to the family of Polycomb Group of proteins (PcG). The PcG proteins form two distinct Polycomb-repressive multiprotein complexes: PRC1 and PRC2. Trimethylation of H3K27 is mainly catalyzed by EZH2, which is the catalytic subunit of the PRC2 core complex. It also contains two other PcG proteins, SUZ12 and EED, which are critical proteins required for the process of transcriptional modulation by PRC2 [7, 23]. The precise mechanisms that facilitate miRNA-induced trimethylation of H3K4 or H3K27 on histone proteins located at the promoter sites are still unknown. In light of the current research on histone methylation and analysis of multiple histone marks by small RNAs, activating the active histone marks such as H3K4 is a dynamic process when compared to the loss of methylation of H3K27. Apart from the histone methylation, there is a paucity of data on the role of DNA methylation during the process of RNAa. The AGO proteins discussed earlier for miRNA transport from the cytoplasm to the nucleus also may play a role in DNA methylation. For example, AGO4 is functionally distinct from AGO1–AGO3 and incompatible with splicing abilities [30], and is not directly involved in the siRNA/miRNA-triggered RNA degradation process. Based on AGO4's structure and its interaction with other AGO family proteins and methylase enzymes in particular, we hypothesize that AGO4 has important roles in RNA-mediated DNA methylation or other epigenetic regulation processes. Thus the RNAa mechanism is a fertile field rich with scientific questions that need answers.

6.5 Conclusion

Recent studies demonstrate that RNAa is an endogenous mechanism of gene regulation, wherein small duplex RNAs including miRNAs activate the gene expression through their interaction with the promoter. Evidence suggests that promoter-guiding effects of small duplex RNAs are primarily regulated through the Argonaute family proteins. It is currently accepted that once small RNAs get bound to the AGO complex, they modify histones and recruit transcription factors to form RNA polymerase II complex on gene promoters. In the past 10 years, RNAa has emerged as an important topic in miRNA research. However, our understanding of RNAa is still limited with many questions still unanswered. These include questions related to the sequence of events that occur during the process of RNAa such as (1) whether the number of seed sequences on the promoters will have different effects on transcription and (2) the number of mechanisms such as histone modifications and recruitment of TFs and/or RNA Pol II that associate with the miRNA binding on the promoter. We expect that the RNA immune precipitation experiments followed by mass spectrometry as well as pulse chase experiments will identify most of the critical proteins associated with conformational changes that lead to the transcriptional activation. Most of the studies on RNAa are employing overexpression methods to increase the levels in miRNA in cells to achieve RNAa. Therefore, it still remains unexplored whether RNAa is an important operational mechanism for transcriptional regulation under normal physiological conditions. Thus more research in this area is required in order to unravel the complexities in RNAa mechanisms, in particular those associated with AGO. Finally, whether this mechanism is found in other cell types and plays a role in diseases other than cancer needs further exploration.

Acknowledgments P. Chaluvally-Raghavan is supported by the Liz-Tilberis Career Development Award and the Ann Schreiber Scholar Award from the Ovarian Cancer Research Foundation and funds from the Marsha Rivkin Center for Ovarian Cancer Research. Ramani Ramchandran is supported by endowment funds from the Department of OBGYN at MCW, programmatic support funds from the Department of Pediatrics, Children's Research Institute, and NIH grants HL123338, HL112639, HL033833, and HL128374.

Conflict of Interest The authors declare that there is no conflict of interest regarding the publication of this paper.

References

1. Ambros V (2001) microRNAs: tiny regulators with great potential. Cell 107(7):823–826
2. Ambros V, Lee RC, Lavanway A, Williams PT, Jewell D (2003) MicroRNAs and other tiny endogenous RNAs in *C. elegans*. Curr Biol 13(10):807–818
3. Azlan A, Dzaki N, Azzam G (2016) Argonaute: the executor of small RNA function. J Genet Genomics 43(8):481–494. doi:10.1016/j.jgg.2016.06.002
4. Barski A, Cuddapah S, Cui K, Roh TY, Schones DE, Wang Z, Wei G, Chepelev I, Zhao K (2007) High-resolution profiling of histone methylations in the human genome. Cell 129 (4):823–837. doi:10.1016/j.cell.2007.05.009

5. Bartel DP (2004) MicroRNAs: genomics, biogenesis, mechanism, and function. Cell 116 (2):281–297
6. Britten RJ, Davidson EH (1969) Gene regulation for higher cells: a theory. Science 165 (3891):349–357
7. Cao R, Wang L, Wang H, Xia L, Erdjument-Bromage H, Tempst P, Jones RS, Zhang Y (2002) Role of histone H3 lysine 27 methylation in polycomb-group silencing. Science 298 (5595):1039–1043. doi:10.1126/science.1076997
8. Chaluvally-Raghavan P, Jeong KJ, Pradeep S, Silva AM, Yu S, Liu W, Moss T, Rodriguez-Aguayo C, Zhang D, Ram P, Liu J, Lu Y, Lopez-Berestein G, Calin GA, Sood AK, Mills GB (2016) Direct upregulation of STAT3 by MicroRNA-551b-3p deregulates growth and metastasis of ovarian cancer. Cell Rep 15(7):1493–1504. doi:10.1016/j.celrep.2016.04.034
9. Dykxhoorn DM, Novina CD, Sharp PA (2003) Killing the messenger: short RNAs that silence gene expression. Nat Rev Mol Cell Biol 4(6):457–467. doi:10.1038/nrm1129
10. Fire A, Xu S, Montgomery MK, Kostas SA, Driver SE, Mello CC (1998) Potent and specific genetic interference by double-stranded RNA in *Caenorhabditis elegans*. Nature 391 (6669):806–811. doi:10.1038/35888
11. Huang V, Qin Y, Wang J, Wang XL, Place RF, Lin GT, Lue TF, Li LC (2010) RNAa is conserved in mammalian cells. Plos One 5(1). doi:ARTN e8848 10.1371/journal.pone.0008848
12. Huang V, Place RF, Portnoy V, Wang J, Qi Z, Jia Z, Yu A, Shuman M, Yu J, Li LC (2012) Upregulation of cyclin B1 by miRNA and its implications in cancer. Nucleic Acids Res 40 (4):1695–1707. doi:10.1093/nar/gkr934
13. Huang V, Zheng J, Qi Z, Wang J, Place RF, Yu J, Li H, Li LC (2013) Ago1 interacts with RNA polymerase II and binds to the promoters of actively transcribed genes in human cancer cells. PLoS Genet 9(9):e1003821. doi:10.1371/journal.pgen.1003821
14. Hutvagner G, Simard MJ (2008) Argonaute proteins: key players in RNA silencing. Nat Rev Mol Cell Biol 9(1):22–32. doi:10.1038/nrm2321
15. Jain R, Iglesias N, Moazed D (2016) Distinct functions of argonaute slicer in siRNA maturation and heterochromatin formation. Mol Cell 63(2):191–205. doi:10.1016/j.molcel.2016.05.039
16. Janowski BA, Younger ST, Hardy DB, Ram R, Huffman KE, Corey DR (2007) Activating gene expression in mammalian cells with promoter-targeted duplex RNAs. Nat Chem Biol 3 (3):166–173. doi:10.1038/nchembio860
17. Li LC, Okino ST, Zhao H, Pookot D, Place RF, Urakami S, Enokida H, Dahiya R (2006) Small dsRNAs induce transcriptional activation in human cells. Proc Natl Acad Sci U S A 103 (46):17337–17342. doi:10.1073/pnas.0607015103
18. Lopez P, Wagner KD, Hofman P, Van Obberghen E (2016) RNA activation of the vascular endothelial growth factor gene (VEGF) promoter by double-stranded RNA and hypoxia: role of noncoding VEGF promoter transcripts. Mol Cell Biol 36(10):1480–1493. doi:10.1128/MCB.01096-15
19. Matsui M, Chu Y, Zhang H, Gagnon KT, Shaikh S, Kuchimanchi S, Manoharan M, Corey DR, Janowski BA (2013) Promoter RNA links transcriptional regulation of inflammatory pathway genes. Nucleic Acids Res 41(22):10086–10109. doi:10.1093/nar/gkt777
20. Matsui M, Li L, Janowski BA, Corey DR (2015) Reduced expression of argonaute 1, argonaute 2, and TRBP changes levels and intracellular distribution of RNAi factors. Sci Rep 5:12855. doi:10.1038/srep12855
21. Matsui M, Prakash TP, Corey DR (2016) Argonaute 2-dependent regulation of gene expression by single-stranded miRNA mimics. Mol Ther 24(5):946–955. doi:10.1038/mt.2016.39
22. Meister G, Landthaler M, Patkaniowska A, Dorsett Y, Teng G, Tuschl T (2004) Human Argonaute2 mediates RNA cleavage targeted by miRNAs and siRNAs. Mol Cell 15 (2):185–197. doi:10.1016/j.molcel.2004.07.007
23. Pasini D, Bracken AP, Hansen JB, Capillo M, Helin K (2007) The polycomb group protein Suz12 is required for embryonic stem cell differentiation. Mol Cell Biol 27(10):3769–3779. doi:10.1128/MCB.01432-06

24. Pasquinelli AE, Reinhart BJ, Slack F, Martindale MQ, Kuroda MI, Maller B, Hayward DC, Ball EE, Degnan B, Muller P, Spring J, Srinivasan A, Fishman M, Finnerty J, Corbo J, Levine M, Leahy P, Davidson E, Ruvkun G (2000) Conservation of the sequence and temporal expression of let-7 heterochronic regulatory RNA. Nature 408(6808):86–89. doi:10.1038/35040556
25. Place RF, Li LC, Pookot D, Noonan EJ, Dahiya R (2008) MicroRNA-373 induces expression of genes with complementary promoter sequences. Proc Natl Acad Sci U S A 105 (5):1608–1613. doi:10.1073/pnas.0707594105
26. Place RF, Noonan EJ, Foldes-Papp Z, Li LC (2010) Defining features and exploring chemical modifications to manipulate RNAa activity. Curr Pharm Biotechnol 11(5):518–526
27. Portnoy V, Huang V, Place RF, Li LC (2011) Small RNA and transcriptional upregulation. Wiley Interdiscip Rev RNA 2(5):748–760. doi:10.1002/wrna.90
28. Reebye V, Saetrom P, Mintz PJ, Huang KW, Swiderski P, Peng L, Liu C, Liu X, Lindkaer-Jensen S, Zacharoulis D, Kostomitsopoulos N, Kasahara N, Nicholls JP, Jiao LR, Pai M, Spalding DR, Mizandari M, Chikovani T, Emara MM, Haoudi A, Tomalia DA, Rossi JJ, Habib NA (2014) Novel RNA oligonucleotide improves liver function and inhibits liver carcinogenesis in vivo. Hepatology 59(1):216–227. doi:10.1002/hep.26669
29. Saini HK, Griffiths-Jones S, Enright AJ (2007) Genomic analysis of human microRNA transcripts. Proc Natl Acad Sci U S A 104(45):17719–17724. doi:10.1073/pnas.0703890104
30. Schurmann N, Trabuco LG, Bender C, Russell RB, Grimm D (2013) Molecular dissection of human argonaute proteins by DNA shuffling. Nat Struct Mol Biol 20(7):818–826. doi:10.1038/nsmb.2607
31. Turunen MP, Lehtola T, Heinonen SE, Assefa GS, Korpisalo P, Girnary R, Glass CK, Vaisanen S, Yla-Herttuala S (2009) Efficient regulation of VEGF expression by promoter-targeted lentiviral shRNAs based on epigenetic mechanism: a novel example of epigenetherapy. Circ Res 105(6):604–609. doi:10.1161/CIRCRESAHA.109.200774
32. Vaucheret H, Vazquez F, Crete P, Bartel DP (2004) The action of ARGONAUTE1 in the miRNA pathway and its regulation by the miRNA pathway are crucial for plant development. Genes Dev 18(10):1187–1197. doi:10.1101/gad.1201404
33. Wang X, Wang J, Huang V, Place RF, Li LC (2012) Induction of NANOG expression by targeting promoter sequence with small activating RNA antagonizes retinoic acid-induced differentiation. Biochem J 443(3):821–828. doi:10.1042/BJ20111491
34. Wang T, Li M, Yuan H, Zhan Y, Xu H, Wang S, Yang W, Liu J, Ye Z, Li LC (2013) saRNA guided iNOS up-regulation improves erectile function of diabetic rats. J Urol 190(2):790–798. doi:10.1016/j.juro.2013.03.043
35. Wei Y, Li L, Wang D, Zhang CY, Zen K (2014) Importin 8 regulates the transport of mature microRNAs into the cell nucleus. J Biol Chem 289(15):10270–10275. doi:10.1074/jbc.C113.541417
36. Yue X, Schwartz JC, Chu Y, Younger ST, Gagnon KT, Elbashir S, Janowski BA, Corey DR (2010) Transcriptional regulation by small RNAs at sequences downstream from 3′ gene termini. Nat Chem Biol 6(8):621–629. doi:10.1038/nchembio.400

Chapter 7
RNAa Induced by TATA Box-Targeting MicroRNAs

Yijun Zhang and Hui Zhang

Abstract Recent studies reveal that some nuclear microRNAs (miRNA) and synthesized siRNAs target gene promoters to activate gene transcription (RNAa). Interestingly, our group identified a novel HIV-1-encoded miRNA, miR-H3, which targets specifically the core promoter TATA box of HIV-1 and activates viral gene expression. Depletion of miR-H3 significantly impaired the replication of HIV-1. miR-H3 mimics could activate viruses from $CD4^{+}$ T cells isolated from patients receiving suppressive highly active antiretroviral therapy, which is very intriguing for reducing HIV-1 latent reservoir. Further study revealed that many cellular miRNAs also function like miR-H3. For instance, let-7i targets the TATA box of the *interleukin-2* (*IL-2*) promoter and upregulates IL-2 expression in T-lymphocytes. In RNAa induced by TATA box-targeting miRNAs, Argonaute (AGO) proteins are needed, but there is no evidence for the involvement of promoter-associated transcripts or epigenetic modifications. We propose that the binding of small RNA-AGO complex to TATA box could facilitate the assembly of RNA Polymerase II transcription preinitiation complex. In addition, synthesized small RNAs targeting TATA box can also efficiently activate transcription of interested genes, such as *insulin*, *IL-2*, and *c-Myc*. The discovery of RNAa induced by TATA box-targeting miRNA provides an easy-to-use tool for activating gene expression.

Y. Zhang
Institute of Human Virology, Zhongshan School of Medicine, Sun Yat-sen University, Guangzhou, Guangdong 510080, China

Key Laboratory of Tropical Disease Control of Ministry of Education, Zhongshan School of Medicine, Sun Yat-sen University, Guangzhou, Guangdong 510080, China

Section of Infectious Diseases, Department of Internal Medicine, Yale University School of Medicine, New Haven, CT 06520, USA

H. Zhang (✉)
Institute of Human Virology, Zhongshan School of Medicine, Sun Yat-sen University, Guangzhou, Guangdong 510080, China

Key Laboratory of Tropical Disease Control of Ministry of Education, Zhongshan School of Medicine, Sun Yat-sen University, Guangzhou, Guangdong 510080, China
e-mail: zhangh92@mail.sysu.edu.cn

L.-C. Li (ed.), *RNA Activation*, Advances in Experimental Medicine and Biology 983, DOI 10.1007/978-981-10-4310-9_7

Keywords RNAa • TATA box • microRNA • Core promoter • Transcription initiation

7.1 Introduction

MicroRNAs (miRNAs) represent an important class of regulators in many crucial biological processes in animals, plants, fungi, and viruses [3, 7]. In the nucleus, primary miRNAs (pri-miRNAs) are transcribed and processed by Drosha and its cofactor DGCR8 to generate precursor miRNAs (pre-miRNAs) [61]. The pre-miRNAs are then exported to the cytoplasm [119] wherein they are further sliced by Dicer to generate ~22 nt mature miRNAs [40] and bound by Argonaute (AGO) proteins and loaded into the RNA-induced silence complexes (RISCs) [27]. It is widely accepted that most miRNAs function through guiding the cytoplasmic RISCs to 3′ UTR of mRNAs to suppress translation or degrade mRNAs [4, 7, 39, 70]. The regulatory mode of miRNA has been extended by the observation that 5′ UTR [68] or the coding region [104] of mRNA could also be the target for miRNA-mediated translation repression. Moreover, miRNAs could switch from suppressor to activator of translation by targeting 5′ UTR [81] or 3′ UTR [109] of mRNAs. Nevertheless, the role of miRNA as a posttranscriptional activator is reported only in a few studies and is not broadly applicable now. These findings reveal that the mechanism of miRNA-mediated gene regulation is quite complicated in metazoans.

Interestingly, a fraction of cellular miRNAs is enriched in the nucleus [41, 65, 102], although their functions are largely unknown. Several studies have suggested that miRNAs could modulate gene expression at transcriptional level. miR-373 can readily induce the expression of E-cadherin and cold-shock domain-containing protein C2 (CSDC2) through targeting a site in their promoters [88]. miR-423-5p induces transcriptional silencing by targeting a highly conserved region in the promoter of the progesterone receptor (PR) gene [122]. miR-744 and miR-1186 induce Ccnb1 expression and manipulate mouse cell proliferation with putative binding sites in the gene promoter [35]. The target sites of these miRNAs are located randomly on gene promoter without a unique feature. The underlying molecular mechanism is considered to be through altering epigenetic modifications of the promoter, including acetylation and/or methylation at histones [35, 64, 91], and promoter-associated RNA (paRNA) transcripts overlapped with the promoter are possible mediators of these modifications, but they are not reported in all RNAa cases [95, 122]. These findings suggest that the mechanism underlying miRNA-mediated transcriptional regulation is sophisticated.

The key RNA interfering (RNAi) components such as Argonaute proteins have been found in the nucleus [23, 77, 103, 113] and associated with RNA polymerase II (RNA Pol II) core transcription machinery [8, 29, 30]. In *Schizosaccharomyces pombe*, centromeric repeats region-derived siRNAs associate with Argonaute 1 and form the RNA-induced transcriptional silencing (RITS) complex to induce the heterochromatin assembly [75, 110, 111]. In *Caenorhabditis elegans*, siRNAs and piRNAs can enter the nucleus and trigger co-transcriptional silencing of genes [29, 30, 58]. SiRNA-directed recruitment of the nuclear RNAi components, nuclear RNAi defective (NRDE) factors, could inhibit RNA Pol II during the elongation phase of transcription [30]. A recent report suggests that the key protein components of the RNAi pathway, i.e., Dicer 2 (DCR2) and AGO2, could directly interact with the transcription machinery and control the processivity of RNA Pol II in *Drosophila* [8]. These findings raise the possibility that small noncoding RNAs could directly interact with the Pol II core transcription machinery and affect gene transcription in metazoans.

TATA box is a widespread gene core promoter motif in archaea and eukaryotes. Approximately 24% of human genes contain a TATA box within the core promoter [117]. The eukaryotic TATA box has a consensus sequence of TATAWAAR [46], with the upstream T nucleotide most commonly located from −33 to −28 (particularly, at −31 or −30) relative to the mRNA start site (+1). TATA box is the position where the RNA polymerase II transcription complex starts to assemble and initiate transcription [93]. The TATA box-binding protein (TBP), a basic subunit of transcription factor II D (TFIID), first binds to the TATA box motif, then recruits other general transcription factors including TFIIB, TFIIA, TFIIE, and TFIIH as well as RNA polymerase II to form a transcription preinitiation complex (PIC). It is noteworthy that TATA-containing promoters belong to the class of sharp, or focused, core promoters, which often control tissue-specific or viral genes [106]. And the transcription from TATA-containing promoters is often highly regulated [37, 93]. Thus, TATA box represents the most conserved and widespread core promoter, which is enriched in the genes of tissue-specific expression or viral origin.

The RNA polymerase II transcription initiation is quite a complicated process. Assembly of the preinitiation complex and its postassembly control are critical early steps in the transcription of eukaryotic genes. The binding of TBP to TATA box is the first step of the assembly of preinitiation complex. The TBP displays ~1000 lower affinity for nonspecific DNA compared with TATA sequences in vitro [14]. Two α-β domains of core TBP form a saddle-like structure, which displays an approximate twofold symmetry [6]. The structure of TBP-TATA complex shows that the large concave and hydrophobic surface of the TBP saddle mediates minor groove binding to TATA box [106]. When entering the complex, the minor groove of B-form DNA becomes widened over the 8 bp of the TATA box. Two pairs of highly conserved phenylalanines insert into the first T-A step and between the last 2 bps. These insertions result in two sharp kinks at either end of TATA and induce bending of ~90° toward the major groove [6]. Bending is reduced to ~30° by single substitutions in the canonical TATAAAAG sequence [18]. The TBP turnover on TATA-containing promoters is significantly higher than that on non-TATA promoters in yeast [114], indicating that it is a highly regulated process. The TATA box-TBP binding is negatively regulated by cellular factors such as nucleosome, NC2, and BTAF1 (Mot1p) [105]. On the contrary, the TATA box-TBP binding could be enhanced by TFIIA, TFIIB, and SWI/SNF [105]. All these regulatory factors are proteins, and other regulators of this step are still waiting to be identified.

7.2 Discovery and Overview of RNAa Induced by TATA Box-Targeting miRNAs

7.2.1 Discovery of HIV-1-Encoded miRNA, miR-H3, Which Targets HIV-1 TATA Box

Virus-encoded miRNAs were initially identified from Epstein-Barr viruses (EBV) [86]. Since then, increasing virus-encoded miRNAs have been identified [87, 108]. Most of these miRNAs were reported in DNA viruses such as herpes and polyoma viruses, but rarely in RNA viruses [108]. Because of the rapid developments of next-generation sequencing technology that is much more sensitive and quantitative than the conventional cDNA clone sequencing method, increasing RNA virus-derived miRNAs have been discovered, especially from human immunodeficiency virus type 1 (HIV-1), West Nile virus (WNV), and bovine leukemia virus [38, 49, 52, 94].

It has been reported that HIV-1 encodes miRNAs and other small RNAs. Bennasser et al. first performed a computational prediction on HIV-1-encoded miRNAs and found five pre-miRNAs candidates [5]. Subsequently, several groups identified HIV-1-encoded miRNAs from the *Nef* and the *TAR* element [50, 51, 80, 82]. Through the next-generation sequencing method, a number of HIV-1-encoded small RNAs were discovered, some of which exhibit the features of miRNA or small interfering RNA (siRNA) [32, 94, 118]. Interestingly, to prevent cleavage of their RNA genome, retroviruses use an alternative RNA source as miRNA precursor. Harwig et al. revealed that nonprocessive transcription from the HIV-1 LTR promoter results in the production of TAR-encoded miRNA-like small RNAs [32]. These TAR RNAs are cleaved by Dicer and the processing is stimulated by the viral Tat protein. Through this special biogenesis pathway HIV-1 produces the TAR-derived miRNA without cleavage of the RNA genome. The HIV-1-derived small RNAs have been shown to modulate the cellular and/or viral gene expression. A Nef-derived miRNA, miR-N367, could block HIV-1 Nef expression in vitro [80]. The expression of the TAR-derived miRNA could protect the infected cells from apoptosis by downregulating cellular genes involved in apoptosis [51, 83].

Nevertheless, the low abundance of HIV-1-derived small RNAs has fueled debates on the existence and function of HIV-1-encoded miRNAs [108]. Through a strategy combining computational prediction and next-generation sequencing, our group identified a novel HIV-1-encoded miRNA, miR-H3 [125]. miR-H3 locates in the mRNA region encoding the active center of reverse transcriptase and exhibits high sequence conservation among different subtypes of HIV-1. Overexpression of miR-H3 increases viral production, and the mutations in the miR-H3 sequence significantly impair the viral replication of wildtype HIV-1_{NL43}, suggesting that it is a viral replication-enhancing miRNA. Interestingly, miR-H3 targets HIV-1 5′ LTR TATA box and activates the viral promoter in a sequence-specific manner to upregulate HIV-1 RNA and protein expression. This is the first report that miRNA specifically targets the core promoter TATA box and activates gene transcription. In contrast to the cytoplasmic miRNAs that require a considerable abundance to bind their massive mRNA targets, the required abundance for the TATA box-targeting miRNAs to fulfill their function has significantly reduced, given that their target has very limited copy number in the nucleus.

7.2.2 Systematic Identification of Cellular TATA Box-Targeting miRNAs

The discovery of miR-H3 evoked the question whether there are cellular miRNAs that also target the TATA box motif of endogenous promoters to activate transcription. The investigation was initiated by searching cellular miRNAs that directly interact with general transcription factors, including RNA Pol II and TBP. Through an RNA-ChIP (Chromatin immunoprecipitation) method, our group found that a fraction of small RNAs of ~22 nt length was significantly enriched by an anti-Pol II antibody in human peripheral blood mononuclear cells (PBMCs), suggesting an association between small RNAs and Pol II [126]. The RNA Pol II-associated small RNAs were further analyzed by miRNA microarray and a number of miRNAs were identified. The RNA-ChIP followed by quantitative RT-PCR (qRT-PCR) was used to confirm the association of Pol II/TBP with some miRNAs, including let-7i, miR-145, and miR-16. Moreover, when treated with DNase I, the association of miRNAs with both Pol II and TBP was significantly reduced, suggesting that this association is DNA dependent. We further investigated whether the miRNA-binding proteins, Argonautes (the key RNAi components), are associated with the core transcription machinery in human cells. Through co-immunoprecipitation assay, we found that both AGO1 and AGO2 bound to Pol II and TBP in HEK293T cells. These results suggest that cellular miRNAs and their binding proteins are associated with the RNA Pol II core transcription machinery in human cells [126].

Computer prediction of miRNA-binding site(s) on the gene core promoter region was then performed with the RNA-hybrid web server [56]. Interestingly, a nearly perfect binding between *interleukin-2* (*IL-2*) core promoter region and let-7i was predicted with a remarkable minimum free energy (MFE) value (−27.6 kcal/mol) [126], suggesting that let-7i might interact with the core promoter and be involved in IL-2 transcription regulation. To determine whether the binding of miRNA to TATA box core promoter is a universal phenomenon, our group further performed a comprehensive computational screen of the putative bindings between 354 TATA box-containing human gene promoters and 1223 human miRNAs from miRBase ver. 16 [54]. When the cutoff of MFE was set to −27 kcal/mol, 13,323 putative bindings were found in the ± 20-bp region centered with the TATA box motif. From 16 gene promoters experimentally examined, four miRNAs, miR-138, miR-92a, let-7c, and miR-181d, were identified to enhance the promoter activity of *insulin*, *calcitonin*, *histone H4-A1*, or *c-myc*, respectively in HEK293T cells. Among these miRNAs, let-7c and miR-138 had a nearly equal distribution in the nucleus and the cytoplasm, while miR-92a and miR-181d were relatively enriched in the cytoplasm [65, 126].

7.3 Physiological Significance of RNAa Induced by TATA Box-Targeting miRNAs

7.3.1 RNAa Induced by TATA Box-Targeting miRNA and HIV-1 Replication

miRNAs play important roles in the interaction between viruses and their hosts. Cellular miRNAs could affect the viral replication and latency and mediate antiviral defense [31]. For example, miR-122 that is enriched in the liver plays a key role in the accumulation of viral RNAs of hepatitis C viruses [45]. A cellular miRNA effectively restricts the accumulation of the retrovirus primate foamy virus type 1 (PFV-1) in human cells [59]. Our group reported that several miRNAs from resting human $CD4^+$ T cells repress the translation of viral proteins and contribute to the latency of HIV-1 [34]. Conversely, viral miRNAs could facilitate viral propagation through reducing viral antigens or impairing the host antiviral immune response. For instance, SV40 miR-S1 downmodulates the production of the viral T antigen (TAg), an early protein which is not required during late infection [100]. An EBV-encoded miRNA, miR-BART5, targets the proapoptotic factor PUMA to promote host cell survival [11]. An hCMV miRNA, miR-UL112-1, was reported to inhibit the expression of the stress-induced ligand MICB and enable hCMV to escape from the immune surveillance by NK cells [99].

Since TATA box represents the most conserved and widespread core promoter, which is enriched in the genes of tissue-specific expression or viral origin [46, 93], the TATA box-targeting miRNAs are expected to modulate such highly regulated or viral gene expression. Our study found that the HIV-1-derived miRNA, miR-H3, targets the TATA box motif of HIV-1 5′ LTR and activates viral transcription. The RNAa effects by miR-H3 were verified in several ways, including HIV-1 5′ LTR-driven luciferase activity in TZM-bl cells, viral RNA accumulation in virus-producing cells, virus-like particle (VLP) production from Env-defective HIV-1 clone pNL4-3-deltaE-EGFP, and replication of wildtype or miR-H3-mutated HIV-1_{NL43} in activated human $CD4^+$ T-lymphocytes. Mutations in the viral genomic sequence of miR-H3, without altering the codon and activity of reverse transcriptase, impaired the miRNA processing and the normal replication of HIV-$1_{NL4–3}$ in primary human $CD4^+$ T cells, suggesting that miR-H3 is a positive regulator for viral replication in physiological conditions [125].

HIV-1 latency in resting $CD4^+$ T cells is the major obstacle for the eradication of viruses from HIV-1-infected patients on suppressive highly active antiretroviral therapy (HAART). As miR-H3 exhibits enhancement activity on HIV-1 transcription, we examined whether this TATA box-targeting miRNA or its mimic small activating RNA (saRNA) could be used to activate HIV-1 latency. One saRNA designed according to miR-H3 sequence, si-HIV-TATA-msig, enhanced the HIV-1 promoter activity and VLP production efficiently in HEK293T cells [125]. When treating the resting $CD4^+$ T cells isolated from HIV-1-infected individuals receiving suppressive HAART with these small RNAs, the viral particles generated from these resting $CD4^+$ T cells were significantly more than those treated with a negative control small RNA. These data indicate that an saRNA complementary to HIV-1 TATA box alone is able to activate HIV-1 transcription in the latently infected resting $CD4^+$ T-lymphocytes [125].

Several approaches have been developed to activate latent viruses transcription for killing, including by activating T-lymphocytes with IL-2 or IL-2 plus anti-CD3/anti-CD28 antibody [13, 69], protein kinase C (PKC) activators [e.g., prostatin [57]], or activating transcription with histone deacetylases inhibitors (HDACi) without inducing host cell activation (such as, valproic acid and suberoylanilide hydroxamic acid) [1, 15, 120]. However, the first approach has been shown to cause serious cytotoxic effects, and the PKC agonists and HDACi are speculated for causing global gene expression activation with unpredictable side effects. Thus, an HIV-1 provirus-specific activating reagent is ideal for purging the latent reservoir. Our study demonstrated that an HIV-1-encoded miRNA, miR-H3, could activate HIV-1 transcription in a sequence-specific manner [125]. Together with our previous finding that some cellular miRNAs contribute to the latency of HIV-1 by inhibiting HIV-1 production [34], a combination of the small RNA(s) targeting HIV-1 TATA box and the inhibitors of these cellular miRNAs will provide an HIV-1-specific and much safer approach for eradicating HIV-1 latent reservoir [124].

7.3.2 RNAa Induced by TATA Box-Targeting miRNA and $CD4^+$ T Cell Death in HIV-1 Infection

The causes of $CD4^+$ T cell depletion in acquired immunodeficiency syndrome patients have not been fully elucidated. Several predisposing factors have been reported to contribute to HIV-1-induced $CD4^+$ T cell death [26]. For example, viral proteins, such as Tat, Nef, Vpr, and Env, can induce cell death [2, 42, 63, 115]. The integration of proviral DNA into the host chromosome is also a trigger of cell death [16]. Recently, Doitsh et al. reported that most $CD4^+$ T cell death during HIV-1 infection is caused by caspase-1-mediated pyroptosis triggered by abortive viral infection [19].

Interleukin-2 is a key cytokine that regulates the proliferation, differentiation, and survival of T cells [22]. By promoting the differentiation of T cells into effector T cells, memory T cells and T helper cells, following stimulation with an antigen, IL-2 activates immune responses to help the host counteract the invasion of pathogens [66]. It has been reported that HIV-1-infection of $CD4^+$ T cells leads to abnormal expression of the IL-2 gene and disturbs the efficient antiviral immune responses mediated by IL-2 [20, 47, 96, 98, 121]. The gradual loss of IL-2 secretion and proliferation is an early sign of T cell exhaustion in HIV-1 infection [90]. IL-2 is also a key cytokine for maintaining the viability of activated $CD4^+$ T cells by inducing *bcl-2*, *c-myc*, and other genes [67, 73]. The administration of IL-2 to HIV-1-infected individuals could significantly increase $CD4^+$ T cell counts compared with antiretroviral therapy alone [28, 53, 84]. Our work demonstrated that cellular miRNA let-7i activates IL-2 gene transcription through targeting the promoter TATA box region and functions as a positive regulator of IL-2 gene expression [126]. And the impaired expressions of several let-7 family members have been observed in chronic HIV-1-infected patients [101]. Thus, it is intriguing to investigate the possible connection among them in the context of HIV-1 infection.

In a recent study, our group revealed that HIV-1 infection decreased the expression of let-7i in $CD4^+$ T cells by attenuating its promoter activity, which led to a decline in IL-2 levels [127]. A let-7i mimic could increase IL-2 expression and subsequently enhance the resistance of $CD4^+$ T cells to HIV-1-induced apoptosis. By contrast, the blockage of let-7i with a specific inhibitor resulted in elevated $CD4^+$ T cell apoptosis during HIV-1 infection. Furthermore, by blocking the IL-2 pathway, we found that the let-7i-mediated $CD4^+$ T cell resistance to apoptosis during HIV-1 infection was dependent on IL-2 signaling rather than an alternative CD95-mediated cell death pathway. Taken together, our findings reveal a novel pathway for HIV-1-induced dysregulation of IL-2 cytokines and depletion of $CD4^+$ T-lymphocytes. This work reveals a novel mode of HIV-1 and host interaction, which is centered on miRNA. First, the HIV-1 infection impairs the biogenesis of host miRNA let-7i. Next, as let-7i is an activator of IL-2 expression through RNAa effect, the dysregulation of the miRNA leads to abnormal expression of the important cytokine that modulates immune responses during viral infection. These findings indicate that HIV-1 has profoundly hijacked the gene regulation network of its host.

Since the binding between miRNA and TATA box motif is sequence specific, the regulation of transcription by miRNAs is more specific and accurate than that by protein transcription factors. Accumulating evidence has demonstrated that the biogenesis and function of miRNAs are also regulated by many signal transduction pathways [17, 55, 79]. Therefore, our findings indicate a novel signal pathway to specifically regulate gene expression at transcriptional level. This is in accordance with the observation that a number of miRNAs are found in the nucleus. Moreover, since the binding between let-7i miRNA and IL-2 promoter TATA box was also predicted in other mammals and validated in mice, it is likely that the enhancement of gene transcription by cellular miRNAs via targeting the TATA box motif is conserved during evolution, at least in mammals and viruses [126].

7.4 Mechanism of RNAa Induced by TATA Box-Targeting miRNAs

7.4.1 Current Model of Promoter-Targeting RNAa Mediated by Small RNAs

The mechanism(s) of RNAa induced by promoter-targeting small RNAs is still largely elusive, although recent studies have provided insights into several important aspects of RNAa, such as the target molecule of RNAa, dependence of RNAa on AGO proteins, and epigenetic changes associated with transcriptional activation [91]. Several studies reported that the noncoding transcript overlapping with the promoter sequence, or promoter-associated RNA (paRNA), is the target of saRNA. For example, noncoding antisense transcript (i.e., synthesized in the opposite direction of target gene transcription) that runs through the promoter has been shown to be involved in the activation of the progesterone receptor (PR) and low-density lipoprotein receptor (LDLR) genes by promoter-targeting saRNA [71, 95]. In another case, a noncoding sense transcript (i.e., synthesized in the same direction of target gene transcription) recruits antigene RNA (agRNA) to activate PR expression [123]. In these cases, the sense or antisense noncoding transcripts might serve as docking sites for the complementary saRNAs and AGO proteins. It is noteworthy that the binding of agRNAs does not affect the abundance of these paRNAs, suggesting that an RNAi cleavage mechanism is not involved [12].

Both RNA and DNA can be the target molecule of activating miRNAs. Since all RNAa events rely on the sequence complementary between saRNAs and their target promoter, there should be base pair interactions between them, whether through a direct interaction between saRNA and promoter DNA or an indirect interaction mediated by noncoding transcripts overlapping the promoter. In the condition of noncoding transcript(s)-mediated interaction, the noncoding transcript(s) might serve as a scaffold for the assembly of AGO-saRNA complex, and then they interact with the promoter to affect its transcription [95]. While in the condition without noncoding transcript, saRNA might directly interact with the target promoter to form DNA:RNA hybrids. In plants, RNA-directed DNA methylation (RdDM) only occurs to the cytosines along RNA-DNA duplexes, indicating a direct RNA-DNA interaction that provides a strong and specific signal for de novo DNA methylation [72, 85]. Thus, it is possible that, with the help of certain proteins such as AGO and general transcription factors, saRNA can interact directly with promoter DNA without the need of paRNA as a docking molecule. This viewpoint has been further supported by our recent observations [126].

RNAa is dependent on AGO proteins, especially AGO2. When AGO2 was knocked down, RNAa-mediated induction of E-cadherin and p21 was abolished, whereas knockdown of the other AGO members (AGO1, 3, and 4) did not significantly impair RNAa [64, 89]. Using ChIP assays, AGO2 has been shown to associate with the antisense transcripts of the PR and LDLR promoter in an agRNA-dependent manner [12, 71]. Moreover, our work showed that AGO1 and AGO2 are associated with Pol II and TBP, the core factors of RNA Pol II transcription machinery [126].

Several studies have shown an increase in RNA Pol II association with the core promoter, indicating that RNAa is a transcriptional mechanism. At the same time, various types of chromatin modification changes have also been identified at different promoters associated with RNAa (e.g., loss of di- and tri-methylation at histone H3K9 at the E-cadherin promoter, reduced acetylation at H3K9 and H3K4 together with increased di- and tri-methylation at H4K4 at the PR promoter, increased H3K4me2 and H3K4me3 levels, as well as decreased H3K9me2 at the mouse VEGF promoter [44, 64, 107]). It seems that there is not a unique pattern of chromatin modification change that is associated with RNAa up to now, which makes it difficult to clarify a causal relationship between them.

7.4.2 No Evidence of paRNA or Epigenetic Modification Change Is Involved in RNAa Induced by TATA Box-Targeting miRNAs

Unlike other studies of the promoter-targeting RNAa [44, 64, 88], the TATA box-targeting miRNAs target a quite unique site of the gene promoter. The current data demonstrate that this gene activation is through direct interaction between miRNA and the TATA box core promoter, which facilitates the PIC assembly and transcription initiation [126]. Mutation assay in both the IL-2 promoter and let-7i miRNA indicated that their interaction is sequence specific [126]. Although some reports suggest that noncoding RNAs derived from the promoter region are the mediator of small RNA-induced chromatin epigenetic modifications [12, 76, 95], the attempt to detect IL-2 promoter-associated noncoding transcript did not yield any positive signal and overexpression of the speculated paRNA did not affect IL-2 promoter activity either. In addition, the TATA box-targeting miRNAs did not induce the change of histone epigenetic modifications of the target gene promoter, such as histone acetylation and methylation. However, similar to other RNAa, the AGO proteins are involved in the TATA box-targeting miRNA-induced RNAa. For instance, the miRNAs binding proteins AGO1 and AGO2 are associated with the general transcription factors such as Pol II and TBP, suggesting an interaction between the "nuclear RISC" and the RNA Pol II core transcription machinery. The knockdown of AGO1 and AGO2 proteins significantly impaired the activating effect of the IL-2 promoter by let-7i, indicating that both AGO1 and AGO2 may be recruited to the IL-2 promoter by let-7i for its regulation [126].

7.4.3 TATA Box-Targeting miRNAs Facilitate Pol II Pre-initiation Complex Assembly and Transcription Initiation

The RNA polymerase II transcription initiation is quite a complicated process. Assembly of the preinitiation complex and its postassembly control are critical early steps in the transcription of eukaryotic genes. The TBP turnover on TATA-containing promoters is significantly higher than that on non-TATA promoters in yeast [114], indicating that it is a highly regulated process. Our results reveal that miRNAs could directly interact with the TATA box motifs and associate with TBP [126] to enhance transcription, implying that the miRNAs are involved in the TBP turnover on TATA-containing promoter and affect the assembly of PICs. The formation of TFIID-TFIIA-promoter DNA complex (D-A complex) is important for transcription activation [112]. It is also reported that the isomerized D-A complex is sufficient for gene activation [10]. Certain gene activators and coactivators could facilitate the formation of D-A complex, but the mechanism has not been fully clarified [9, 24, 97, 116]. The TATA box-targeting miRNAs might have facilitated the D-A complex formation to activate transcription via binding to the core promoters.

Although several studies suggest that small RNAs-induced transcriptional gene silence or activation is through epigenetic programming in the promoters, the details on how these small RNAs guide these modifications in a sequence-specific manner are still elusive [35, 36, 48, 74, 88, 92, 122]. In any model, the RNA:DNA interaction seems to be an unavoidable issue to ensure the specificity of the epigenetic modifications on the target promoter. However, the way in which they interact needs to be further discussed. Given that the fully assembled PIC is over two million daltons, it is likely that a small RNA could be included within such a large complex. During transcription initiation, the TBP binds to the TATA box motif and unwinds the DNA double helix [6], which could enable the pairing between a complementary small RNA and its single-strand DNA target in the nucleus. The interaction of small noncoding RNA with the single-stranded DNA may stabilize the PICs under certain physiological conditions. It is also possible that miRNA function as a "wedge" to maintain the "open" status of the transcription initiation site and to improve the processivity of Pol II by facilitating the loading of new Pol II units to the vicinity of the transcription start site (Fig. 7.1). Moreover, small RNAs might interact with the double-strand promoter DNAs as suggested before [36]. The binding of small RNA-AGO complexes to the TATA box motif could recruit TBP and other general transcription factors to the core promoter [126] and facilitate the assembly of PICs. After TBP has bound to the TATA box, the small RNAs may be released from this special site. However, these hypotheses need to be clarified by further studies. A crystal structure analysis will be quite useful to verify whether small RNAs are incorporated into the PICs assembled on the TATA box core promoter.

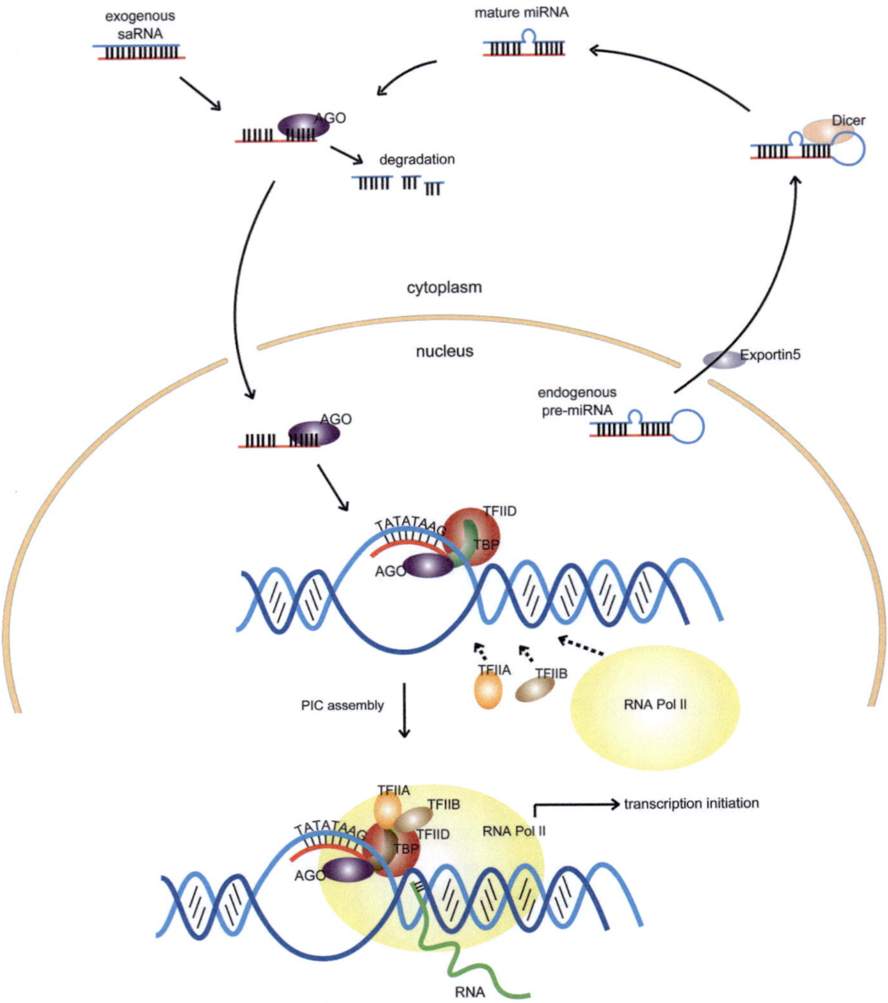

Fig. 7.1 Mechanism of RNAa induced by TATA box-targeting small RNAs. In cell cytoplasm, the guide strand of exogenous saRNA or endogenous mature miRNA is loaded into AGO protein (e.g., AGO2) to form an AGO-RNA complex. Then the AGO-RNA complex is imported into the cell nucleus by free diffusion or with the help of unidentified factor(s). In the nucleus, the small RNA guides the AGO-RNA complex to the TATA box of the target gene, which has been opened by TBP (TFIID). The binding of AGO-RNA complex stabilizes the open status of TATA box, which facilitates the assembly of preinitiation complex and transcription initiation of RNA Pol II. AGO, Argonaut proteins; saRNA, small activating RNA; miRNA, microRNA. TATA box is represented by a "TATATAAG" sequence

A recent report suggested that the key protein components of the RNAi pathway, i.e., DICER2 and AGO2, are associated with RNA Pol II core transcription machinery [8], while our study further reveals that cellular miRNAs could directly interact with the core promoter. These findings together suggest that the miRNA-

Argonaute complexes in the nucleus (nuclear RISCs) could interact with the core transcription machine and participate in different steps of the transcription process.

7.5 Rational Design of TATA Box-Targeting saRNAs to Activate Gene Expression

siRNAs are widely used to repress gene expression by targeting mRNAs [25, 43, 60, 78]. However, so far, there is not an easy-to-use method to upregulate gene expression specifically. Increasing evidence shows that cellular miRNAs and synthetic small RNAs could inhibit or induce gene expression through targeting gene promoters. However, the information about the unique feature of the activating small RNAs is still lacking. And it is also difficult to predict whether a promoter-targeting small RNA will either up- or downregulate gene expression. Thus, there is a need for a simple and efficient method to design small activating RNAs.

To mimic the RNAa effect induced by TATA box-targeting miRNA, the saRNAs targeting the TATA box region were designed and optimized to activate gene expression [21]. These TATA box-targeting saRNAs effectively activated 14 out of 16 gene promoters. Furthermore, we find the following features can improve the activation efficiency of the TATA box-targeting saRNAs: (a) complementary to the TATA box-centered region; (b) UA (uracil and adenine) usage at the first two bases of the antisense strand; (c) 23 nts in length; (d) 2′-O-Methyl (2′-OMe) modification at the 3′ terminus of the antisense strand; and (e) avoiding mismatches at the 3′ end of the antisense strand.

The simplified design procedure of activating small RNAs promises its good potential in basic research and therapeutic applications. For instance, it is well known that it is very difficult to transfect some primary cells such as peripheral blood mononuclear cell with plasmid DNA [33, 62]. When the expression level of a certain gene needs to be elevated, the specific TATA box-targeting saRNAs instead of plasmid DNA could be transfected into the primary cells to activate gene expression, which would be much easier and more efficient. Moreover, the application of TATA box-targeting saRNAs on certain highly interested target genes, such as *insulin*, tumor repressor genes (*p21*), and DNA repair genes (*BRCA2*), may provide a novel treatment strategy for diseases such as diabetes and cancer. Further studies should be conducted to find the target position(s) and the sequence characteristics of saRNAs that could enhance the promoter activities of the genes without the TATA box motif.

7.6 Summary

Recent studies revealed that some nuclear miRNAs and synthesized siRNAs bind to gene promoter and activate gene transcription (the RNAa phenomenon). A paradigm of how these small RNAs work is that they target dispersed positions of gene promoter, and interact with the sense or antisense transcript of promoter as well as AGO proteins, to change the epigenetic modifications of target promoter and activate gene expression. Interestingly, we identified a novel HIV-1-encoded miRNA, miR-H3, that binds specifically to the core promoter TATA box to activate viral gene expression. The depletion of this miRNA significantly impaired the viral replication, suggesting that it is a positive regulator of viral propagation. The treatment of miR-H3 mimics could activate viruses from the $CD4^+$ T cells isolated from the patients receiving HAART, which is very intriguing for reducing the HIV-1 latent reservoir as a specific latency reversing reagent.

We further revealed that many cellular miRNAs also function as miR-H3 by targeting gene core promoter TATA box to induce gene expression. For instance, let-7i targets the TATA box region of the IL-2 gene promoter and upregulates its expression in T-lymphocytes. Given that IL-2 is an important cytokine controlling the proliferation, differentiation, and survival of T cells, let-7i becomes a part of the regulatory network of T cell homeostasis. Especially during HIV-1 infection, the expression levels of both IL-2 and let-7 miRNA family are decreased, but whether they were correlated was unknown. Our findings bridged this gap by showing that let-7i is a positive regulator of IL-2 gene expression, and the suppression of let-7i/IL-2 axis in HIV-1 infection contributes to the depletion of $CD4^+$ T cells.

Unlike many other reports of RNAa induced by miRNAs, no evidence of the promoter-associated transcript is involved in the RNAa induced by TATA box-targeting miRNAs, nor were significant change of epigenetic modifications observed on the promoter of the target gene. AGO proteins, e.g., AGO 1 and AGO 2, are still indispensable for the activation induced by TATA box-targeting miRNAs. Thus, we proposed a model wherein the binding of small RNA-AGO complexes to the TATA box motif could facilitate the assembly of RNA Polymerase II PIC and transcription initiation. The synthesized small RNAs that target the TATA box can efficiently activate transcription of interested genes, such as *insulin*, *IL-2*, and *c-Myc*. Several features of the TATA box-targeting siRNAs are optimized to improve their activation efficiency. The discovery of RNAa induced by TATA box-targeting miRNA provides us an easy-to-use tool for activating gene expression.

References

1. Archin NM, Espeseth A, Parker D, Cheema M, Hazuda D, Margolis DM (2009) Expression of latent HIV induced by the potent HDAC inhibitor suberoylanilide hydroxamic acid. AIDS Res Hum Retrovir 25:207–212

2. Badley AD, Dockrell D, Simpson M, Schut R, Lynch DH, Leibson P, Paya CV (1997) Macrophage-dependent apoptosis of CD4+ T lymphocytes from HIV-infected individuals is mediated by FasL and tumor necrosis factor. J Exp Med 185:55–64
3. Bartel DP (2004) MicroRNAs: genomics, biogenesis, mechanism, and function. Cell 116:281–297
4. Bartel DP (2009) MicroRNAs: target recognition and regulatory functions. Cell 136:215–233
5. Bennasser Y, Le SY, Yeung ML, Jeang KT (2004) HIV-1 encoded candidate micro-RNAs and their cellular targets. Retrovirology 1:43
6. Burley SK, Roeder RG (1996) Biochemistry and structural biology of transcription factor IID (TFIID). Annu Rev Biochem 65:769–799
7. Bushati N, Cohen SM (2007) microRNA functions. Annu Rev Cell Dev Biol 23:175–205
8. Cernilogar FM, Onorati MC, Kothe GO, Burroughs AM, Parsi KM, Breiling A, Lo Sardo F, Saxena A, Miyoshi K, Siomi H, Siomi MC, Carninci P, Gilmour DS, Corona DF, Orlando V (2011) Chromatin-associated RNA interference components contribute to transcriptional regulation in drosophila. Nature 480:391–395
9. Chi T, Carey M (1993) The ZEBRA activation domain: modular organization and mechanism of action. Mol Cell Biol 13:7045–7055
10. Chi T, Carey M (1996) Assembly of the isomerized TFIIA – TFIID – TATA ternary complex is necessary and sufficient for gene activation. Genes Dev 10:2540–2550
11. Choy EY, Siu KL, Kok KH, Lung RW, Tsang CM, To KF, Kwong DL, Tsao SW, Jin DY (2008) An Epstein-Barr virus-encoded microRNA targets PUMA to promote host cell survival. J Exp Med 205:2551–2560
12. Chu Y, Yue X, Younger ST, Janowski BA, Corey DR (2010) Involvement of argonaute proteins in gene silencing and activation by RNAs complementary to a non-coding transcript at the progesterone receptor promoter. Nucleic Acids Res 38:7736–7748
13. Chun TW, Engel D, Mizell SB, Hallahan CW, Fischette M, Park S, Davey RT Jr, Dybul M, Kovacs JA, Metcalf JA, Mican JM, Berrey MM, Corey L, Lane HC, Fauci AS (1999) Effect of interleukin-2 on the pool of latently infected, resting CD4+ T cells in HIV-1-infected patients receiving highly active anti-retroviral therapy. Nat Med 5:651–655
14. Coleman RA, Pugh BF (1995) Evidence for functional binding and stable sliding of the TATA binding protein on nonspecific DNA. J Biol Chem 270:13850–13859
15. Contreras X, Schweneker M, Chen CS, Mccune JM, Deeks SG, Martin J, Peterlin BM (2009) Suberoylanilide hydroxamic acid reactivates HIV from latently infected cells. J Biol Chem 284:6782–6789
16. Cooper A, Garcia M, Petrovas C, Yamamoto T, Koup RA, Nabel GJ (2013) HIV-1 causes CD4 cell death through DNA-dependent protein kinase during viral integration. Nature 498:376–379
17. Cullen BR (2004) Transcription and processing of human microRNA precursors. Mol Cell 16:861–865
18. Delgadillo RF, Whittington JE, Parkhurst LK, Parkhurst LJ (2009) The TATA-binding protein core domain in solution variably bends TATA sequences via a three-step binding mechanism. Biochemistry 48:1801–1809
19. Doitsh G, Galloway NL, Geng X, Yang Z, Monroe KM, Zepeda O, Hunt PW, Hatano H, Sowinski S, Munoz-Arias I, Greene WC (2014) Cell death by pyroptosis drives CD4 T-cell depletion in HIV-1 infection. Nature 505:509–514
20. Fan J, Bass HZ, Fahey JL (1993) Elevated IFN-gamma and decreased IL-2 gene expression are associated with HIV infection. J Immunol 151:5031–5040
21. Fan M, Zhang Y, Huang Z, Liu J, Guo X, Zhang H, Luo H (2014) Optimizations of siRNA design for the activation of gene transcription by targeting the TATA-box motif. PLoS One 9: e108253
22. Gaffen SL, Liu KD (2004) Overview of interleukin-2 function, production and clinical applications. Cytokine 28:109–123

23. Gagnon KT, Li L, Chu Y, Janowski BA, Corey DR (2014) RNAi factors are present and active in human cell nuclei. Cell Rep 6:211–221
24. Ge H, Roeder RG (1994) Purification, cloning, and characterization of a human coactivator, PC4, that mediates transcriptional activation of class II genes. Cell 78:513–523
25. Gitlin L, Karelsky S, Andino R (2002) Short interfering RNA confers intracellular antiviral immunity in human cells. Nature 418:430–434
26. Gougeon ML (2003) Apoptosis as an HIV strategy to escape immune attack. Nat Rev Immunol 3:392–404
27. Gregory RI, Chendrimada TP, Cooch N, Shiekhattar R (2005) Human RISC couples microRNA biogenesis and posttranscriptional gene silencing. Cell 123:631–640
28. Group IES, Committee SS, Abrams D, Levy Y, Losso MH, Babiker A, Collins G, Cooper DA, Darbyshire J, Emery S, Fox L, Gordin F, Lane HC, Lundgren JD, Mitsuyasu R, Neaton JD, Phillips A, Routy JP, Tambussi G, Wentworth D (2009) Interleukin-2 therapy in patients with HIV infection. N Engl J Med 361:1548–1559
29. Guang S, Bochner AF, Pavelec DM, Burkhart KB, Harding S, Lachowiec J, Kennedy S (2008) An Argonaute transports siRNAs from the cytoplasm to the nucleus. Science 321:537–541
30. Guang S, Bochner AF, Burkhart KB, Burton N, Pavelec DM, Kennedy S (2010) Small regulatory RNAs inhibit RNA polymerase II during the elongation phase of transcription. Nature 465:1097–1101
31. Harwig A, Das AT, Berkhout B (2014) Retroviral microRNAs. Curr Opin Virol 7:47–54
32. Harwig A, Jongejan A, VAN Kampen AH, Berkhout B, DAS AT (2016) Tat-dependent production of an HIV-1 TAR-encoded miRNA-like small RNA. Nucleic Acids Res 44 (9):4340–4353
33. Herweijer H, Wolff JA (2003) Progress and prospects: naked DNA gene transfer and therapy. Gene Ther 10:453–458
34. Huang J, Wang F, Argyris E, Chen K, Liang Z, Tian H, Huang W, Squires K, Verlinghieri G, Zhang H (2007) Cellular microRNAs contribute to HIV-1 latency in resting primary CD4+ T lymphocytes. Nat Med 13:1241–1247
35. Huang V, Place RF, Portnoy V, Wang J, Qi Z, Jia Z, Yu A, Shuman M, Yu J, Li LC (2012) Upregulation of Cyclin B1 by miRNA and its implications in cancer. Nucleic Acids Res 40:1695–1707
36. Huang XA, Yin H, Sweeney S, Raha D, Snyder M, Lin H (2013) A major epigenetic programming mechanism guided by piRNAs. Dev Cell 24:502–516
37. Huisinga KL, Pugh BF (2004) A genome-wide housekeeping role for TFIID and a highly regulated stress-related role for SAGA in Saccharomyces cerevisiae. Mol Cell 13:573–585
38. Hussain M, Torres S, Schnettler E, Funk A, Grundhoff A, Pijlman GP, Khromykh AA, Asgari S (2012) West Nile virus encodes a microRNA-like small RNA in the 3′ untranslated region which up-regulates GATA4 mRNA and facilitates virus replication in mosquito cells. Nucleic Acids Res 40:2210–2223
39. Hutvagner G, Zamore PD (2002) A microRNA in a multiple-turnover RNAi enzyme complex. Science 297:2056–2060
40. Hutvagner G, Mclachlan J, Pasquinelli AE, Balint E, Tuschl T, Zamore PD (2001) A cellular function for the RNA-interference enzyme Dicer in the maturation of the let-7 small temporal RNA. Science 293:834–838
41. Hwang HW, Wentzel EA, Mendell JT (2007) A hexanucleotide element directs microRNA nuclear import. Science 315:97–100
42. Jacotot E, Ravagnan L, Loeffler M, Ferri KF, Vieira HL, Zamzami N, Costantini P, Druillennec S, Hoebeke J, Briand JP, Irinopoulou T, Daugas E, Susin SA, Cointe D, Xie ZH, Reed JC, Roques BP, Kroemer G (2000) The HIV-1 viral protein R induces apoptosis via a direct effect on the mitochondrial permeability transition pore. J Exp Med 191:33–46
43. Jacque JM, Triques K, Stevenson M (2002) Modulation of HIV-1 replication by RNA interference. Nature 418:435–438

44. Janowski BA, Younger ST, Hardy DB, Ram R, Huffman KE, Corey DR (2007) Activating gene expression in mammalian cells with promoter-targeted duplex RNAs. Nat Chem Biol 3:166–173
45. Jopling CL, Yi M, Lancaster AM, Lemon SM, Sarnow P (2005) Modulation of hepatitis C virus RNA abundance by a liver-specific MicroRNA. Science 309:1577–1581
46. Juven-Gershon T, Hsu JY, Theisen JW, Kadonaga JT (2008) The RNA polymerase II core promoter – the gateway to transcription. Curr Opin Cell Biol 20:253–259
47. Kassu A, D'souza M, O'connor BP, Kelly-Mcknight E, Akkina R, Fontenot AP, Palmer BE (2009) Decreased 4-1BB expression on HIV-specific CD4+ T cells is associated with sustained viral replication and reduced IL-2 production. Clin Immunol 132:234–245
48. Kawasaki H, Taira K (2004) Induction of DNA methylation and gene silencing by short interfering RNAs in human cells. Nature 431:211–217
49. Kincaid RP, Burke JM, Sullivan CS (2012) RNA virus microRNA that mimics a B-cell oncomiR. Proc Natl Acad Sci U S A 109:3077–3082
50. Klase Z, Kale P, Winograd R, Gupta MV, Heydarian M, BERRO R, Mccaffrey T, Kashanchi F (2007) HIV-1 TAR element is processed by Dicer to yield a viral micro-RNA involved in chromatin remodeling of the viral LTR. BMC Mol Biol 8:63
51. Klase Z, Winograd R, Davis J, Carpio L, Hildreth R, Heydarian M, Fu S, Mccaffrey T, Meiri E, Ayash-Rashkovsky M, Gilad S, Bentwich Z, Kashanchi F (2009) HIV-1 TAR miRNA protects against apoptosis by altering cellular gene expression. Retrovirology 6:18
52. Klase ZA, Sampey GC, Kashanchi F (2013) Retrovirus infected cells contain viral microRNAs. Retrovirology 10:15
53. Kovacs JA, Baseler M, Dewar RJ, Vogel S, Davey RT Jr, Falloon J, Polis MA, Walker RE, Stevens R, Salzman NP et al (1995) Increases in CD4 T lymphocytes with intermittent courses of interleukin-2 in patients with human immunodeficiency virus infection. A preliminary study. N Engl J Med 332:567–575
54. Kozomara A, Griffiths-Jones S (2011) miRBase: integrating microRNA annotation and deep-sequencing data. Nucleic Acids Res 39:D152–D157
55. Krol J, Loedige I, Filipowicz W (2010) The widespread regulation of microRNA biogenesis, function and decay. Nat Rev Genet 11:597–610
56. Kruger J, Rehmsmeier M (2006) RNAhybrid: microRNA target prediction easy, fast and flexible. Nucleic Acids Res 34:W451–W454
57. Kulkosky J, Culnan DM, Roman J, Dornadula G, Schnell M, Boyd MR, Pomerantz RJ (2001) Prostratin: activation of latent HIV-1 expression suggests a potential inductive adjuvant therapy for HAART. Blood 98:3006–3015
58. Le Thomas A, Rogers AK, Webster A, Marinov GK, Liao SE, Perkins EM, Hur JK, Aravin AA, Toth KF (2013) Piwi induces piRNA-guided transcriptional silencing and establishment of a repressive chromatin state. Genes Dev 27:390–399
59. Lecellier CH, Dunoyer P, Arar K, Lehmann-Che J, Eyquem S, Himber C, Saib A, Voinnet O (2005) A cellular microRNA mediates antiviral defense in human cells. Science 308:557–560
60. Lee NS, Dohjima T, Bauer G, Li H, Li MJ, Ehsani A, Salvaterra P, Rossi J (2002) Expression of small interfering RNAs targeted against HIV-1 rev transcripts in human cells. Nat Biotechnol 20:500–505
61. Lee Y, Ahn C, Han J, Choi H, Kim J, Yim J, Lee J, Provost P, Radmark O, Kim S, Kim VN (2003) The nuclear RNase III Drosha initiates microRNA processing. Nature 425:415–419
62. Li S, Huang L (2000) Nonviral gene therapy: promises and challenges. Gene Ther 7:31–34
63. Li CJ, Friedman DJ, Wang C, Metelev V, Pardee AB (1995) Induction of apoptosis in uninfected lymphocytes by HIV-1 Tat protein. Science 268:429–431
64. Li LC, Okino ST, Zhao H, Pookot D, Place RF, Urakami S, Enokida H, Dahiya R (2006) Small dsRNAs induce transcriptional activation in human cells. Proc Natl Acad Sci U S A 103:17337–17342

65. Liao JY, Ma LM, Guo YH, Zhang YC, Zhou H, Shao P, Chen YQ, Qu LH (2010) Deep sequencing of human nuclear and cytoplasmic small RNAs reveals an unexpectedly complex subcellular distribution of miRNAs and tRNA 3′ trailers. PLoS One 5:e10563
66. Liao W, Lin JX, Leonard WJ (2011) IL-2 family cytokines: new insights into the complex roles of IL-2 as a broad regulator of T helper cell differentiation. Curr Opin Immunol 23:598–604
67. Lord JD, Mcintosh BC, Greenberg PD, Nelson BH (2000) The IL-2 receptor promotes lymphocyte proliferation and induction of the c-myc, bcl-2, and bcl-x genes through the trans-activation domain of Stat5. J Immunol 164:2533–2541
68. Lytle JR, Yario TA, Steitz JA (2007) Target mRNAs are repressed as efficiently by microRNA-binding sites in the 5′ UTR as in the 3′ UTR. Proc Natl Acad Sci U S A 104:9667–9672
69. Manninen A, Renkema GH, Saksela K (2000) Synergistic activation of NFAT by HIV-1 nef and the Ras/MAPK pathway. J Biol Chem 275:16513–16517
70. Martinez J, Patkaniowska A, Urlaub H, Luhrmann R, Tuschl T (2002) Single-stranded antisense siRNAs guide target RNA cleavage in RNAi. Cell 110:563–574
71. Matsui M, Sakurai F, Elbashir S, Foster DJ, Manoharan M, Corey DR (2010) Activation of LDL receptor expression by small RNAs complementary to a noncoding transcript that overlaps the LDLR promoter. Chem Biol 17:1344–1355
72. Matzke M, Aufsatz W, Kanno T, Daxinger L, Papp I, Mette MF, Matzke AJ (2004) Genetic analysis of RNA-mediated transcriptional gene silencing. Biochim Biophys Acta 1677:129–141
73. Miyazaki T, Liu ZJ, Kawahara A, Minami Y, Yamada K, Tsujimoto Y, Barsoumian EL, Permutter RM, Taniguchi T (1995) Three distinct IL-2 signaling pathways mediated by bcl-2, c-myc, and lck cooperate in hematopoietic cell proliferation. Cell 81:223–231
74. Morris KV, Chan SW, Jacobsen SE, Looney DJ (2004) Small interfering RNA-induced transcriptional gene silencing in human cells. Science 305:1289–1292
75. Motamedi MR, Verdel A, Colmenares SU, Gerber SA, Gygi SP, Moazed D (2004) Two RNAi complexes, RITS and RDRC, physically interact and localize to noncoding centromeric RNAs. Cell 119:789–802
76. Napoli S, Pastori C, Magistri M, Carbone GM, Catapano CV (2009) Promoter-specific transcriptional interference and c-myc gene silencing by siRNAs in human cells. EMBO J 28:1708–1719
77. Nishi K, Nishi A, Nagasawa T, Ui-Tei K (2013) Human TNRC6A is an Argonaute-navigator protein for microRNA-mediated gene silencing in the nucleus. RNA 19:17–35
78. Novina CD, Murray MF, Dykxhoorn DM, Beresford PJ, Riess J, Lee SK, Collman RG, Lieberman J, Shankar P, Sharp PA (2002) siRNA-directed inhibition of HIV-1 infection. Nat Med 8:681–686
79. O'Donnell KA, Wentzel EA, Zeller KI, Dang CV, Mendell JT (2005) c-Myc-regulated microRNAs modulate E2F1 expression. Nature 435:839–843
80. Omoto S, Ito M, Tsutsumi Y, Ichikawa Y, Okuyama H, Brisibe EA, Saksena NK, Fujii YR (2004) HIV-1 nef suppression by virally encoded microRNA. Retrovirology 1:44
81. Orom UA, Nielsen FC, Lund AH (2008) MicroRNA-10a binds the 5′UTR of ribosomal protein mRNAs and enhances their translation. Mol Cell 30:460–471
82. Ouellet DL, Plante I, Landry P, Barat C, Janelle ME, Flamand L, Tremblay MJ, Provost P (2008) Identification of functional microRNAs released through asymmetrical processing of HIV-1 TAR element. Nucleic Acids Res 36:2353–2365
83. Ouellet DL, Vigneault-Edwards J, Letourneau K, Gobeil LA, Plante I, Burnett JC, rossi JJ, Provost P (2013) Regulation of host gene expression by HIV-1 TAR microRNAs. Retrovirology 10:86
84. Pandolfi F, Pierdominici M, Marziali M, Livia Bernardi M, Antonelli G, Galati V, D'Offizi G, Aiuti F (2000) Low-dose IL-2 reduces lymphocyte apoptosis and increases naive CD4 cells in HIV-1 patients treated with HAART. Clin Immunol 94:153–159

85. Pelissier T, Thalmeir S, Kempe D, Sanger HL, Wassenegger M (1999) Heavy de novo methylation at symmetrical and non-symmetrical sites is a hallmark of RNA-directed DNA methylation. Nucleic Acids Res 27:1625–1634
86. Pfeffer S, Zavolan M, Grasser FA, Chien M, Russo JJ, Ju J, John B, Enright AJ, Marks D, Sander C, Tuschl T (2004) Identification of virus-encoded microRNAs. Science 304:734–736
87. Pfeffer S, Sewer A, Lagos-Quintana M, Sheridan R, Sander C, Grasser FA, van Dyk LF, Ho CK, Shuman S, Chien M, Russo JJ, Ju J, Randall G, Lindenbach BD, Rice CM, Simon V, Ho DD, Zavolan M, Tuschl T (2005) Identification of microRNAs of the herpesvirus family. Nat Methods 2:269–276
88. Place RF, Li LC, Pookot D, Noonan EJ, Dahiya R (2008) MicroRNA-373 induces expression of genes with complementary promoter sequences. Proc Natl Acad Sci U S A 105:1608–1613
89. Place RF, Noonan EJ, Foldes-Papp Z, Li LC (2010) Defining features and exploring chemical modifications to manipulate RNAa activity. Curr Pharm Biotechnol 11:518–526
90. Porichis F, Kaufmann DE (2011) HIV-specific CD4 T cells and immune control of viral replication. Curr Opin HIV AIDS 6:174–180
91. Portnoy V, Huang V, Place RF, Li LC (2011) Small RNA and transcriptional upregulation. Wiley Interdiscip Rev RNA 2:748–760
92. Rajasethupathy P, Antonov I, Sheridan R, Frey S, Sander C, Tuschl T, Kandel ER (2012) A role for neuronal piRNAs in the epigenetic control of memory-related synaptic plasticity. Cell 149:693–707
93. Sandelin A, Carninci P, Lenhard B, Ponjavic J, Hayashizaki Y, Hume DA (2007) Mammalian RNA polymerase II core promoters: insights from genome-wide studies. Nat Rev Genet 8:424–436
94. Schopman NC, Willemsen M, Liu YP, Bradley T, van Kampen A, Baas F, Berkhout B, Haasnoot J (2011) Deep sequencing of virus-infected cells reveals HIV-encoded small RNAs. Nucleic Acids Res 40:414–427
95. Schwartz JC, Younger ST, Nguyen NB, Hardy DB, Monia BP, Corey DR, Janowski BA (2008) Antisense transcripts are targets for activating small RNAs. Nat Struct Mol Biol 15:842–848
96. Seddiki N, Phetsouphanh C, Swaminathan S, Xu Y, Rao S, Li J, Sutcliffe EL, Denyer G, Finlayson R, Gelgor L, Cooper DA, Zaunders J, Kelleher AD (2013) The microRNA-9/B-lymphocyte-induced maturation protein-1/IL-2 axis is differentially regulated in progressive HIV infection. Eur J Immunol 43:510–520
97. Shykind BM, Kim J, Sharp PA (1995) Activation of the TFIID-TFIIA complex with HMG-2. Genes Dev 9:1354–1365
98. Sieg SF, Bazdar DA, Harding CV, Lederman MM (2001) Differential expression of interleukin-2 and gamma interferon in human immunodeficiency virus disease. J Virol 75:9983–9985
99. Stern-Ginossar N, Elefant N, Zimmermann A, Wolf DG, Saleh N, Biton M, Horwitz E, Prokocimer Z, Prichard M, Hahn G, Goldman-Wohl D, Greenfield C, Yagel S, Hengel H, Altuvia Y, Margalit H, Mandelboim O (2007) Host immune system gene targeting by a viral miRNA. Science 317:376–381
100. Sullivan CS, Grundhoff AT, Tevethia S, Pipas JM, Ganem D (2005) SV40-encoded microRNAs regulate viral gene expression and reduce susceptibility to cytotoxic T cells. Nature 435:682–686
101. Swaminathan S, Suzuki K, Seddiki N, Kaplan W, Cowley MJ, Hood CL, Clancy JL, Murray DD, Mendez C, Gelgor L, Anderson B, Roth N, Cooper DA, Kelleher AD (2012) Differential regulation of the Let-7 family of microRNAs in CD4+ T cells alters IL-10 expression. J Immunol 188:6238–6246
102. Taft RJ, Simons C, Nahkuri S, Oey H, Korbie DJ, Mercer TR, Holst J, Ritchie W, Wong JJ, Rasko JE, Rokhsar DS, Degnan BM, Mattick JS (2010) Nuclear-localized tiny RNAs are associated with transcription initiation and splice sites in metazoans. Nat Struct Mol Biol 17:1030–1034

103. Tan GS, Garchow BG, Liu X, Yeung J, Morris JPT, Cuellar TL, Mcmanus MT, Kiriakidou M (2009) Expanded RNA-binding activities of mammalian Argonaute 2. Nucleic Acids Res 37:7533–7545
104. Tay Y, Zhang J, Thomson AM, Lim B, Rigoutsos I (2008) MicroRNAs to Nanog, Oct4 and Sox2 coding regions modulate embryonic stem cell differentiation. Nature 455:1124–1128
105. Thomas MC, Chiang CM (2006) The general transcription machinery and general cofactors. Crit Rev Biochem Mol Biol 41:105–178
106. Tora L, Timmers HT (2010) The TATA box regulates TATA-binding protein (TBP) dynamics in vivo. Trends Biochem Sci 35:309–314
107. Turunen MP, Lehtola T, Heinonen SE, Assefa GS, Korpisalo P, Girnary R, Glass CK, Vaisanen S, Yla-Herttuala S (2009) Efficient regulation of VEGF expression by promoter-targeted lentiviral shRNAs based on epigenetic mechanism: a novel example of epigenetherapy. Circ Res 105:604–609
108. Umbach JL, Cullen BR (2009) The role of RNAi and microRNAs in animal virus replication and antiviral immunity. Genes Dev 23:1151–1164
109. Vasudevan S, Tong Y, Steitz JA (2007) Switching from repression to activation: microRNAs can up-regulate translation. Science 318:1931–1934
110. Verdel A, Jia S, Gerber S, Sugiyama T, Gygi S, Grewal SI, Moazed D (2004) RNAi-mediated targeting of heterochromatin by the RITS complex. Science 303:672–676
111. Volpe TA, Kidner C, Hall IM, Teng G, Grewal SI, Martienssen RA (2002) Regulation of heterochromatic silencing and histone H3 lysine-9 methylation by RNAi. Science 297:1833–1837
112. Wang W, Gralla JD, Carey M (1992) The acidic activator GAL4-AH can stimulate polymerase II transcription by promoting assembly of a closed complex requiring TFIID and TFIIA. Genes Dev 6:1716–1727
113. Weinmann L, Hock J, Ivacevic T, Ohrt T, Mutze J, Schwille P, Kremmer E, Benes V, Urlaub H, Meister G (2009) Importin 8 is a gene silencing factor that targets argonaute proteins to distinct mRNAs. Cell 136:496–507
114. van Werven FJ, van Teeffelen HA, Holstege FC, Timmers HT (2009) Distinct promoter dynamics of the basal transcription factor TBP across the yeast genome. Nat Struct Mol Biol 16:1043–1048
115. Westendorp MO, Frank R, Ochsenbauer C, Stricker K, Dhein J, Walczak H, Debatin KM, Krammer PH (1995) Sensitization of T cells to CD95-mediated apoptosis by HIV-1 Tat and gp120. Nature 375:497–500
116. White J, Brou C, Wu J, Lutz Y, Moncollin V, Chambon P (1992) The acidic transcriptional activator GAL-VP16 acts on preformed template-committed complexes. EMBO J 11:2229–2240
117. Yang C, Bolotin E, Jiang T, Sladek FM, Martinez E (2007) Prevalence of the initiator over the TATA box in human and yeast genes and identification of DNA motifs enriched in human TATA-less core promoters. Gene 389:52–65
118. Yeung ML, Bennasser Y, Watashi K, Le SY, Houzet L, Jeang KT (2009) Pyrosequencing of small non-coding RNAs in HIV-1 infected cells: evidence for the processing of a viral-cellular double-stranded RNA hybrid. Nucleic Acids Res 37:6575–6586
119. Yi R, Qin Y, Macara IG, Cullen BR (2003) Exportin-5 mediates the nuclear export of pre-microRNAs and short hairpin RNAs. Genes Dev 17:3011–3016
120. Ylisastigui L, Archin NM, Lehrman G, Bosch RJ, Margolis DM (2004) Coaxing HIV-1 from resting CD4 T cells: histone deacetylase inhibition allows latent viral expression. AIDS 18:1101–1108
121. Younes SA, Yassine-Diab B, Dumont AR, Boulassel MR, Grossman Z, Routy JP, Sekaly RP (2003) HIV-1 viremia prevents the establishment of interleukin 2-producing HIV-specific memory CD4+ T cells endowed with proliferative capacity. J Exp Med 198:1909–1922
122. Younger ST, Corey DR (2011) Transcriptional gene silencing in mammalian cells by miRNA mimics that target gene promoters. Nucleic Acids Res 39:5682–5691

123. Yue X, Schwartz JC, Chu Y, Younger ST, Gagnon KT, Elbashir S, Janowski BA, Corey DR (2010) Transcriptional regulation by small RNAs at sequences downstream from 3′ gene termini. Nat Chem Biol 6:621–629
124. Zhang H (2009) Reversal of HIV-1 latency with anti-microRNA inhibitors. Int J Biochem Cell Biol 41:451–454
125. Zhang Y, Fan M, Geng G, Liu B, Huang Z, Luo H, Zhou J, Guo X, Cai W, Zhang H (2014a) A novel HIV-1-encoded microRNA enhances its viral replication by targeting the TATA box region. Retrovirology 11:23
126. Zhang Y, Fan M, Zhang X, Huang F, Wu K, Zhang J, Liu J, Huang Z, Luo H, Tao L, Zhang H (2014b) Cellular microRNAs up-regulate transcription via interaction with promoter TATA-box motifs. RNA 20:1878–1889
127. Zhang Y, Yin Y, Zhang S, Luo H, Zhang H (2016) HIV-1 infection-induced suppression of the Let-7i/IL-2 axis contributes to CD4(+) T cell death. Sci Rep 6:25341

Chapter 8
miRNA-Mediated RNAa by Targeting Enhancers

Qingping Zou, Ying Liang, Huaibing Luo, and Wenqiang Yu

Abstract MicroRNAs (miRNAs) are small noncoding RNAs that act as negative regulators of gene expression in the cytoplasm. Yet, emerging evidence has shown that miRNAs are also distributed in the nucleus, with its function largely undetermined. At the same time, while miRNAs and enhancers show obvious tissue specificity, the interaction between miRNAs and enhancers in gene regulation remains unknown. By screening miRNA databases, we have identified a subset of miRNAs, called nuclear activating miRNAs (NamiRNAs). As enhancer regulators, NamiRNAs are able to activate gene expression at the transcriptional level. In addition, we found that the regulation of enhancers *mediated* by NamiRNAs depends on the presence of intact enhancers and AGO2 protein. More interesting is that NamiRNAs promote global gene transcription through the binding and activation of their targeted enhancers. Our results demonstrate a novel role for miRNA as an enhancer trigger for transcriptional gene activation. Further study of the function and molecular mechanism for NamiRNAs in tumorigenesis and development is of great significance.

Keywords microRNA • NamiRNA • Enhancer • Activation • AGO2 • Tissue specificity

8.1 Introduction

MicroRNAs (miRNAs) are a group of small noncoding RNAs of approximately 22 nucleotides (nts) in length that are endogenous and single-stranded. The first miRNA discovered by Victor Ambros [14] downregulates the expression of lin-14

Qingping Zou and Ying Liang contributed equally to this work.

Q. Zou • Y. Liang • H. Luo • W. Yu (✉)
Laboratory of RNA Epigenetics, Institutes of Biomedical Sciences & Department of General Surgery, Huashan Hospital, Cancer Metastasis Institute, Fudan University, 130 Dong-An Road, Shanghai 200032, China

Department of Biochemistry and Molecular Biology, Shanghai Medical College, Key Laboratory of Ministry of Education, Department of Molecular Biology, Fudan University, 130 Dong-An Road, Shanghai 200032, China
e-mail: wenqiangyu@fudan.edu.cn

L.-C. Li (ed.), *RNA Activation*, Advances in Experimental Medicine and Biology 983, DOI 10.1007/978-981-10-4310-9_8

by targeting the 3′ UTR (3′ untranslated region). Since then, a growing number of miRNAs have been recounted to function as a regulator of gene expression. Canonically, miRNAs affect gene expression through posttranslational gene silencing, that is, they play a negative role in the regulation of translation or degradation of target mRNAs by binding to the complementary sequences that are often located in the 3′ UTR in the cytoplasm [1]. However, an increasing body of evidence shows that many miRNAs are also located within the nucleus. As such, it is difficult to explain the mechanism of nuclear miRNAs and establish characteristics of gene activation if one was to use the conventional negative-regulation theory of miRNAs in the cytoplasm.

8.2 Localization and Function of miRNAs

8.2.1 Localization of miRNAs in the Nucleus

To our knowledge, miRNAs positioned in the cytoplasm usually play the role of a negative regulator in gene expression. For instance, miR-17-92, as a classical cancer-related miRNA family, were able to manipulate cell growth by inhibiting the expression of E2F1 [19]. miR-19b also downregulates the expression of PTEN to activate AKT signaling pathway, thereby promoting an EMT transformation of A549 cells, allowing the invasion and migration ability of the A549 cells to be enhanced [16]. Most studies of miRNA so far, as briefly described above, have focused on those in the cytoplasm. But what about miRNAs located in other cell organelles? It should not be neglected that researchers have found that miRNAs in mitochondria may have a positive regulation role as well. A recent study reported that miR-1 specifically expressed during myocyte formation can promote the translation of specific target genes in mitochondria, which is different from cytosolic miRNAs [37]. Another report showed that miR-29b, which has a sequence similar to the nuclear localization signal, was mainly concentrated in the nucleus of Hela and NIH3T3 cells [11]. Clearly, it was not unique. Liao et al. have showed that miR-29b is located in the nucleus even in the earlier years. The distribution of miRNAs in the nucleus was further confirmed by analyzing the distribution of multiple miRNAs in the cytoplasm and nucleus using deep sequencing techniques [17]. It is indicated that the different miRNA localization may lead to contrasting functions.

On one hand, the association between miRNA localization and its function has always been overlooked in today's miRNAs research. Given this, the mechanism of miRNAs in the nucleus remain elusive. On the other hand, with the development of deep sequencing technology, increasing evidence shows that miRNAs are localized not only in the cytoplasm but also in the nucleus [11, 17]. The nucleus is the control center of cells as well as the main storage location of genetic information. It is natural to think that the cellular localization of miRNAs affects their biological functions, and miRNAs in the nucleus can play a pivotal role in gene regulation. Then what role do miRNAs in the nucleus play?

8.2.2 Positive Regulation of Gene Expression by miRNAs

With increasing advancement of miRNA research and sequencing technology, growing research has reported that miRNA can promote gene expression as well as translation in certain special cases [4, 21, 27, 28, 33]. Other than the classical theory that miRNAs negatively regulate target genes by binding to the 3′ UTR of mRNA, it has also been proved that miRNAs can bind to the promoter of the target genes to facilitate gene transcriptions. As demonstrated by Orom et al., miR-10a has been shown to interact with ribosomal proteins in the 5′ UTR of mRNA to promote gene translation [20]. Vasudevan et al. found that miRNAs promote transcription through binding proteins AGO2 and FXR1, revealing the two faces of miRNAs [28]. Another evidence to support this idea is that Place et al. showed that miRNA-373 targets E-cadherin and cold-shock domain-containing protein C2 (CSDC2) to promote gene expression [21]. This phenomenon was termed in an earlier study as RNAa (RNA activation) to explain that small activating RNAs are capable of stimulating gene expression at the transcriptional level via binding the specific promoter regions [15]. Furthermore, saRNAs (small activating RNAa) have been found as a widespread phenomenon in many mammal cells, such as in mouse, rat, and nonhuman primate cells [13]. Taken together, these lines of evidence show that besides negative regulation, miRNAs also display a pattern of positive regulation of gene expression. However, its precise molecular mechanism in the nucleus requires further investigation.

8.2.3 Tissue-Specific miRNAs

As small noncoding RNAs, miRNAs play an important role in regulating the expression of many key genes during physiological processes, such as cell differentiation or tumorigenesis, as well as the development and metastasis of cancer and other pathological processes. Another noteworthy feature of miRNAs is that its expression exhibits a characteristic of strong tissue specificity. For example, miR-122 shows a concentrated enrichment in the liver as an important regulator for biological functions [18]; miR-1 is mainly concentrated in myocardium and skeletal muscle cells [5]; miR-124 has a high level of expression in the central nervous system. Moreover, tissue-specific expression of miRNA is closely related to the function of the organs. Knockout of miR-1 leads to death of Drosophila due to differentiation disorders of muscle cells. In mice, knockout of miR-1-2 results in death for 50% of mice because of cardiac developmental disorders [38].

However, the regulatory mechanism of tissue-specific expression of miRNAs remains unclear. Studies have shown that a transduction of neuron-specific miR-124 into Hela cells can significantly inhibit the expression of nonneuronal genes, which is believed to induce a drift for genes in Hela cells toward those in neurons [26]. Here we wonder whether miR-124 itself is able to lead to this phenomenon. We believe that heterologous expression of miR-124 is able to promote the expression of neuron-specific genes, which is more helpful in explaining why miRNA exhibits specific expressions in different tissue cells. Unfortunately, it goes against the currently acknowledged negative regulation of miRNA and, even more, there is no applicable theory and mechanism to annotate it. To this end, we hope to explore whether miRNA completes this function independently or together with additional factors to cope with tissue-specific matters.

Interestingly, enhancers also exhibit a characteristic of distinct tissue specificity in distribution. Enhancers with tissue-specific distribution are closely related to many physiological and pathological processes, such as embryonic development and disease occurrence. As an important regulator of tissue-specific gene expression, how then does enhancer manipulate its function in the nucleus? Moreover, does miRNA in the nucleus interact with enhancer to cooperate in the regulation of tissue-specific functions?

8.3 Characteristics of Enhancers

8.3.1 Definition of Enhancers

Enhancers are cis-regulatory elements that could exist at long distances from their target genes and regulate their expression by a precise spatiotemporal pattern during development [24]. Enhancer was firstly named by Schaffner and colleagues to describe a 72-bp sequence of the SV40 virus genome about 30 years ago. This sequence could enhance the transcription of a reporter gene in HeLa cells [2]. Enhancers hold their own unique characteristics, including histone modification (such as H3K4me1, H3K27me3, and H3K27ac), p300/CBP binding, DNase I hypersensitivity, and production of eRNAs. In general, enhancer regions show dramatic chromatin accessibility and DNase I hypersensitivity. These chromosomal states are benefited for RNAPII binding and eRNA production. At the same time, p300/CBP can interact with transcriptional activators or general transcription machinery, including RNAPII. As an acetyltransferase, p300/CBP can catalyze the acetylation of H3K27 when it binds to the enhancer regions. Along with the occurrences of other histone modifications, a series of adjacent genes could be activated by enhancers.

According to chromatin states, enhancers can be classified into three types: inactive or "poised" enhancers, active enhancers, and overactive enhancers.

Generally, inactive enhancers displayed active H3K4me1 and repressive Polycomb protein-associated H3K27me3 marks, and they can become active enhancers when H3K27 is acetylated by p300/CBP. Active enhancers could achieve an overactive status that displayed distinctly H3K27ac, H3K4me1, and H3K4me3 [34].

8.3.2 Tissue-Specific Enhancers

Multicellular organisms require different kinds of cells to establish a complete system. But all cells from one organism have the same genomic DNAs, so they require establishment of spatiotemporal patterns of gene expression during development [30]. Enhancers are the key regulators for determining gene expression in different cell types. Distribution of enhancers in various cell types showed significant differences [9]. Although mammalian genomes contain millions of potential enhancers, only a small subset of them is active in a given cell type. Here, cell type-specific enhancer selection involves the binding of lineage-determining transcription factors [8]. Therefore, tissue-specific enhancers are pivotal elements when realizing gene expression in a tissue-specific manner. Super-enhancer is a typical tissue-specific enhancer. And as its name suggests, super-enhancer is defined as a class of regulatory regions in genomic DNAs with a high density of enhancer-associated chromatin marks or transcriptional coactivators, especially for the mediator complex [32]. In any given cell type, super-enhancers are highly enriched in the biological processes that define the identities of the cell types [10].

It has now been 30 years since enhancers were first found, yet we still hardly understand their precise regulatory mechanism on gene expression. Given that miRNAs are short chains of noncoding RNAs which display remarkable tissue specificity, similar to enhancers, it seems logical to ask whether there is a link between miRNAs and enhancers. Could miRNA regulate the activation of enhancers as an enhancer trigger? These questions require our attention.

8.4 NamiRNAs-Mediated RNAa by Targeting Enhancers

8.4.1 Overlap of miRNAs and Enhancers

Taking all the above-mentioned analyses into consideration, we can suppose that miRNAs are in fact associated with enhancers to function in the nucleus. Four years ago, when our lab studied the epigenetic regulation of miRNAs, we found that more than 400 miRNA precursors in the genome by which data obtained from the University of California Santa Cruz (UCSC) Genome Browser were highly overlapped with histone modification markers H3K4me1 or H3K27ac. This finding pushed us to stick to our idea that tissue-specific miRNAs and enhancers are two important biological events that may crosstalk during the regulatory process. For example, hsa-miR-3179, hsa-miR-26a-1, and hsa-miR-24 are highly overlapped to the enhancer regions in the genome. In addition, the enrichment of enhancer markers, including H3K27ac, P300/CBP and DNase I high-sensitivity (DHS) on 1594 annotated miRNA precursors loci, was performed in seven different kinds of cell lines. We surprisingly found that many miRNAs are highly overlapped to the enhancer regions in the genome and the expression of nuclear miRNAs exhibits a positive correlation with their neighboring genes. Moreover, H3K27ac-enriched miRNAs are inclined to be present in the nucleus.

8.4.2 Enhancer Activation by NamiRNAs

Importantly, enhancers are able to be activated by miRNAs and thus play crucial roles in gene expression. We termed these miRNAs as NamiRNAs (Nuclear activating miRNAs). These miRNAs loci are enriched with epigenetic markers that display enhancer activity such as histone H3K27ac, P300/CBP, and DNase I high-sensitivity loci. Both miR-26a-1 and miR-339 are examples of NamiRNAs to explore their functions. Their enhancer activities tested by the reporter gene analysis showed that, when overexpressed, they could activate nearby enhancers. miR-26a-1, surrounded by protein-coding genes *ITGA9*, *CTDSPL*, *VILL*, and *PLCD1* in a 400 kb window, can activate the transcription of the neighboring genes *ITGA9* and *VILL* when overexpressed. Similarly, miR-339 surrounded by *GPER* is equally upregulated when compared with other genes neighboring *GPER*. Therefore, these miRNA-activated enhancers may play certain roles on gene regulation and represent a novel mechanism of epigenetic regulation.

More information needs to be added to demonstrate the mechanism of enhancer-mediated NamiRNAs activation. When miR-24 is overexpressed in HEK293T cells, the expression of miR-24 is significantly upregulated. Using ChIP-seq, we found that miR-24 can change the chromatin state of the enhancer loci, causing H3K27ac to be enriched in the miR-24 targeted enhancer region. When the integrity of miR-24's targeted enhancer region is disrupted by gene editing techniques, the function of gene activation by exogenous miRNA becomes entirely blocked, which not only implies that the activation process of NamiRNAs is accomplished by interaction with enhancers but also shows that this process depends on the integrity of the enhancers. AGO2 has been reported to exist and play certain roles in the nucleus [35, 39]. Our work gives evidence that AGO2 is associated with miRNA activation [35], although more detailed information needs to be obtained to further demonstrate the correlation between AGO2 and miRNA activation.

8.4.3 Gene Activation by Enhancers

Since our previous work showed that miR-24 can activate each of their neighboring genes, we wondered whether miRNA may also be able to activate more gene expressions on a genome-wide level. Fortunately, we received a satisfactory answer through microarray data and bioinformatics analysis by miRanda, which, on one hand, showed that miR-24-1 could either increase or decrease gene expression and play its conventional roles in the cytoplasm. On the other hand, similar motifs like the seed sequence of miR-24-3p were found in enriched H3K27ac regions and miR-24-1 appeared more frequently in activated enhancer regions than those sequences that were randomly selected. These phenomena are powerful evidence in support of our idea that NamiRNA can activate gene expression by targeting enhancers in the nucleus, while also repressing gene expression in the cytoplasm. Moreover, a great enrichment of H3K27ac was determined in 3282 enhancer regions corresponding to miR-24-1 overexpression, revealing that upregulated genes are inclined to reside near activated enhancers. Interestingly, the upregulated neighboring genes failed to be activated when the predicted enhancer locus was disrupted by gene editing methods. Taking what we have learned above, we believe that miRNAs can activate global gene expressions by targeting the integral enhancer regions.

It is reasonable to harbor the view that miRNAs-mediated RNAa by targeting enhancers is a general phenomenon. Interestingly, when we compared the above-mentioned 1594 enhancer-associated miRNAs based on H3K27ac enrichment of miRNA loci across seven cell lines, a total of 303 miRNAs were identified within the peaks of H3K27ac modification, which indicated that 303 miRNAs loci were overlapped within the enhancer markers H3K27ac. According to the specific distribution of the enhancers on the genome in different cell types, we classified miRNAs into three categories: enhancers distributed in all cell lines, enhancers distributed in some cell lines, and finally tissue-specific enhancers. There are various examples to demonstrate the fact that miRNAs can bind enhancers to realize the function of activation. Taking GM12878 cell line (UCSC) for example, there always exist some miRNAs with activation-related enhancer markers. As such, miRNA-24 is not a lone example to illustrate our theory that miRNAs display a pattern of activation by targeting associated enhancers.

8.4.4 Two Faces of miRNAs

miRNAs are involved in almost all physiological and pathological processes [3]. However, with the discovery of NamiRNAs, we may have to reevaluate the function of miRNAs. What is the core function of miRNA? Xiao et al. showed that miRNAs distributed in the nucleus could activate enhancers, thus leading to increase transcriptions of adjacent genes [35]. The function of transcriptional activation and translational inhibition may be analogous to the relationship between water and fire, that is, miRNAs mainly play a negative regulatory role, similar to the case of water putting out a flame in the cell cytoplasm. They target the 3′ UTR of mRNAs and inhibit the translation of their target genes. Conversely, miRNAs can be compared to a fire in the cell nucleus, igniting its target enhancers, and then inducing a series of gene activation involved in cell physiological and biochemical processes [35]. These two regulatory mechanisms may coexist, mutually coordinate, and together maintain the normal physiology of cells.

8.5 Mechanism of miRNA-Mediated RNAa

8.5.1 Original Source of NamiRNAs

The classical theory of miRNAs is that genomic DNAs are firstly transcribed into pri-miRNAs. Thereafter, pri-miRNAs are cleaved into pre-miRNAs by Drosha in the nucleus. Then, pre-miRNAs are transported into the cytoplasm via Exportin5, and are cut again by Dicer to form mature miRNAs [22]. However, this theory cannot explain some of the existing problems surrounding miRNAs, including distribution of miRNAs in the nucleus [16, 19]. One plausible explanation is that miRNAs in the cytoplasm can be carried back into the nucleus through some transport proteins. But there is no crucial evidence to support this. Another reasonable explanation is that the pre-miRNAs could mature directly in the nucleus by Dicer cleavage (Fig. 8.1). This is a simpler and more energy-efficient way. More interesting is that emerging evidence has already supported this explanation. Dicer and AGO2 proteins have already been showed to exist in the nucleus [23, 31]. They are the key proteins for miRNAs maturation and RNA-induced silencing complex (RISC) formation.

8.5.2 Is It Possible That miRNAs Act as an Enhancer Trigger by Binding to Enhancers?

Another question that draws our attention is the regulation of enhancers by miRNAs. As known, miRNA can exert its function by directly binding to the 3′ UTR of mRNA. However, RNA could also bind to the genomic DNA and form an RNA-DNA hybrid, such as gRNA in the CRISPR/Case9 system [29]. This suggests that miRNAs have the potential abilities to bind to the complementary DNA sequences. Our results have demonstrated that miRNAs are unable to activate gene transcription when enhancer sequences have been destroyed or the seed sequences mutation of miRNAs [35]. While more evidence is needed to support this, it is nevertheless a reasonable and possible hypothesis to explain the regulation of miRNAs on enhancers (Fig. 8.2).

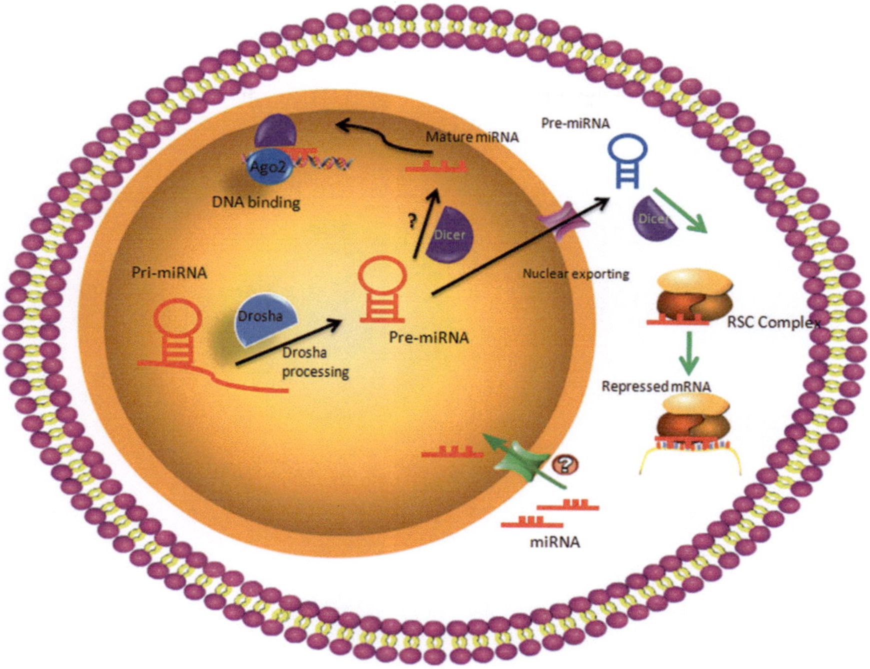

Fig. 8.1 *Two possible pathways for the origins of nuclear miRNAs.* The classical theory is that genomic DNAs are first transcribed to produce pri-miRNAs, which are then cleaved into pre-miRNAs by Drosha in the nucleus. After being exported into the cytoplasm by Exportin 5, pre-miRNAs are further processed by Dicer into mature miRNAs which can then be imported back into the nucleus;the other hypothesis is that pre-miRNAs could become mature miRNAs in the nucleus processed by Dicer

8.5.3 What Is the Function of AGO2 Protein on NamiRNA-Mediated RNAa?

Argonaute (AGO) proteins are able to form RISC complexes with miRNAs or siRNAs, and regulate their target genes expression. Among the four AGO variants in human cells (AGO1–4), AGO2 is the only AGO protein with catalytic activity. So we doubt that AGO2 could be involved in NamiRNA-mediated RNAa. Xiao's experimental data showed that shAGO2 could significantly interfere with the regulation of NamiRNAs on gene activation through enhancers [35]. These results indicate that the activation of gene transcription by miRNAs depends on AGO2 protein. In addition, the formation of RNA-DNA hybrid may also need AGO2 protein, since AGO2 is the key protein for recognizing miRNAs during the process of regulating gene expression in cell nucleus [7, 12]. According to our results then, we draw a model to illuminate the function of AGO2 during gene activation by miRNAs (Fig. 8.2).

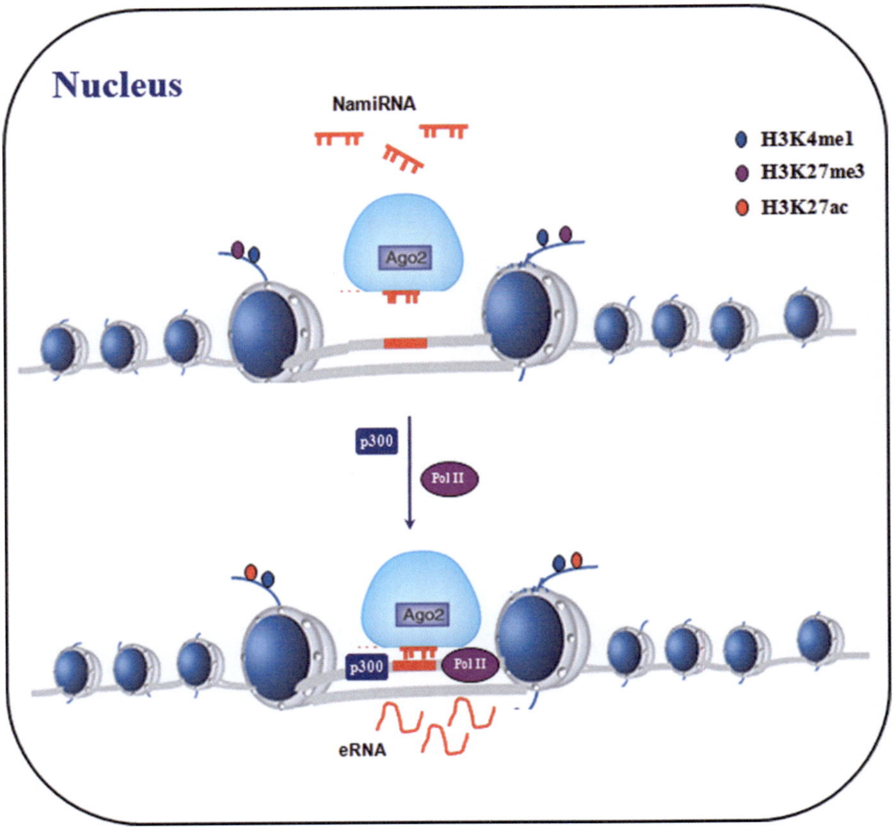

Fig. 8.2 *Schematic diagram shows enhancer activation by NamiRNAs in an AGO2-dependent manner*. Generally, inactive or "poised" enhancers display active H3K4me1 and repressive Polycomb protein-associated H3K27me3 marks. NamiRNAs in association with AGO2 bind to enhancer sequences through an incomplete base-pairing. The complex can recruit p300 and RNAPII to produce a large number of eRNAs. Furthermore, p300 can catalyze the acetylation of H3K27, thereby transforming the inactive enhancer into an active one

8.5.4 Interaction of miRNAs with Enhancers

In addition to enhancers activation, miRNAs expression regulation is also a critical process. miRNAs are produced primarily by RNA polymerase II or polymerase III, which are further matured by the cleavage of Drosha and Dicer. However, the regulation of miRNAs expression is still unknown. Here, Xiao's results provide a good explanation for genes activation by NamiRNAs. They found that the majority of NamiRNAs are located in the enhancer regions, and there is a dramatic overlap between NamiRNAs and enhancers [35]. In addition, Sharp PA and his colleagues recently have also studied the regulation between miRNAs and enhancers. Interestingly, their results showed that super-enhancers could in deed modulate the miRNA expression profiles and even affect the expression of multiple tumor-associated miRNAs [25]. This undoubtedly affirms our studies and supports our hypothesis. In turn, activation of miRNAs could further activate additional enhancers, and promote a series of genes transcription. It can be said that NamiRNAs and enhancers mutually coordinate with each other and together achieve the regulation of gene expression.

8.6 Prospects and Challenges of NamiRNA Research

8.6.1 Biological Significance of NamiRNA-Mediated RNAa

Tumor metastasis is a common obstacle in the course of cancer therapy. The way to overcome tumor metastasis has become the key to curing cancer. We found a highly abnormal expression of miRNAs in breast cancer and osteosarcoma, as well as other high metastatic tumor tissues in our recent studies (unpublished results). More importantly, most of these miRNAs are enhancer-targeting NamiRNAs which activate the expression of many tumor-related genes. In addition, multiple miRNAs have been reported to be involved in the pathogenesis of systemic lupus erythematosus (SLE) [6, 36]. Our analysis showed that most of these miRNAs have the characteristics of NamiRNAs, indicating that NamiRNAs may be involved in tumor metastasis, SLE, and other important biological processes.

8.6.2 The Dual Functions of miRNAs: Activation and Repression

The two faces of miRNAs can also be displayed in the process of embryonic development, which is a gradual differentiation from embryonic stem cells, as well as a process of inhibiting genes expression. In contrast, it is a dedifferentiation process and a series of genes are gradually activated during the induced pluripotent stem cells (IPS cells) production. Thus, the activated and repressed functions of miRNAs are always mutually kept in balance at different stages. During cell differentiation, most of the cytoplasmic miRNAs play an inhibitory role through translational inhibition; while in the process of dedifferentiation, the nuclear NamiRNAs could promote the transcription of large numbers of genes through enhancers. Undoubtedly, these two interactive systems are essential for maintaining the physiological activity of cells. In fact, since miRNAs have been found to be involved in almost all of the biological and biochemical processes of organisms, we believe that miRNAs regulation is also pivotal in embryonic development.

8.6.3 *Prospects for Future Research*

The current research on NamiRNA is only the tip of the iceberg. Many questions have yet to be answered. How does AGO2 participate in NamiRNA transcriptional regulation? How do miRNAs balance the regulation of transcriptional activation and translational inhibition in cells? Our results showed that NamiRNAs may be involved in a variety of physiological and pathological processes. So it is necessary to clarify the regulatory mechanism of NamiRNAs. NamiRNAs exploration has a long way to go, and we look forward to pushing the development of miRNAs research in the future.

References

1. Ambros V (2004) The functions of animal microRNAs. Nature 431(7006):350–355
2. Banerji J, Rusconi S, Schaffner W (1981) Expression of a beta-globin gene is enhanced by remote SV40 DNA sequences. Cell 27:299–308
3. Beermann J, Piccoli MT, Viereck J et al (2016) Non-coding RNAs in development and disease: background, mechanisms, and therapeutic approaches. Physiol Rev 96(4):1297–1325
4. Buchan JR, Parker R (2007) The two faces of miRNA. Science 318(5858):1931–1934
5. Chen JF, Mandel EM, Thomson JM et al (2006) The role of microRNA-1 and microRNA-133 in skeletal muscle proliferation and differentiation. Nat Genet 38(2):228–233
6. Divekar AA, Dubey S, Gangalum PR et al (2011) Dicer insufficiency and miR-155 overexpression in lupus Treg cells: an apparent paradox in the setting of an inflammatory milieu. J Immunol 186(2):924–930
7. Gagnon KT, Li L, Chu Y et al (2014) RNAi factors are present and active in human cell nuclei. Cell Rep 6(1):211–221
8. Heinz S, Benner C, Span N et al (2010) Simple combinations of lineage determining transcription factors prime cis-regulatory elements required for macrophage and B cell identities. Mol Cell 38:576–589
9. Heinz S, Romanoski CE, Benner C et al (2015) The selection and function of cell type-specific enhancers. Nat Rev Mol Cell Biol 16(3):144–154
10. Hnisz D, Abraham BJ, Lee TI et al (2013) Super-enhancers in the control of cell identity and disease. Cell 155:934–947
11. Hwang HW, Wentzel EA, Mendell JT (2007) A hexanucleotide element directs microRNA nuclear import. Science 315:97–100
12. Janowski BA, Huffman KE, Schwartz JC et al (2006) Involvement of AGO1 and AGO2 in mammalian transcriptional silencing. Nat Struct Mol Biol 13(9):787–792
13. Jiao AL, Slack FJ (2014) RNA-mediated gene activation. Epigenetics 9(1):27–36
14. Lee RC, Feinbaum RL, Ambros V (1993) The C.elegans heterochronic gene lin-4 encodes small RNAs with antisense complementarity to lin-14. Cell 75(5):843–854
15. Li LC, Okino ST, Zhao H et al (2006) Small dsRNAs induce transcriptional activation in human cells. Proc Natl Acad Sci U S A 103(46):17337–17342
16. Jing Li, Yang S, Yan W et al (2015) MicroRNA-19 triggers epithelial–mesenchymal transition of lung cancer cells accompanied by growth inhibition. Lab Investig 95(9):1056–1070
17. Liao JY, Ma LM, Guo YH et al (2010) Deep sequencing of human nuclear and cytoplasmic small RNAs reveals an unexpectedly complex subcellular distribution of miRNAs and tRNA 3′ trailers. PLoS One 5(5):e10563

18. Luna JM, Scheel TK, Danino T et al (2015) Hepatitis C virus RNA functionally sequesters miR-122. Cell 160(6):1099–1110
19. O'Donnell KA, Wentzel EA, Zeller KI et al (2005) c-Myc regulated microRNAs modulate E2F1expression [J]. Nature 435(7043):839–843
20. Orom UA, Nielsen FC, Lund AH (2008) MicroRNA-10a binds the 5′ UTR of ribosomal protein mRNAs and enhances their translation. Mol Cell 30:460–471
21. Place RF, Li LC, Pookot D et al (2008) MicroRNA-373 induces expression of genes with complementary promoter sequences. Proc Natl Acad Sci U S A 105(5):1608–1613
22. Romero-Cordoba SL, Salido-Guadarrama I, Rodriguez-Dorantes M et al (2014) miRNA biogenesis: biological impact in the development of cancer. Cancer Biol Ther 15 (11):1444–1455
23. Sharma NR, Wang X, Majerciak V et al (2016) Cell type- and tissue context-dependent nuclear distribution of human ago2. J Biol Chem 291(5):2302–2309
24. Smith E, Shilatifard A (2014) Enhancer biology and enhanceropathies. Nat Struct Mol Biol 21 (3):210–219
25. Suzuki HI, Young RA, Sharp PA (2017) Super-enhancer-mediated RNA processing revealed by integrative microRNA network analysis. Cell 168(6):1000–1014.e15
26. Sonntag KC, Woo TU, Krichevsky AM (2012) Converging miRNA functions in diverse brain disorders: a case for miR-124 and miR-126. Exp Neurol 235(2):427–435
27. Vasudevan S, Steitz JA (2007) AU-rich-element-mediated upregulation of translation by FXR1 and Argonaute 2. Cell 128(6):1105–1118
28. Vasudevan S, Tong Y, Steitz JA (2007) Switching from repression to activation: microRNAs can up-regulate translation. Science 318(5858):1877–1878
29. Wang FY, Qi LS (2016) Applications of CRISPR genome engineering in cell biology. Trend Cell Biol 26(11):875–888
30. Wamstad JA, Wang X, Demuren OO et al (2014) Distal enhancers: new insights into heart development and disease. Trends Cell Biol 24(5):294–302
31. White E, Schlackow M, Kamieniarz-Gdula K et al (2014) Human nuclear Dicer restricts the deleterious accumulation of endogenous doublestranded RNA. Nat Struct Mol Biol 21(6):552–559
32. Whyte WA, Orlando DA, Hnisz D et al (2013) Master transcription factors and mediator establish super-enhancers at key cell identity genes. Cell 153:307–319
33. Wu LG, Belasco JG (2008) Let me count the ways: mechanisms of gene regulation by miRNAs and siRNAs. Mol Cell 29(1):1–7
34. Wu F, Shi YG, Shi Y et al (2016) Suppression of enhancer overactivation by a RACK7-Histone demethylase complex. Cell 165(2):331–342
35. Xiao M, Li J, Li W et al (2016) MicroRNAs activate gene transcription epigenetically as an enhancer trigger. RNA Biol 6:0
36. Yuan Y, Kasar S, Underbayev C et al (2012) Role of microRNA-15a in autoantibody production in interferon-augmented murine model of lupus. Mol Immunol 52(2):10
37. Zhang X, Zuo X, Yang B et al (2014) MicroRNA directly enhances mitochondrial translation during muscle differentiation. Cell 158(3):607–619
38. Zhao Y, Ransom JF, Li A et al (2007) Dysregulation of cardiogenesis, cardiac conduction, and cell cycle in mice lacking miRNA-1-2. Cell 129:303–317
39. Zisoulis DG, Kai ZS, Chang RK et al (2012) Autoregulation of microRNA biogenesis bylet-7 and Argonaute. Nature 486:541–544

Part III
RNA Activation Guided by Other Small RNAs

Chapter 9
Specific Increase of Protein Levels by Enhancing Translation Using Antisense Oligonucleotides Targeting Upstream Open Frames

Xue-Hai Liang, Wen Shen, and Stanley T. Crooke

Abstract A number of diseases are caused by low levels of key proteins; therefore, increasing the amount of specific proteins in human bodies is of therapeutic interest. Protein expression is downregulated by some structural or sequence elements present in the 5′ UTR of mRNAs, such as upstream open reading frames (uORF). Translation initiation from uORF(s) reduces translation from the downstream primary ORF encoding the main protein product in the same mRNA, leading to a less efficient protein expression. Therefore, it is possible to use antisense oligonucleotides (ASOs) to specifically inhibit translation of the uORF by base-pairing with the uAUG region of the mRNA, redirecting translation machinery to initiate from the primary AUG site. Here we review the recent findings that translation of specific mRNAs can be enhanced using ASOs targeting uORF regions. Appropriately designed and optimized ASOs are highly specific, and they act in a sequence- and position-dependent manner, with very minor off-target effects. Protein levels can be increased using this approach in different types of human and mouse cells, and, importantly, also in mice. Since uORFs are present in around half of human mRNAs, the uORF-targeting ASOs may thus have valuable potential as research tools and as therapeutics to increase the levels of proteins for a variety of genes.

Keywords Translation • uORF • Regulation • Antisense • Oligonucleotides

9.1 Introduction

Abnormal expression of proteins can cause diseases due to the essential roles that proteins play in various biological processes. Regulating protein expression or function is thus an important opportunity for therapeutics. Downregulation of

X.-H. Liang (✉) • W. Shen • S.T. Crooke
Department of Core Antisense Research, Ionis Pharmaceuticals, Inc., 2855 Gazelle Court, Carlsbad, CA 92010, USA
e-mail: Lliang@ionisph.com

L.-C. Li (ed.), *RNA Activation*, Advances in Experimental Medicine and Biology 983, DOI 10.1007/978-981-10-4310-9_9

specific protein expression via different mechanisms, especially using antisense technology, has been developed both as a research tool and for therapeutics [7, 14, 61, 77, 98]. The pharmacokinetics and pharmacodynamics of antisense oligonucleotide (ASO) drugs have been well studied and efficient delivery of ASOs *in vivo* to different organs has been achieved via many different routes, including subcutaneous injection [24–26, 38]. Importantly, ASO drugs have been approved by FDA, for example, for the treatment of patients with homozygous familial hypercholesterolemia [90]. However, in many cases a disease can be caused by low levels of functional proteins [16, 79]; therefore, specific increases in the amount of a protein may be beneficial. Multiple approaches have been developed to achieve this purpose, such as delivery of expression constructs for gene therapy, downregulation of repressor proteins or antisense transcripts, inhibition of nonsense-mediated mRNA decay or modulation of pre-mRNA splicing to increase mRNA levels [30, 31, 67, 69]. However, these approaches have certain limitations, for example, the *in vivo* delivery of large DNA molecules remains challenging and reduction of antisense transcripts may result in nonspecific effects and is applicable only to a limited number of target proteins [30, 51, 65, 70]. New methodology is thus in high demand for efficient and selective increase of the levels of proteins, especially *in vivo*, for a large number of targetable genes.

Regulation of gene expression occurs at many steps between transcription and translation. Often several mechanisms act synergistically to control the level of a protein [27, 36]. Translation, as one of the most active biological events in cells, is an important step that controls the levels of proteins. Translational regulation has been well studied and is responsive to many stimuli including stress, cell growth rate, and physiological conditions [12, 34, 68, 92].

9.2 Translation Initiation

In eukaryotic cells, translation occurs mainly in a cap-dependent manner, although cap-independent translation, which is often mediated by internal ribosome entry sites (IRES), also operates for a subset of mRNAs [3, 5, 33]. During translation, the eIF4F complex, containing the cap-binding protein eIF4E, scaffold protein eIF4G, and a helicase eIF4A, binds to the 5′ m^7G cap of mRNAs. The binding of eIF4A to the mRNA may help to resolve the structure formed within the 5′ end of the mRNA, facilitating the subsequent recruitment of other protein factors [2, 34, 85]. Next, the preformed 43S complex, which contains the 40S ribosome small subunit (SSU), eIF2-GTP-tRNAiMet ternary complex, and eIF1 and eIF3 initiation factors, joins the mRNA, generating the 48S preinitiation complex (PIC) [34]. The PIC then searches for the AUG start codon within the mRNA in a 5′→3′ direction by a scanning process [43, 48]. Upon identifying the start codon, the anticodon of the initiator Met-tRNAi base-pairs with the mRNA start codon, leading to conformational changes and release of protein factors including eIF2 proteins, and the recruitment of the 60S ribosomal large subunit (LSU). This results in the formation of the 80S ribosome complex, which translates the open reading frame (ORF) of the mRNA [85].

The commonly accepted model for the finding of the start codon is the scanning model in which the PIC searches for the first AUG to initiate translation, as proposed by Kozak more than 30 years ago [43, 46, 48]. However, it was found later that in a number of mRNAs, the AUG codon leading the main ORF is not always the AUG nearest to the 5′ cap [17]. Many mRNAs contain one or more upstream AUG (uAUG) trinucleotides prior to the primary AUG (pAUG) [96]. This observation led to a revised "Scanning model," which takes into consideration of the context sequence of the AUG codon – the kozak sequence [48]. If an AUG codon is present in an appropriate context sequence, the AUG codon is preferentially used for translation initiation. The consensus Kozak sequence was identified to be 5′-GCCGCC(A/G)CC**AUG**G-3′, with the underlined nucleotides, −3 and +4 relative to the start codon, being more important in influencing start codon utilization [44, 45, 47]. The A at −3 position was shown to contact eIF2α subunit and the +4 G interacts with Helix 44 in the 18S rRNA present in the small subunit, and these interactions may contribute to the recognition of an AUG as a start codon [72]. The PIC can bypass an AUG trinucleotide with a weaker consensus sequence, preferring the downstream AUG with a relatively strong Kozak sequence. This is known as the leaky scanning model [29, 48].

Although translational regulation can occur in many different ways, such as by altering the level or modification of translation factors that act in trans, cis-acting elements present in the mRNAs are also involved in modulating protein production [22, 34, 68, 85]. Due to the nature of directional PIC scanning, it is not surprising that some elements present in the 5′ UTR of mRNAs can affect translation. For example, strong secondary structures in the 5′ UTR, especially when close to the 5′ cap, can inhibit translation [4, 17, 49]. The presence of IRES in the 5′ UTR of mRNAs can recruit ribosome in a cap-independent manner, leading to increased translation under certain stress conditions [3, 18, 86]. Some RNA-binding proteins may bind to the 5′ UTR region of certain mRNAs to modulate translation activity [71]. In additional to the 5′ UTR, the 3′ UTR also contains elements that regulate protein levels. For instance, certain sequences or structural elements in 3′ UTR can interact with miRNAs or proteins, which regulate protein production either by affecting mRNA stability or translatability [9, 22, 94].

9.3 Many mRNAs Contain Upstream Open Reading Frames That Can Inhibit Translation of the Main Proteins

As described above, many mRNAs have one or more AUGs in the 5′ UTRs [6], which may encode upstream ORFs (uORFs) (Fig. 9.1a). uORFs are defined as those 5′ UTRs containing an AUG and at least one additional amino acid in addition to a

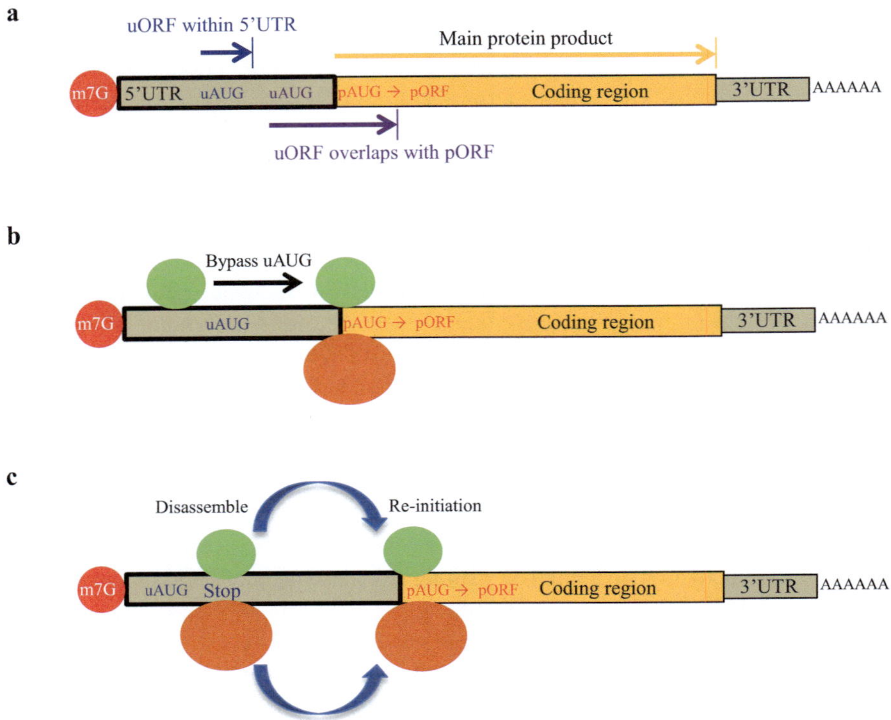

Fig. 9.1 Many mammalian mRNAs contain uORFs. (**a**) Depiction of the structure of an mRNA containing uORFs. The 5′ m^{7}G cap, 5′ UTR, coding region, 3′ UTR, and the poly(A) tail are indicated. The primary start codon (pAUG) for the primary ORF (pORF) is shown. A uORF present in the 5′ UTR may stop within 5′ UTR, or overlaps with pORFs but in a different frame. (**b**) Translation of the pORF encoded by a uORF-containing mRNA may occur via a leaky scanning mechanism. Although some mRNAs contain uORFs, the preinitiation complex may bypass the uAUG codon, for example, due to unfavorite context sequence or structure. In this case, the scanning machinery will recognize the downstream pAUG and initiate translation of the pORF. (**c**) Translation of pORF through reinitiation. When a uORF is present entirely within 5′ UTR, the subunits of the ribosome dissociate at the stop codon of the uORF, and may rejoin at the downstream pAUG, leading to translation reinitiation. The *green* and *orange circles* represent small and large subunits of the ribosome, respectively

stop codon. An uORF-mediated translational regulation is present in different eukaryotic cells. Genome-wide analysis revealed that around 13%, 44%, and 49% of mRNAs contain one or more uORFs in yeast, mouse, and human, respectively [8, 50, 96]. Around two third of oncogenes harbor uORFs [66], implying a pathway-preferred translation regulation through this mechanism. Among the human uORF-containing mRNAs, approximately 70% are conserved between human and mouse or human and rat, either at the sequence level or at the position level [32]. This conserved feature suggests that translation regulation by uORFs is not random and may thus have biological importance [10]. Indeed, it was found that uORFs can cause widespread reduction in expression of proteins encoded by downstream pORFs [8, 35, 96]. The presence of uORF tends to inhibit translation of downstream pORFs through different mechanisms. For example, translation initiation at the uAUG site can reduce translation initiation at the pAUG site, thus leading to reduced expression of the main protein product encoded by the pORF [6, 96, 97]. In addition, the peptide translated from the uORF may interact with the translation machinery and promotes ribosome blockage [40]. In certain cases, the termination codon of the uORF within 5′ UTR can be recognized as an NMD signal, resulting in decreased mRNA stability [62].

Some uORFs are present entirely in the 5′ UTR, that is, the uORF stops within the 5′ UTR. In this case, translation of downstream pORF may be mediated either by leaky scanning or translation reinitiation, that is, the disassembled ribosome subunits may rejoin at the downstream pAUG sites (Fig. 9.1b–c). There are also many uORFs that overlap with pORF, either in the same frame as the pORF, or in different frames. When uORF and pORF are present in the same frame, alternative translation initiation may occur. This is more prevalent for mitochondria targeted proteins [41, 91]. This situation will not be discussed further here. When a uORF overlaps with pORF but in different frames, translation of pORF is most likely mediated by a leaky scanning process, since translation termination of uORF occurs downstream from the pAUG site. The effect of uORFs on pORF expression can be influenced by many factors (Fig. 9.2). For example, the position of uORFs within the 5′ UTR, the number of uORFs per mRNA, the length of encoded amino acids by the uORF (or the translation time), the Kozak consensus sequence of uAUG, and whether or not the uORF overlaps with pORF, all can contribute to the inhibitory effects of uORFs [6, 96, 97]. Greater inhibition has been observed with more uORFs per mRNA, longer peptides encoded by the uORF, or stronger Kozak sequence at the uAUG sites [6, 96, 97]. In addition, longer distance between a uORF and the 5′ cap or shorter distance between the uORF and the pAUG site may have stronger inhibitory effect.

Since uORFs inhibit translation of pORFs mainly due to translation initiation at the uAUG site, it seemed possible to selectively inhibit translation from uORF and that this would enhance translation initiation at the downstream pAUG site, leading to specific increase in the level of the protein encoded by the pORF. Previous studies have already shown that translation of a particular mRNA could be inhibited using antisense oligonucleotides (ASO) that base-pair with the target mRNA without triggering mRNA degradation [11, 52], making ASOs a good candidate to specifically increase protein level by inhibiting translation from uORF.

9.4 Antisense Oligonucleotides

ASOs, either as research tools or as drugs, have been well studied and developed for more than 25 years [14]. ASOs act through Watson-Crick base-pairing with target RNAs, thus ensuring high sequence-specificity. ASOs are designed to have different configurations based on the desired mechanisms of action post hybridization.

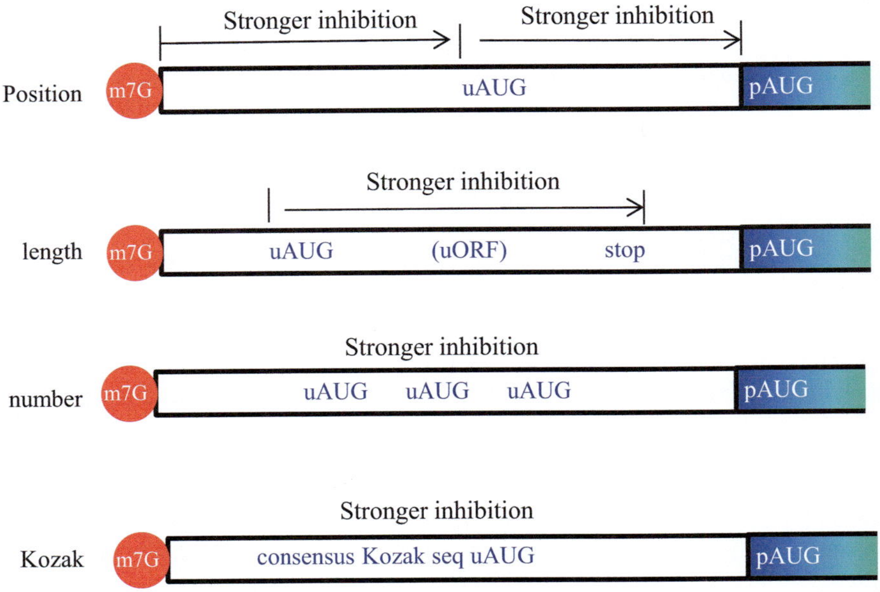

Fig. 9.2 Many factors can affect the inhibitory effects of uORFs. These factors include, but not limited to: the position of a uORF relative to the 5′ cap and to the start codon of the pORF; the length of the amino acid sequence encoded by the uORF (or the time needed to translate the uORF); the number of uORFs within a 5′ UTR; and the Kozak consensus sequence around the uAUG codon. In addition, the structure of the 5′ UTR may also affect the utilization of uAUG

One type of the commonly used ASOs, known as gapmer ASOs, has a chimeric design that contains deoxynucleotides (in most cases 10 nts) in the middle portion, and 3-5 ribonucleotides at both 5′ and 3′ ends [1, 13, 20]. These gapmer ASOs can trigger target RNA degradation by directing RNase H1 cleavage of the RNA strand that base-pairs with the DNA portion of the gapmer ASO [59, 99]. Another type of ASO is usually modified at the 2′-position of the ribose and used to modulate pre-mRNA splicing [31, 42, 64, 75]. To increase nuclease resistance and to enhance ASO pharmacology, the phosphodiester backbone (PO) of the ASOs is often substituted with phosphorothioate (PS) backbone [21, 26, 84, 89], which replaces a nonbridging oxygen with a sulfur (Fig. 9.3a). Compared with PO-ASOs, PS-ASO can be delivered *in vivo* without the need of formulation, and can be readily taken up by cells via endocytic pathways [26, 37]. These enhanced properties of PS-ASOs in delivery and trafficking are believed to be a result of increased protein binding. Indeed, PS-ASOs bind proteins much more promiscuously and avidly than PO-ASOs [28, 55–57, 82, 95]. The binding by proteins prevents PS-ASOs from being rapidly discarded through urinary excretion, and enhances interaction with cell surface and thus internalization [15, 26, 39].

In addition to modifications at the backbones, different chemical modifications at the 2′-position of the ribose have also been used to enhance the hybridization affinity to the RNA target as well as the pharmacological properties. The commonly used 2′ modifications include, but are not limited to, 2′-*O*-methyl (Me), 2′-*O*-

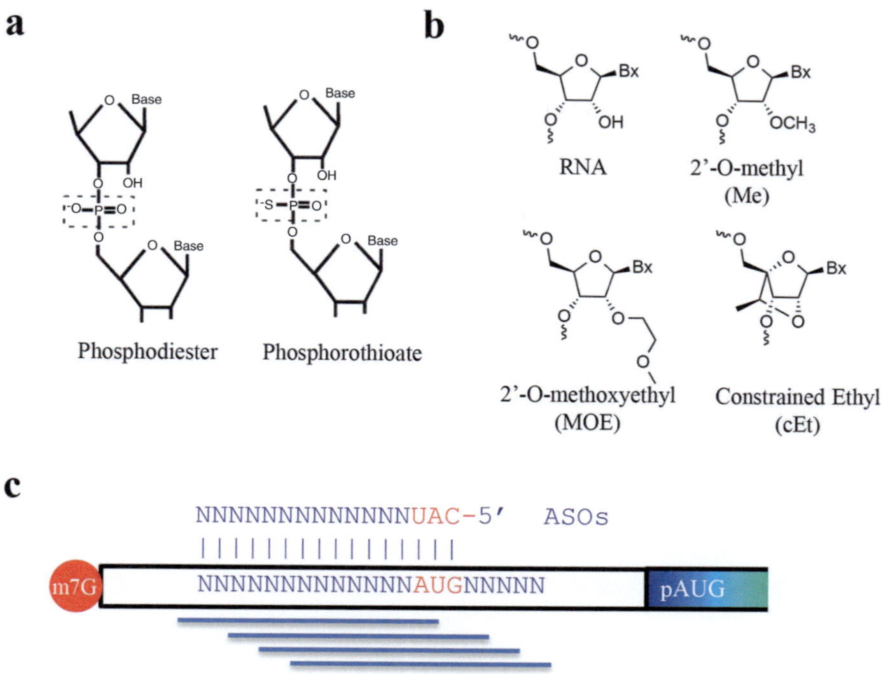

Fig. 9.3 The commonly used chemical modifications for oligonucleotides. (**a**) The backbone of ASOs can be a phosphodiester (PO), which is present in natural DNA or RNA, or a phosphorothioate (PS) linkage, where a nonbridging oxygen is replaced by a sulfur, as indicated by a *dashed box*. (**b**) Commonly used modifications in the 2′ position of the ribose. The 2′-OH group can be replaced with a 2′-*O*-methyl (Me), a 2′-*O*-methoxyethyl (MOE), or a constrained Ethyl (cEt), which forms a cyclic linkage with the 4′ carbon within the ribose. *Bx*, the base of a nucleotide. (**c**) Schematic representation of ASOs targeting the uORF region. N depicts any nucleotides. The suggested binding region of an ASO to inhibit uORF translation is indicated by lines. The best ASO binding sites for each particular mRNA target needs to be experimentally determined. The uAUG is shown in *red*

methoxyethyl (MOE), 2′-F, and 2′-constrained ethyl (cEt) (Fig. 9.3b), which can increase the melting temperature (Tm) by approximately 0.5, 0.5, 3–5, and 3–5 °C/modification, respectively, based on the sequence context [23, 80]. In addition, we have also found that different 2′-modifications of ASOs can significantly affect the binding to intracellular proteins, with ASOs containing more hydrophobic 2′-modifications showing more promiscuous and more tight protein binding [55–57, 81, 82, 93]. Thus, by changing ASO chemical modifications, together with ASO length and chemical configuration, different Tm and protein binding properties can be achieved for ASOs to fulfill different purposes.

To increase levels of specific protein levels by inhibiting translation from uORFs, the ASOs need to have the following characteristics: (1) the targeted mRNA level should not be reduced by the ASOs, that is, the ASOs should not activate RISC or RNase H1 pathways; (2) the ASOs should bind to the uORF region to inhibit translation initiation from the uAUG; and (3) the ASOs should not block the ribosome scanning process so that the translation machinery can reach the downstream pORF. Therefore, the Tm of the ASO should be properly designed to allow removal of the ASO from the target mRNA by the PIC machinery. Finally, for therapeutics, the ASOs should be efficiently delivered *in vivo* with relative ease. Fortunately, we have shown that these requirements can be met by designing ASOs with appropriate length and chemical modifications.

9.5 Protein Levels Can Be Increased Using ASOs Targeting uORFs Both *in vitro* and *in vivo*

The first protein that was increased using uORF-targeting ASOs is RNase H1, the key enzyme required for ASO-directed RNA degradation [99]. Although the mRNA level of RNase H1 is fairly abundant, its protein level is relatively low [14]. It has been shown that RNase H1 mRNA contains a uORF in both human and mouse, and that it inhibits translation of the pORF [87]. The uORF of RNase H1 overlaps with the pORF, and a point mutation to convert the uAUG to an uUUG dramatically increased translation activity by 5–7 fold, as evaluated using a reporter assay [58]. To inhibit translation initiation from the uORF, we designed ASOs to base-pair with the uAUG region of human RNase H1 mRNA, at different positions (Fig. 9.3c). The ASOs were initially designed as 16-mer oligonucleotides modified with 2′-*O*-methyl in the ribose of all nucleotides that are linked with phosphodiester backbones (PO/Me). The 2′-*O*-methyl modification was included to increase ASO stability and binding affinity to the target mRNA. This type of ASO does not trigger RNase H1 or RISC activity [11, 78, 83], and, consistently, transfection of these ASOs into human cells did not alter the level of RNase H1 mRNA [58].

Transfection of the uORF-targeting ASOs in HeLa or HEK293 cells significantly increased the level of RNase H1 protein as early as 6–8 h after transfection. The protein level could be increased to 170–250%, in an ASO concentration dependent manner [58]. The increased RNase H1 protein is obviously functional, since enhanced target reduction by RNase H1-dependent gapmer ASOs was observed in cells treated with the uORF ASO, but not with control ASOs. This and our additional observations suggest that even modest increase in the levels of endogenous proteins can lead to phenotypical changes. These findings provide evidence that ASOs targeting the uORF region of an mRNA can specifically increase the level of the encoded protein.

ASOs targeting the uORF region can increase protein levels for different transcripts. The effects of uORF ASOs on several additional transcripts, including human SFXN3, mouse LRPPRC, and MRPL11 were also tested. The mRNAs of these genes contain uORFs that have been experimentally demonstrated to exhibit inhibitory effects to different extents (20–80% reduction) [8]. ASOs targeting the uORF region of human SFXN3 mRNA sequence specifically increased the protein level to ~135%, in an ASO dose dependent manner, as tested in different human cell lines [58]. Importantly, ASOs targeting uORFs can increase protein levels in different species. PO/Me ASOs targeting the uORF regions of mouse LRPPRC and MRPL11 mRNAs increased the levels of the corresponding proteins to 150–250% in mouse MHT cells and bEnd' cells [58]. These results confirmed that it is applicable to use ASOs targeting uORF regions of different genes to increase the levels of proteins in different cell types, and in different species.

An important step to use ASOs as therapeutic agents for treatment of diseases is that the ASOs need to be delivered and function *in vivo*. To test this, we designed 2′-*O*-methylated ASOs linked with phosphorothioate backbones. This type of ASO can be delivered *in vivo* by nearly all routes including systemic administration, without the need of formulation [26]. PS modification of ASOs provides strong nuclease resistance, thus increasing the stability of the ASOs [21, 26, 84, 89]. 2′-MOE gapmer PS-ASOs have tissue elimination half-lives ranging from 2 to 4 weeks [24]. In addition, PS-ASOs bind more proteins than PO-ASOs. This property helps to prevent ASOs from being rapidly discarded and enhances cellular uptake both *in vitro* and *in vivo* [15, 26, 39]. Although PS modification slightly reduces the Tm, with ~0.5 °C per modification, this affinity loss can be easily compensated by employing 2′ modifications in the ASOs.

Similar to PO/Me ASOs, the PS/Me ASOs targeting uORFs also increased protein levels in both human and mouse cells [58], indicating that this type of modification supports the ASO activity in increasing protein levels. Importantly, when a PS/Me ASO targeting mouse LRPPRC uORF region was administrated into mice by subcutaneous injection, the level of LRPPRC protein was specifically increased by around 80% in mouse liver after 96 h of ASO treatment [58]. In addition, increased protein levels in mice treated with uORF-targeting ASOs were also achieved for other targets with therapeutic interests (our unpublished data). These observations, together with the fact that PS-ASOs have been well developed as drugs for both delivery and pharmacology, suggest that ASOs targeting uORFs can be potentially used as therapeutic agents to increase protein levels.

9.6 The uORF ASOs Are Highly Specific in Increasing the Levels of Proteins

Although potential off-target effects have been a concern for antisense-mediated techniques [19, 54, 88], the uORF ASOs appears to be highly specific, with very minor off-target effects [58]. This is largely due to the fact that the effect of uORF ASOs is not only sequence-specific but also highly position-dependent. Indeed, not all ASOs targeting the RNase H1 5′ UTR could increase the level of the protein, although all ASOs have perfect complementarity to the mRNA [58]. The ASOs targeting the uAUG region increased the protein level; however, shifting the target position even a few nucleotides might significantly reduce the ASO activity in increasing protein levels [58]. In addition, control ASOs that do not base-pair with the RNase H1 mRNA had no effect on the level of RNase H1 protein. Furthermore, it appears that the uORF ASOs effect in a hybridization-dependent manner, as mismatches introduced in the ASOs dramatically reduced the ASO activity, and the upregulation effect of the uORF ASO could be attenuated by subsequent transfection of an oligonucleotide complementary to the uORF ASO.

The high specificity was further demonstrated by the observation that, although the ASO targeting RNase H1 uORF could potentially base-pair with the mRNAs of SRPL2B (11 bp) and FGFR1 (13 bp), neither their mRNA nor the protein levels were affected by the treatment of the uORF ASO [58]. The potential base-pairing of the ASO to the off-targeted mRNAs, even if the mRNA site is accessible to the ASO, might be effectively disrupted by the translation machinery. This view is supported by the findings that the on-target ASO/mRNA interaction, which is a 100% match with the highest possible affinity, could be disrupted by the ribosome during translation or scanning along the mRNA molecules (see below). The minor off-target effect was also confirmed by evaluating the global gene expression profile using microarray analysis. Only dozens of genes exhibited altered expression in the uORF ASO treated cells, compared with mock treated cells, with very modest level changes in either direction (mostly with absolute log2 value less than 2) [58]. These results indicate that the ASOs targeting the uORF regions increased specific protein levels, with very minor off-target effects.

9.7 uORF-Targeting ASOs can Increase Protein Levels by Enhancing Translation

The uORF ASOs increase protein levels by enhancing translation from the pORF. This is supported by the observations that the nascent protein level of LRPPRC was increased by the treatment of cells with an ASO targeting the uORF, as demonstrated by pulse-chase labeling using S^{35}-methionine, followed by immunoprecipitation [58]. However, global translation was not affected by the uORF ASO. In agreement with enhanced translation, treatment with the ASO targeting RNase H1 uORF shifted the mRNA toward polysomes, a signature of more active translation. In addition, the half-life of the targeted protein was not affected by uORF ASO treatment, suggesting that the increased protein level is not resulted from increased protein stability.

The uORF ASOs are designed to base-pair with mRNA (and pre-mRNA) sequence without triggering RNA degradation via mechanisms, such as RISC or RNase H1-mdeiated cleavage. Consistently, the levels of mature mRNAs were not affected by the uORF ASOs for different targeted genes. Although ASOs that bind to certain regions of pre-mRNAs could influence splicing, the uORF ASO did not affect the splicing pattern of nearby exons, as exemplified for the RNase H1 pre-mRNA [58]. This is in agreement with the observations that splicing modulating ASOs are highly position-dependent [31, 74, 75]. In addition, uORF ASO treatment did not alter the nuclear/cytoplasmic distribution and the length of the poly(A) tail of the targeted mature mRNA. These results are consistent with the view that the uORF ASOs enhance translation of the targeted mRNAs, most likely by redirecting translation initiation from the uAUG to the downstream pAUG. This hypothesis is supported by the observation that the effect of uORF ASO on protein

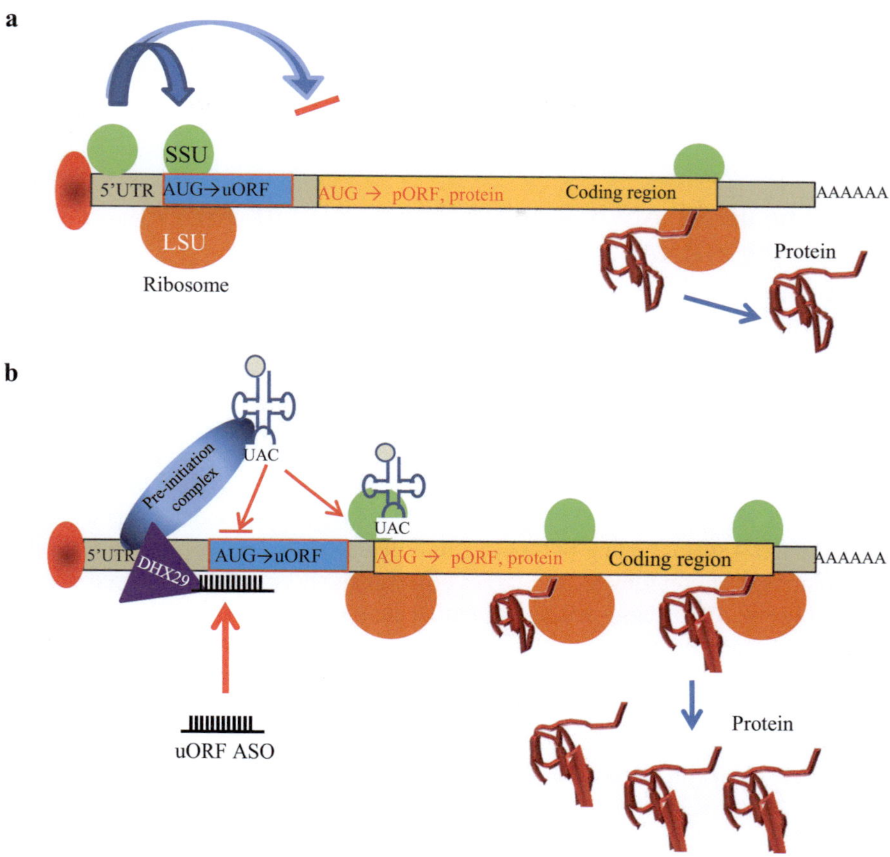

Fig. 9.4 ASOs targeting the uORF region of an mRNA can enhance translation of the downstream pORF. (**a**) Translation of uORF can inhibit translation of pORF. Translation initiation at the uORF region reduces the accessibility of pAUG to the translation initiation machinery, leading to less efficient production of the protein encoded by the pORF. (**b**) ASOs that base-pair with the uORF region can reduce the recognition of uAUG by the PIC, leading to increased utilization of the pAUG start codon, therefore, more efficient protein synthesis. However, the bound ASO needs to be removed from the mRNA by helicases, such as DHX29, enabling further scanning and translation. SSU, small subunit; LSU, large subunit

level is helicase-dependent, since reduction of DHX29, a helicase required for solving the 5′ UTR structure [73], significantly diminished the ASO effect on translation activation. It seems that the uORF ASO needs to transiently bind to the uAUG region, thus to inhibit recognition of this position as a start codon. It is not clear how the uORF ASOs "cheat" the PIC to bypass the uAUG codon and continue to scan the downstream sequence. However, it appears that the bound ASOs need to be removed from the mRNA by the helicase(s) present in the scanning machinery to allow the PIC to reach and recognize the downstream pAUG site (Fig. 9.4). In alignment with this hypothesis, transfection of the uORF ASOs did not affect the structure of the 5′ UTR region of RNase H1 mRNA that was targeted by the uORF ASOs, as demonstrated by structural probing using DMS modification [58]. Thus, these results suggest that the ASOs transiently interact with the target mRNAs to alter the start codon utilization.

9.8 The uORF ASOs Require Proper Affinity to the mRNAs to Increase Protein Levels

As described above, the uORF ASOs should be able to bind to the uAUG region and inhibit recognition of the uAUG codon by the PIC. However, the ASO/mRNA interaction should not be too strong to be dissociated by the helicase(s) in the PIC machinery during scanning. This feature requires a proper affinity of the ASOs, that is, the Tm. ASOs with Tm ranging from 55 to 85 °C might be active in increasing the level of RNase H1 protein [58] (and our unpublished data), however, the best Tm of an ASO for a particular target should be experimentally determined.

Although ASOs targeting the uAUG and the upstream sequence within 5′ UTR could increase protein levels, the target position should also be optimized experimentally using ASOs base-pairing with different positions of the uORF region. The position-dependent feature of uORF ASOs may limit the selection of available sequences; however, the proper Tm of the ASO can be obtained by changing the length of ASOs, based on the mRNA sequence. ASOs with 14–20 nt could be functional in many cases, although longer ASOs have not been tested [58]. In addition, chemical modification of ASOs contributes greatly to the affinity of the ASOs to target RNAs. For example, PS backbone modification reduces Tm by 0.5 °C per modification, whereas different 2′ modifications can increase Tm to a different degree [23, 80]. These modifications can be used in combination to optimize the ASO design, together with altering ASO length.

Interestingly, structure-activity relationship studies showed that in addition to the Tm, the type of modification and the position of the modified nucleotides in the ASO also influence the activity of uORF ASOs in increasing protein levels [58]. For example, although both Me and MOE could increase Tm to a similar level, PO/MOE ASOs tend to reduce protein levels by inhibiting translation, likely because MOE modification may inhibit helicase activity [76], which is required for translation of mRNAs containing structures in the 5′ UTR [29, 73]. In addition, we also tested the effects of cEt modification on ASO activity. Surprisingly, adding cEt modification to three nucleotides at the 3′, but not 5′, of the ASOs increased ASO potency in activating translation [58]. Since the Tm values are comparable for the ASOs containing 5′ or 3′ cEt modified nucleotides, these observations suggest that the uORF ASO activity is related not only to the Tm of the ASO but also to some other biophysical/biochemical features of the modified nucleotides. Thus, the design of ASOs for this purpose needs further optimization.

9.9 Additional Notes

Although around half of human mRNAs contain uORFs, some uORFs may not be translated and do not inhibit the translation of pORFs. Thus it is important to determine whether a given uORF is inhibitory using a reporter assay, ideally by converting the uAUG to a nonstart codon. In addition, some uORFs may only play roles in regulation of pORF translation under stress conditions [96, 97]; therefore, their inhibitory effects under normal condition may not be obvious. Since translation is tightly regulated, physiological changes could significantly affect the translational activity of the ribosome and, in turn, the effects of uORF ASOs. For example, cell confluence affects not only the transfection efficiency of ASOs but also the translation rate of cells. It is known that in fast growing cells (such as those in exponential growth) translation activity is robust, whereas in cells at the stationary phase (e.g., confluent cells) the translation rate is reduced [60]. Since the uORF ASOs act to increase protein levels by modulating translation, the effect of uORF ASOs on protein levels might be affected by the cell status. In addition, as described above, the uORF ASO needs to have a proper affinity to sufficiently inhibit translation initiation at the uAUG but to allow being unwound by the helicases. Thus, a proper Tm of an ASO also needs to be experimentally optimized, by changing the length or chemical modifications of the ASO. Although the target site is relatively limited to the uAUG region, the best target position for a particular mRNA should also be experimentally determined.

9.10 Perspectives

Many diseases are caused by reduced levels or insufficient function of proteins; therefore, increasing the levels of specific proteins is of therapeutic interest. The approach described here for ASO-mediated increase of specific proteins by targeting uORFs provides a novel opportunity for therapeutics as well as a research tool. ASO drugs have been well developed for over 25 years. Their delivery, stability, pharmacokinetics, and the mechanisms of action have been well studied [14, 26]. These advances make uORF ASOs an ideal candidate as therapeutic agents to increase protein levels in patients. uORFs are found in around half of human mRNAs, and new uORFs can be created by mutations that may reduce protein levels and cause diseases, such as mutation in HBB gene that is implicated in beta thalassemia and sickle-cell disease [63]. In addition, recent studies have found that translation initiation can occur at non-AUG sites within 5′ UTR that could also inhibit translation of pORFs [53]. These non-AUG-mediated uORFs might be targetable as well by ASOs. Taken together, ASOs will find broad utility to increase specific protein levels for a large number of genes.

Acknowledgments This work has been supported by an internal funding from Ionis Pharmaceutics, Inc.

References

1. Agrawal S (1999) Importance of nucleotide sequence and chemical modifications of antisense oligonucleotides. Biochim Biophys Acta 1489(1):53–68
2. Alekhina OM, Vassilenko KS (2012) Translation initiation in eukaryotes: versatility of the scanning model. Biochemistry (Mosc) 77(13):1465–1477
3. Araujo PR, Yoon K, Ko D, Smith AD, Qiao M, Suresh U, Burns SC, Penalva LO (2012) Before it gets started: regulating translation at the 5′ UTR. Comp Funct Genomics 2012:475731
4. Babendure JR, Babendure JL, Ding JH, Tsien RY (2006) Control of mammalian translation by mRNA structure near caps. RNA 12(5):851–861
5. Baird SD, Turcotte M, Korneluk RG, Holcik M (2006) Searching for IRES. RNA 12 (10):1755–1785
6. Barbosa C, Peixeiro I, Romao L (2013) Gene expression regulation by upstream open reading frames and human disease. PLoS Genet 9(8):e1003529
7. Bennett CF, Swayze EE (2010) RNA targeting therapeutics: molecular mechanisms of antisense oligonucleotides as a therapeutic platform. Annu Rev Pharmacol Toxicol 50:259–293
8. Calvo SE, Pagliarini DJ, Mootha VK (2009) Upstream open reading frames cause widespread reduction of protein expression and are polymorphic among humans. Proc Natl Acad Sci U S A 106(18):7507–7512
9. Chekulaeva M, Filipowicz W (2009) Mechanisms of miRNA-mediated post-transcriptional regulation in animal cells. Curr Opin Cell Biol 21(3):452–460
10. Chew GL, Pauli A, Schier AF (2016) Conservation of uORF repressiveness and sequence features in mouse, human and zebrafish. Nat Commun 7:11663
11. Chiang MY, Chan H, Zounes MA, Freier SM, Lima WF, Bennett CF (1991) Antisense oligonucleotides inhibit intercellular adhesion molecule 1 expression by two distinct mechanisms. J Biol Chem 266(27):18162–18171
12. Chu D, von der Haar T (2012) The architecture of eukaryotic translation. Nucleic Acids Res 40 (20):10098–10106
13. Crooke ST, Bennett CF (1996) Progress in antisense oligonucleotide therapeutics. Annu Rev Pharmacol Toxicol 36:107–129
14. Crooke ST, Vickers T, Lima W, Wu H-J (2008) Mechanisms of antisense drug action, an introduction. In: Crooke ST (ed) Antisense drug technology – principles, strategies, and application, 2nd edn. CRC Press, Boca Raton, pp 3–46
15. Crooke ST, Wang S, Vickers TA, Shen W, Liang X-H (2016) Cellular uptake and trafficking of antisense oligonucleotides. Nat Biotechnol 35(3):230–237
16. Crosby J, Peloso GM, Auer PL, Crosslin DR, Stitziel NO, Lange LA, Lu Y, Tang ZZ, Zhang H, Hindy G, Masca N, Stirrups K, Kanoni S, Do R, Jun G, Hu Y, Kang HM, Xue C, Goel A, Farrall M, Duga S, Merlini PA, Asselta R, Girelli D, Olivieri O, Martinelli N, Yin W, Reilly D, Speliotes E, Fox CS, Hveem K, Holmen OL, Nikpay M, Farlow DN, Assimes TL, Franceschini N, Robinson J, North KE, Martin LW, DePristo M, Gupta N, Escher SA, Jansson JH, Van Zuydam N, Palmer CN, Wareham N, Koch W, Meitinger T, Peters A, Lieb W, Erbel R, Konig IR, Kruppa J, Degenhardt F, Gottesman O, Bottinger EP, O'Donnell CJ, Psaty BM, Ballantyne CM, Abecasis G, Ordovas JM, Melander O, Watkins H, Orho-Melander M, Ardissino D, Loos RJ, McPherson R, Willer CJ, Erdmann J, Hall AS, Samani NJ, Deloukas P, Schunkert H, Wilson JG, Kooperberg C, Rich SS, Tracy RP, Lin DY, Altshuler D, Gabriel S, Nickerson DA, Jarvik GP, Cupples LA, Reiner AP, Boerwinkle E, Kathiresan S (2014) Loss-of-function mutations in APOC3, triglycerides, and coronary disease. N Engl J Med 371 (1):22–31

17. Davuluri RV, Suzuki Y, Sugano S, Zhang MQ (2000) CART classification of human 5′ UTR sequences. Genome Res 10(11):1807–1816
18. Deforges J, Locker N, Sargueil B (2015) mRNAs that specifically interact with eukaryotic ribosomal subunits. Biochimie 114:48–57
19. Deng Y, Wang CC, Choy KW, Du Q, Chen J, Wang Q, Li L, Chung TK, Tang T (2014) Therapeutic potentials of gene silencing by RNA interference: principles, challenges, and new strategies. Gene 538(2):217–227
20. Dias N, Stein CA (2002) Antisense oligonucleotides: basic concepts and mechanisms. Mol Cancer Ther 1(5):347–355
21. Eckstein F (2000) Phosphorothioate oligodeoxynucleotides: what is their origin and what is unique about them? Antisense Nucleic Acid Drug Dev 10(2):117–121
22. Fabian MR, Sonenberg N, Filipowicz W (2010) Regulation of mRNA translation and stability by microRNAs. Annu Rev Biochem 79:351–379
23. Freier SM, Altmann KH (1997) The ups and downs of nucleic acid duplex stability: structure-stability studies on chemically-modified DNA:RNA duplexes. Nucleic Acids Res 25 (22):4429–4443
24. Geary RS (2009) Antisense oligonucleotide pharmacokinetics and metabolism. Expert Opin Drug Metab Toxicol 5(4):381–391
25. Geary RS, Yu RZ, Siwkoswki A, Leivin AA (2008) Pharmacokinetic/pharmacodynamic properties of phosphorothioate 2′-O-(2-Methoxyethyl)-modified antisense oligonucleotides in animals and man. In: Crooke ST (ed) Antisense drug technology – principles, strategies, and applications. CRC Press, Boca Raton, pp 305–326
26. Geary RS, Norris D, Yu R, Bennett CF (2015) Pharmacokinetics, biodistribution and cell uptake of antisense oligonucleotides. Adv Drug Deliv Rev 87:46–51
27. Gomez C, Esther Ramirez M, Calixto-Galvez M, Medel O, Rodriguez MA (2010) Regulation of gene expression in protozoa parasites. J Biomed Biotechnol 2010:726045
28. Guvakova MA, Yakubov LA, Vlodavsky I, Tonkinson JL, Stein CA (1995) Phosphorothioate oligodeoxynucleotides bind to basic fibroblast growth factor, inhibit its binding to cell surface receptors, and remove it from low affinity binding sites on extracellular matrix. J Biol Chem 270(6):2620–2627
29. Hinnebusch AG (2014) The scanning mechanism of eukaryotic translation initiation. Annu Rev Biochem 83:779–812
30. Hoyng SA, de Winter F, Tannemaat MR, Blits B, Malessy MJ, Verhaagen J (2015) Gene therapy and peripheral nerve repair: a perspective. Front Mol Neurosci 8:32
31. Hua Y, Vickers TA, Baker BF, Bennett CF, Krainer AR (2007) Enhancement of SMN2 exon 7 inclusion by antisense oligonucleotides targeting the exon. PLoS Biol 5(4):e73
32. Iacono M, Mignone F, Pesole G (2005) uAUG and uORFs in human and rodent 5′ untranslated mRNAs. Gene 349:97–105
33. Jackson RJ (2005) Alternative mechanisms of initiating translation of mammalian mRNAs. Biochem Soc Trans 33(Pt 6):1231–1241
34. Jackson RJ, Hellen CU, Pestova TV (2010) The mechanism of eukaryotic translation initiation and principles of its regulation. Nat Rev Mol Cell Biol 11(2):113–127
35. Johnstone TG, Bazzini AA, Giraldez AJ (2016) Upstream ORFs are prevalent translational repressors in vertebrates. EMBO J 35(7):706–723
36. Jones B (2015) Gene expression: layers of gene regulation. Nat Rev Genet 16(3):128–129
37. Juliano RL, Carver K (2015) Cellular uptake and intracellular trafficking of oligonucleotides. Adv Drug Deliv Rev 87:35–45
38. Juliano R, Bauman J, Kang H, Ming X (2009) Biological barriers to therapy with antisense and siRNA oligonucleotides. Mol Pharm 6(3):686–695
39. Juliano RL, Ming X, Carver K, Laing B (2014) Cellular uptake and intracellular trafficking of oligonucleotides: implications for oligonucleotide pharmacology. Nucleic Acid Ther 24 (2):101–113

40. Karagyozov L, Godfrey R, Bohmer SA, Petermann A, Holters S, Ostman A, Bohmer FD (2008) The structure of the 5′-end of the protein-tyrosine phosphatase PTPRJ mRNA reveals a novel mechanism for translation attenuation. Nucleic Acids Res 36(13):4443–4453
41. Kazak L, Reyes A, Duncan AL, Rorbach J, Wood SR, Brea-Calvo G, Gammage PA, Robinson AJ, Minczuk M, Holt IJ (2013) Alternative translation initiation augments the human mitochondrial proteome. Nucleic Acids Res 41(4):2354–2369
42. Kole R, Krieg AM (2015) Exon skipping therapy for Duchenne muscular dystrophy. Adv Drug Deliv Rev 87:104–107
43. Kozak M (1980) Evaluation of the "scanning model" for initiation of protein synthesis in eucaryotes. Cell 22(1 Pt 1):7–8
44. Kozak M (1986) Point mutations define a sequence flanking the AUG initiator codon that modulates translation by eukaryotic ribosomes. Cell 44(2):283–292
45. Kozak M (1987) An analysis of 5′-noncoding sequences from 699 vertebrate messenger RNAs. Nucleic Acids Res 15(20):8125–8148
46. Kozak M (1989) Circumstances and mechanisms of inhibition of translation by secondary structure in eucaryotic mRNAs. Mol Cell Biol 9(11):5134–5142
47. Kozak M (1997) Recognition of AUG and alternative initiator codons is augmented by G in position +4 but is not generally affected by the nucleotides in positions +5 and +6. EMBO J 16 (9):2482–2492
48. Kozak M (2002) Pushing the limits of the scanning mechanism for initiation of translation. Gene 299(1–2):1–34
49. Kozak M (2005) Regulation of translation via mRNA structure in prokaryotes and eukaryotes. Gene 361:13–37
50. Lawless C, Pearson RD, Selley JN, Smirnova JB, Grant CM, Ashe MP, Pavitt GD, Hubbard SJ (2009) Upstream sequence elements direct post-transcriptional regulation of gene expression under stress conditions in yeast. BMC Genomics 10:7
51. Leader B, Baca QJ, Golan DE (2008) Protein therapeutics: a summary and pharmacological classification. Nat Rev Drug Discov 7(1):21–39
52. Lecosnier S, Cordier C, Simon P, Francois JC, Saison-Behmoaras TE (2011) A steric blocker of translation elongation inhibits IGF-1R expression and cell transformation. FASEB J: Off Publ Fed Am Soc Exp Biol 25(7):2201–2210
53. Lee S, Liu B, Lee S, Huang SX, Shen B, Qian SB (2012) Global mapping of translation initiation sites in mammalian cells at single-nucleotide resolution. Proc Natl Acad Sci U S A 109(37):E2424–E2432
54. Liang XH, Hart CE, Crooke ST (2013) Transfection of siRNAs can alter miRNA levels and trigger non-specific protein degradation in mammalian cells. Biochim Biophys Acta 1829 (5):455–468
55. Liang XH, Shen W, Sun H, Prakash TP, Crooke ST (2014) TCP1 complex proteins interact with phosphorothioate oligonucleotides and can co-localize in oligonucleotide-induced nuclear bodies in mammalian cells. Nucleic Acids Res 42(12):7819–7832
56. Liang XH, Sun H, Shen W, Crooke ST (2015) Identification and characterization of intracellular proteins that bind oligonucleotides with phosphorothioate linkages. Nucleic Acids Res 43 (5):2927–2945
57. Liang XH, Shen W, Sun H, Kinberger GA, Prakash TP, Nichols JG, Crooke ST (2016a) Hsp90 protein interacts with phosphorothioate oligonucleotides containing hydrophobic 2′-modifications and enhances antisense activity. Nucleic Acids Res 44(8):3892–3907
58. Liang XH, Shen W, Sun H, Migawa MT, Vickers TA, Crooke ST (2016b) Translation efficiency of mRNAs is increased by antisense oligonucleotides targeting upstream open reading frames. Nat Biotechnol 34(8):875–880
59. Lima W, Wu H, Crooke ST (2008) The RNase H mechanism. In: Crooke ST (ed) Antisense drug technology – principles, strategies, and applications, 2nd edn. CRC Press, Boca Raton, pp 47–74

60. Ling J, Morley SJ, Traugh JA (2005) Inhibition of cap-dependent translation via phosphorylation of eIF4G by protein kinase Pak2. EMBO J 24(23):4094–4105
61. Lu PY, Xie F, Woodle MC (2005) In vivo application of RNA interference: from functional genomics to therapeutics. Adv Genet 54:117–142
62. Lykke-Andersen S, Jensen TH (2015) Nonsense-mediated mRNA decay: an intricate machinery that shapes transcriptomes. Nat Rev Mol Cell Biol 16(11):665–677
63. Mansilla-Soto J, Riviere I, Sadelain M (2011) Genetic strategies for the treatment of sickle cell anaemia. Br J Haematol 154(6):715–727
64. McClorey G, Wood MJ (2015) An overview of the clinical application of antisense oligonucleotides for RNA-targeting therapies. Curr Opin Pharmacol 24:52–58
65. Meng L, Ward AJ, Chun S, Bennett CF, Beaudet AL, Rigo F (2015) Towards a therapy for Angelman syndrome by targeting a long non-coding RNA. Nature 518(7539):409–412
66. Morris DR, Geballe AP (2000) Upstream open reading frames as regulators of mRNA translation. Mol Cell Biol 20(23):8635–8642
67. Mourich DV, Oda SK, Schnell FJ, Crumley SL, Hauck LL, Moentenich CA, Marshall NB, Hinrichs DJ, Iversen PL (2014) Alternative splice forms of CTLA-4 induced by antisense mediated splice-switching influences autoimmune diabetes susceptibility in NOD mice. Nucleic Acid Ther 24(2):114–126
68. Muller PP, Trachsel H (1990) Translation and regulation of translation in the yeast Saccharomyces cerevisiae. Eur J Biochem/FEBS 191(2):257–261
69. Nomakuchi TT, Rigo F, Aznarez I, Krainer AR (2016) Antisense oligonucleotide-directed inhibition of nonsense-mediated mRNA decay. Nat Biotechnol 34(2):164–166
70. Pearson S, Jia H, Kandachi K (2004) China approves first gene therapy. Nat Biotechnol 22 (1):3–4
71. Pichon X, Wilson LA, Stoneley M, Bastide A, King HA, Somers J, Willis AE (2012) RNA binding protein/RNA element interactions and the control of translation. Curr Protein Pept Sci 13(4):294–304
72. Pisarev AV, Kolupaeva VG, Pisareva VP, Merrick WC, Hellen CU, Pestova TV (2006) Specific functional interactions of nucleotides at key −3 and +4 positions flanking the initiation codon with components of the mammalian 48S translation initiation complex. Genes Dev 20(5):624–636
73. Pisareva VP, Pisarev AV, Komar AA, Hellen CU, Pestova TV (2008) Translation initiation on mammalian mRNAs with structured 5′ UTRs requires DExH-box protein DHX29. Cell 135 (7):1237–1250
74. Pramono ZA, Wee KB, Wang JL, Chen YJ, Xiong QB, Lai PS, Yee WC (2012) A prospective study in the rational design of efficient antisense oligonucleotides for exon skipping in the DMD gene. Hum Gene Ther 23(7):781–790
75. Rigo F, Hua Y, Chun SJ, Prakash TP, Krainer AR, Bennett CF (2012) Synthetic oligonucleotides recruit ILF2/3 to RNA transcripts to modulate splicing. Nat Chem Biol 8(6):555–561
76. Rogers GW Jr, Lima WF, Merrick WC (2001) Further characterization of the helicase activity of eIF4A. Substrate specificity. J Biol Chem 276(16):12598–12608
77. Rossbach M (2010) Small non-coding RNAs as novel therapeutics. Curr Mol Med 10 (4):361–368
78. Schneider PN, Olthoff JT, Matthews AJ, Houston DW (2010) Use of fully modified 2′-O-methyl antisense oligos for loss-of-function studies in vertebrate embryos. Genesis 49 (3):117–123
79. Segalat L (2007) Loss-of-function genetic diseases and the concept of pharmaceutical targets. Orphanet J Rare Dis 2:30
80. Seth PP, Siwkowski A, Allerson CR, Vasquez G, Lee S, Prakash TP, Kinberger G, Migawa MT, Gaus H, Bhat B, Swayze EE (2008) Design, synthesis and evaluation of constrained methoxyethyl (cMOE) and constrained ethyl (cEt) nucleoside analogs. Nucleic Acids Symp Ser (Oxf) 52:553–554

81. Shen W, Liang XH, Crooke ST (2014) Phosphorothioate oligonucleotides can displace NEAT1 RNA and form nuclear paraspeckle-like structures. Nucleic Acids Res 42 (13):8648–8662
82. Shen W, Liang XH, Sun H, Crooke ST (2015) 2′-Fluoro-modified phosphorothioate oligonucleotide can cause rapid degradation of P54nrb and PSF. Nucleic Acids Res 43(9):4569–4578
83. Shibahara S, Mukai S, Nishihara T, Inoue H, Ohtsuka E, Morisawa H (1987) Site-directed cleavage of RNA. Nucleic Acids Res 15(11):4403–4415
84. Shimayama T, Nishikawa F, Nishikawa S, Taira K (1993) Nuclease-resistant chimeric ribozymes containing deoxyribonucleotides and phosphorothioate linkages. Nucleic Acids Res 21(11):2605–2611
85. Sonenberg N, Hinnebusch AG (2009) Regulation of translation initiation in eukaryotes: mechanisms and biological targets. Cell 136(4):731–745
86. Spriggs KA, Bushell M, Willis AE (2010) Translational regulation of gene expression during conditions of cell stress. Mol Cell 40(2):228–237
87. Suzuki Y, Holmes JB, Cerritelli SM, Sakhuja K, Minczuk M, Holt IJ, Crouch RJ (2010) An upstream open reading frame and the context of the two AUG codons affect the abundance of mitochondrial and nuclear RNase H1. Mol Cell Biol 30(21):5123–5134
88. Svoboda P (2007) Off-targeting and other non-specific effects of RNAi experiments in mammalian cells. Curr Opin Mol Ther 9(3):248–257
89. Swayze EE, Bhat B (2008) The medicinal chemistry of oligonucleotides. In: Crooke ST (ed) Antisense drug technology – principles, strategies, and applications, 2nd edn. CRC Press, Boca Raton, pp 143–182
90. Toth PP (2013) Emerging LDL therapies: mipomersen-antisense oligonucleotide therapy in the management of hypercholesterolemia. J Clin Lipidol 7(3 Suppl):S6–10
91. Van Damme P, Gawron D, Van Criekinge W, Menschaert G (2014) N-terminal proteomics and ribosome profiling provide a comprehensive view of the alternative translation initiation landscape in mice and men. Mol Cell Proteomics: MCP 13(5):1245–1261
92. van der Velden AW, Thomas AA (1999) The role of the 5′ untranslated region of an mRNA in translation regulation during development. Int J Biochem Cell Biol 31(1):87–106
93. Vickers TA, Crooke ST (2016) Development of a quantitative BRET affinity assay for nucleic acid-protein interactions. PLoS One 11(8):e0161930
94. von Roretz C, Di Marco S, Mazroui R, Gallouzi IE (2011) Turnover of AU-rich-containing mRNAs during stress: a matter of survival. Wiley Interdiscip Rev RNA 2(3):336–347
95. Weidner DA, Valdez BC, Henning D, Greenberg S, Busch H (1995) Phosphorothioate oligonucleotides bind in a non sequence-specific manner to the nucleolar protein C23/nucleolin. FEBS Lett 366(2-3):146–150
96. Wethmar K (2014) The regulatory potential of upstream open reading frames in eukaryotic gene expression. Wiley Interdiscip Rev RNA 5(6):765–778
97. Wethmar K, Smink JJ, Leutz A (2010) Upstream open reading frames: molecular switches in (patho)physiology. Bioessays 32(10):885–893
98. Wilson JA, Richardson CD (2006) Future promise of siRNA and other nucleic acid based therapeutics for the treatment of chronic HCV. Infect Disord Drug Targets 6(1):43–56
99. Wu H, Lima WF, Zhang H, Fan A, Sun H, Crooke ST (2004) Determination of the role of the human RNase H1 in the pharmacology of DNA-like antisense drugs. J Biol Chem 279 (17):17181–17189

Chapter 10
Repurposing CRISPR System for Transcriptional Activation

Meng Chen and Lei Stanley Qi

Abstract In recent years, Clustered Regularly Interspaced Short Palindromic Repeats (CRISPR)/CRISPR-associated protein 9 (Cas9) system has become the most popular one for genome editing. When the nuclease domains of Cas9 protein are mutated into deactivated form (dCas9), CRISPR/dCas9 still retains the ability to bind the targeted DNA sequence, but loses the endonuclease cleavage activity. Taking advantage of the characteristics of this engineered nuclease inactive Cas9, the CRISPR/dCas system has been repurposed into versatile RNA-guided, DNA-targeting platforms, such as genome imaging, gene regulation, and epigenetic modification. Specifically, fusion of dCas9 with activation domains allows specific and efficient transcriptional activation on a genome-wide scale among diverse organisms. The purpose of this chapter is to review most important the recently published literature on CRISPR/dCas9-based transcriptional activation systems. Compared with the conventional approaches for enhancement of the expression of specific genes of interest, CRISPR/Cas9-based system has emerged as a promising technology for genome regulation, allowing specificity, convenience, robustness, and scalability for endogenous gene activation.

Keywords CRISPR/Cas • Gene regulation • Transcriptional activation • Endogenous gene activation

---

M. Chen
Department of Bioengineering, Stanford University, Stanford, CA 94305, USA

Gladstone Institute of Cardiovascular Disease, San Francisco, CA 94158, USA

L.S. Qi (✉)
Department of Bioengineering, Stanford University, Stanford, CA 94305, USA

Department of Chemical and Systems Biology, Stanford University, Stanford, CA 94305, USA

ChEM-H, Stanford University, Stanford, CA 94305, USA
e-mail: stanley.qi@stanford.edu

L.-C. Li (ed.), *RNA Activation*, Advances in Experimental Medicine and Biology 983, DOI 10.1007/978-981-10-4310-9_10

10.1 Introduction

Clustered regularly interspaced short palindromic repeats (CRISPR)/CRISPR-associated (Cas) protein systems provide adaptive immunity against viruses and plasmids in bacteria and archaea. In contrast to type I and III CRISPR/Cas systems which employ a set of Cas proteins for RNA-guided immune surveillance, the type II bacterial CRISPR/Cas system uses only a single Cas protein, known as Cas9, to mediate foreign DNA recognition and cleavage [16]. In the process, CRISPR RNA (crRNA) hybridizes with cognate trans-activating crRNA (tracrRNA) to form a unique dual-RNA structure that directs Cas9 to specific DNA target site that is complementary to the 20-nucleotide (nt), guide-RNA sequence and further introduces site-specific double-stranded breaks (DSBs) in target DNA upon recognition of the protospacer adjacent motif (PAM) sequence [12, 17]. Notably, the single chimeric guide RNA (sgRNA) mimicking the natural dual RNA by fusing crRNA with tracrRNA via a tetraloop is sufficient to guide the endonuclease Cas9 to specific DNA target site for DNA degradation (Fig.10.1a). By changing the 20-nucleotide guide-RNA sequence located on the 5′ end of sgRNA, this simplified two-component CRISPR–Cas9 system can be easily programmed to target virtually any DNA sequence of interest in the genome. In the cells, the generated site-specific DSBs by CRISPR–Cas9 can be further repaired, either by the error-prone nonhomologous end joining (NHEJ) pathway or by high-fidelity-homology-directed repair (HDR) pathway when a repair DNA template is present [11] (Fig. 10.1a). Since the first demonstration of its power for genome editing in mammalian cells [9, 22], CRISPR RNA-guided Cas9 system has drawn worldwide attention due to its simplicity and robustness and quickly become the most common tool for genome engineering in a variety of organisms [11].

Cas9 is a multidomain and multifunctional DNA endonuclease [16]. It contains two distinct nuclease domains responsible for double-stranded (ds) DNA cleavage: the HNH domain of Cas9 cleaves the target DNA strand, while the RuvC-like domain of Cas9 cleaves the nontarget DNA strand [12, 17]. Mutations in both nuclease domains (D10A: Asp10 → Ala, H840A: His840 → Ala) result in an RNA-guided DNA-binding protein without endonuclease activity [17, 24]. This engineered nuclease-deficient Cas9, termed dCas9, when fused to effector domains with distinct regulatory functions, enables the repurposing of the CRISPR–Cas9 system to a general platform for RNA-guided DNA targeting without cleavage activity (Fig. 10.1b), thereby allowing versatile genome modification beyond permanent genome editing [10], such as gene regulation [13, 24], epigenetic modulation [19], and live-cell imaging [7]. Specifically, fusion of dCas9 to activation effector domains allows specific and efficient transcriptional activation on a genome-wide scale in diverse organisms [14].

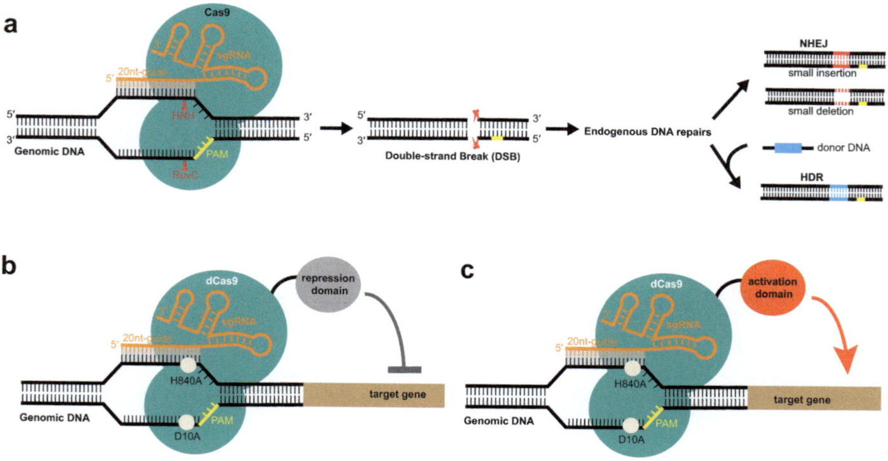

Fig. 10.1 *Schematic diagram showing how CRISPR/Cas9 system is used for genome engineering.* (**a**) Cas9 is guided by an sgRNA to a specific DNA locus, where HNH and RuvC nuclease domains cut the double-strand DNA to form a double-stranded break (DSB). The generated DSB is further repaired, either by the error-prone nonhomologous end joining (NHEJ) pathway or by high-fidelity homology-directed repair (HDR) pathway when a repair DNA template is present. (**b**) Activation or repression domain is fused to the catalytically inactive Cas9 variant for gene regulation

10.2 CRISPRa Activation Systems

CRISPR/dCas9-mediated gene activation (hereafter referred to as CRISPRa) systems, consisting of dCas9-activation domain fusion proteins and sgRNA, can target the specific promoter or enhancer region of the gene of interest. The activation domain can be one or several activation domains, such as VP64 acidic transactivation domain (four copies of *Herpes simplex* virus protein 16), or can be full length or part of an epigenetic modifier, such as the core of histone acetyltransferase p300. Therefore, CRISPRa can function by either directly activating transcription, or modifying the chromatin conformation, or function through recruiting additional transcriptional and/or epigenetic activators to the targeted region. Basically, depending on which component of CRISPR/dCas9 is fused with the activation domain, there are at least three categories of CRISPRa systems: dCas9-activation domain fusion protein, sgRNA-activation domain fusion protein, or combined CRISPRa system with both dCas9- and sgRNA-conjugated with activation domains. For each category, they can also be further classified depending on what the activation domain is. These CRISPRa platforms will be discussed in detail in the following sections (Fig. 10.2).

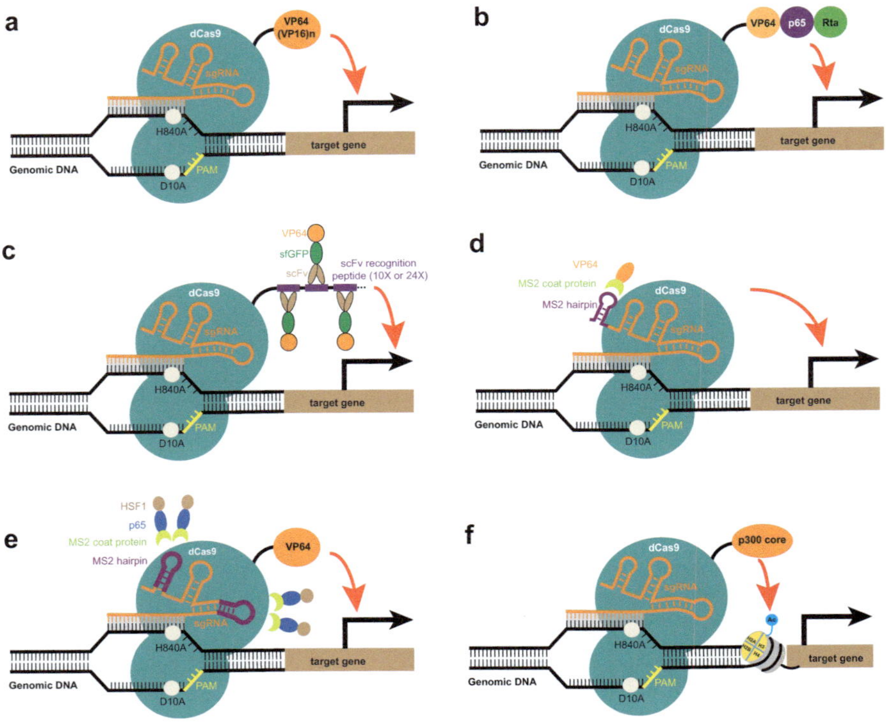

Fig. 10.2 *Different types of CRISPRa activation systems.* (**a**) The dCas9–VP64 (or multiple copies of VP16) system (**b**) The dCas9–VPR system (**c**) The dCas9–SunTag system (**d**) The sgRNA-activation domain system (**e**) The SAM system (**f**) The dCas9–p300 core system

10.2.1 The dCas9–VP64 CRISPRa System

The dCas9–VP64 CRISPRa system, first reported in 2013 as being able to activate targeted endogenous genes, represents the first-generation CRISPRa system, while all the other further improved versions are generally considered as second-generation activation systems. When dCas9 is genetically fused with a C-terminal VP64 acidic transactivation domain (four copies of *Herpes simplex* virus protein 16), it can activate both reporter gene and endogenous genes with a single sgRNA by transient delivery into mammalian cells (Fig. 10.2a). In addition, the use of multiple sgRNAs was able to achieve synergistic activation of a broad range of selected genes (interleukin 1 receptor antagonist, IL1RN), achaete-scute family bHLH transcription factor 1 (ASCL1), nanog homeobox (NANOG), myogenic differentiation 1 (MYOD1), hemoglobin subunit gamma ½ (HBG1/2), vascular endothelial growth factor A (VEGFA), and neurotrophin 3 (NTF3). Furthermore, RNA sequencing demonstrated that targeted gene activation was quite specific with no detectable off-target gene activation [21, 23].

As expected, increasing the number of VP16 repeat domains, such as dCas9–VP96, VP64–dCas9–VP64, dCas9–VP160, and dCas9–VP192 (Fig. 10.2a), has been shown to more efficiently upregulate the expression of endogenous genes, such as interleukin 1 receptor antagonist (IL1RN), SRY-Box2 (SOX2), POU class 5 homeobox 1 (POU5F1 or OCT4), both at mRNA and/or protein levels [1, 4, 8, 18]. Among them, the dCas9–VP192 leads to the highest increase in OCT4 expression levels, up to about 70-fold. Furthermore, human skin fibroblasts can be reprogrammed into inducing pluripotent stem cells (iPSCs) by replacing OCT4 overexpression with dCas9–VP192-mediated activation of endogenous OCT4. The epigenetic changes at OCT4 distal enhancer induced by CRISPRa were investigated and shown to have more active histone mark H3K27Ac, consistent with the previous report that the VP64 transactivation domain recruits the activating complex component p300 and facilitates histone acetylation [1]. In another study, Black et al. found that VP64–dCas9–VP64-mediated endogenous gene activation of mouse neuronal transcription factors Brn2, Ascl1, and Myt1l (BAM factors) directly reprogrammed cultured primary mouse embryonic fibroblasts (PMEFs) to functional induced neuronal cells[3]. Mechanistically, they found that the rapid and sustained elevated levels of endogenous gene expression corresponded to an increase of the epigenetically active markers H3K27ac and H3K4me3 at the target loci. Similar to dCas9–VP64, the efficient activation of endogenous genes also required multiple sgRNAs. In addition, the enhancement of gene activation was also observed with multiple sgRNAs tiling the promoter region, which suggested that recruitment of more activators could be helpful for increasing activation efficiency. This strategy was applied for the development of the second generation of CRISPR activation systems.

10.2.2 The dCas9–SunTag CRISPRa System

SunTag, a protein scaffold with repeating peptide epitope array, that can recruit multiple copies of antibody–activator fusion protein, has been initially developed for imaging of single molecule in living cells. When antibody–VP64 fusion protein was delivered with dCas9-SunTag fusion protein, the system demonstrated strong activation of endogenous gene expression. In one study, Marvin et al. used dCas9 fused with a carboxy-terminal SunTag array consisting of ten copies of a small peptide epitope, and recruited theoretically ten copies of single-chain variable fragment (scFV)–superfolded GFP (sfGFP)–VP64 (scFV–sfGFP–VP64) antibody–activator fusion proteins to a single dCas9 (Fig. 10.2c). Using the dCas9–SunTag–10x (scFV–sfGFP–VP64) system, 10–40-fold activation of the C-X-C Motif Chemokine Receptor 4 (CXCR4) gene was achieved with only one sgRNA, which led to the manipulation of cell migration. Using the SunTag system, Gilbert et al. performed a saturating screen in which they tested the activity of every unique sgRNA broadly tiling around the transcription start sites (TSSs) of 49 genes known to modulate cellular susceptibility to ricin, and observed a peak of active sgRNAs for SunTag CRISPRa system at −400 to −50 bp upstream from the transcription start site (TSS) [26].

10.2.3 The dCas9–VPR CRISPRa System

The tripartite activator domain that consists of VP64, Nuclear Factor NF-κ-B P65 subunit activation domain (p65AD) and Epstein–Barr virus R transactivator (Rta) (VPR) was developed to enhance the CRISPR/dCas9-based activation of endogenous genes (Fig. 10.2b). A set of genes related to cellular reprogramming, development, and gene therapy were activated with three to four gRNAs delivered in concert. When compared to the dCas9–VP64 activator, dCas9–VPR showed significantly (22–320-fold) greater activation of endogenous targets than dCas9–VP64. Furthermore, in accordance with previous studies, we noted an inverse correlation between basal expression level and relative expression gain induced by CRISPR activation systems [5].

10.2.4 The sgRNA-Activation Domain of CRISPRa System

In addition to fusing different transactivation domains to either the amino or carboxy terminus of the dCas9 protein, sgRNA can also be engineered to gain more robust activation. Zalatan et al. first introduced a single-RNA hairpin domain to the end of the sgRNA, connected by a two-base linker. For the recruitment RNA modules, they used the well-characterized viral RNA sequences MS2, PP7, and Com, which are recognized by the MCP, PCP, and Com RNA-binding proteins, respectively. Then they fused the transcriptional activation domain VP64 to each of the corresponding RNA-binding proteins for the purpose of the activation of targeted genes (Fig. 10.2d). On the other hand, when a repression domain KRAB is engineered into RNA-binding proteins, the system is good for transcriptional inhibition. Overall, the successful application of scaffold RNA-mediated transcriptional control in human and yeast cells paves the way for simultaneous ON/OFF gene regulatory switches mediated by orthogonal RNA-binding proteins fused to transcriptional activators (VP64) or repressors (KRAB) [29].

10.2.5 The Combined CRISPRa System

Based on the crystal structure of the *Streptococcus pyogenes* dCas9 (D10A/H840A) in complex with a single-guide RNA (sgRNA) and complementary target DNA, Konermann et al. developed synergistic activation mediator (SAM) system [20]. They selected a minimal hairpin aptamer, which selectively binds dimerized

MS2 bacteriophage coat proteins (MCP), and appended it to the sgRNA tetraloop and stem loop 2 (Fig. 10.2e). Together with MS2-mediated transactivation factors MCP-p65AD-heat shock factor 1 (HSF1), dCas9–VP64 significantly enhanced the efficiency of activation of protein-coding genes and long noncoding RNAs (lincRNA) with one single-guide RNA, and enabled multiplexed activation of ten genes simultaneously. The ability to activate target genes using individual sgRNAs greatly facilitates the development of pooled, genome-wide transcriptional activation screening. Based on the SAM system, they successfully performed a screening for genes that confer resistance to a BRAF inhibitor in melanoma cells [20].

10.2.6 The dCas9–Epigenetic Modifier CRISPRa System

The dCas9 can also be fused with an epigenetic modifier to directly manipulate the epigenetic states at the enhancer region, thereby activating the targeted genes. This system uses different mechanism of action from the dCas9-activating transcription factor fusion protein systems mentioned as above. While the activator domains used in the previous engineered transcriptional factors such as VP64 act as scaffolds for recruiting multiple components of the preinitiation complex including transcriptional and epigenetic factors, and do not enzymatically modulate the chromatin state directly, the dCas9–epigenetic modifier fusion protein directly alter the specific epigenetic marks at specific location.

In one study, fusion of dCas9 to the catalytic core of the transcription activator acetyltransferase p300 (dCas9–p300core), a highly conserved acetyltransferase involved in a wide range of cellular processes, has been demonstrated to activate genes in human cells (Fig. 10.2f). The fusion protein catalyzes acetylation of histone H3 lysine 27 at its target sites, leading to specific and robust transcriptional activation of target genes including IL1RN, Myogenic Differentiation 1 (MYOD) and OCT4 from both promoters and enhancers with an individual guide RNA [15].

With the expansion of the CRISPRa toolbox, it will be necessary to compare the activation by these different systems across many endogenous genes in a variety of cell types, in order to determine which tool is best suited for specific genes and cell types. Recently, Chavez et al. performed a series of experiments in human embryonic kidney (HEK) 293T cells to compare the activation efficiency of many published CRISPRa activation systems, and three from the second generation in particular – VPR, SAM, and SunTag – appeared to be the most potent [6]. For nine selected coding genes and noncoding genes, the activation levels can reach up to several orders of magnitude higher than those of the first-generation dCas9–VP64 activator. Among the three, SAM seems to deliver high levels of gene induction most consistently, although none of the three was obviously superior to the other.

In addition to the application of CRISPRa in mammalian cells, CRISPR/Cas9-based activation system was also tested in the bacteria, *Saccharomyces cerevisiae* and *Drosophila melanogaster* cells for activating endogenous loci [6]. For example, a fusion of dCas9 with the ω-subunit of the *E. coli* polymerase allowed assembly of the holoenzyme for reporter gene activation in *E. coli*. Activation levels depended on the distance between the dCas9 binding sequence and the promoter element. It is possible that activation can be further optimized by changing the protein linker between dCas9 and the activation domain and/or by using different activation domains [2].

10.3 Advantages of CRISPRa System

The dCas9-guide RNA-mediated DNA target recognition requires both the PAM sequence in target DNA and Watson–Crick base pairing between the 20-nt guide RNA sequence and the complementary target DNA sequence. It has been shown that the sequences fully complementary to the guide RNA but lacking a nearby PAM are ignored by CRISPR/Cas9 system [25]. Compared with small activating RNA-mediated gene activation which only depend on Watson–Crick base pairing between mRNAs and saRNAs, the two-component requirement of CRISPR/Cas9 recognition renders more specificity with minimal off-target effects and we need only consider off-targets adjacent to a PAM, because potential targets lacking a PAM are unlikely to be interrogated.

In addition to small activating RNA, customized DNA-binding proteins such as zinc-finger proteins or transcription activator-like effectors (TALEs) have been used as tools for sequence-specific DNA targeting and gene regulation. These proteins robustly target DNA through programmable DNA-binding domains and can recruit effectors for transcription activation in a modular way. However, because each DNA-binding protein needs to be individually designed, the construction and delivery for the purpose of simultaneously regulating multiple loci is technically challenging. In contrast, one of the benefits of dCas9-based transcription effectors over the customized DNA-binding proteins is the ease with which multiple loci can be regulated, with only single-guide RNA (sgRNA) for each additional locus one desires to activate.

The conventional methods for gene overexpression include the use of cDNA overexpression vectors or cDNA libraries. However, cloning large cDNA sequences into viral vectors and manipulating several gene isoforms simultaneously are difficult. Also, the cDNA constructs often do not capture the full complexity of transcript isoforms, and they are independent of the endogenous regulatory context. Additionally, synthesizing large-scale libraries for genome-wide screening is not cost effective. Therefore, CRISPRa system has emerged as an ideal technology for genome regulation, providing specificity, convenience, robustness, and scalability for gene activation.

10.4 Applications and Limitations of CRISPRa System

CRISPR-based activation system could be applied to regulate gene expression in a variety of biological processes, including stem cell differentiation, silenced gene activation, genetic defect compensation, cell fate engineering, and genome-wide screening. To study whether CRISPRa could be used for direct cell reprogramming, Black et al. used a dCas9 with both N-terminal and C-terminal VP64 transactivation domains (VP64–dCas9–VP64) to achieve multiplex activation of the neurogenic factors Brn2, Ascl1, and Mytl1 (BAM factors) and demonstrated direct cellular reprogramming from fibroblast to induced neuronal cells through targeted activation of endogenous genes [3].

Another example is for HIV treatment. Although the combined antiretroviral therapies (cARTs) have had a marked impact on the treatment and progression of HIV/AIDS, the most significant limitation of currently available cARTs is the inability to extinguish the integrated latent HIV reservoirs, resulting in a persistent infection even under lifelong treatment. A promising strategy to eradicate latent HIV reservoirs is to reactivate the dormant virus in the presence of combined antiretroviral therapies (cARTs). Recently, several groups simultaneously reported that CRISPR-based activation is highly effective at inducing transcriptional activation of latent HIV-1 infection specifically in human T cells, providing an exciting new avenue towards latent HIV therapy (Saayman et al. 2016).

Whereas loss-of-function screens can be conducted using RNAi or Cas9-based tools, gain-of-function screens have been confined to cDNA overexpression libraries. Compared with all the limitations with the available cDNA libraries, CRISPRa-based targeted gene regulation on a genome-wide scale is a powerful strategy for interrogating, perturbing, and engineering biological systems. Taking advantage of the robust SAM system, Konermann et al. performed a genome-wide screening for genes that, upon activation, confer resistance to a BRAF inhibitor, using a library consisting of 70,290 guides targeting all human coding isoforms. The screens exhibited a high degree of consistency with 100% validation of the top ten hits. The top hits included genes previously shown to be able to confer resistance, and novel candidates were validated using individual sgRNA and complementary DNA overexpression. Furthermore, gene expression signature based on the top screening hits correlated with markers of BRAF inhibitor resistance in cell lines and patient-derived samples, proving the potential of Cas9-based activators as a powerful genetic perturbation technology [20].

The discovery of the RNA-mediated programmable CRISPR/Cas9 technology, has transformed the field of biology. While CRISPR/dCas9-mediated gene activation represents dramatic advantages over conventional approaches, there are several concerns with its broad application. In addition to the general problems with CRISPR system, such as off-target effects, delivery issue, and potential immunogenicity [28], the major concern is that the activation capability of endogenous genes by CRISPRa system is not as robust as that of cDNA overexpression approach and is heavily dependent on the selection of sgRNAs. This could be a potential concern for biological processes, such as direct reprogramming or transdifferentiation from one mature cell type to another, which might require a large amount of factors in order to overcome the force of gravity on the famous Waddington's epigenetic landscape [27]. Thus how to effectively design the most robust and specific guide RNAs for transcriptional activation needs further exploration.

References

1. Balboa D, Weltner J, Eurola S et al (2015) Conditionally stabilized dCas9 activator for controlling gene expression in human cell reprogramming and differentiation. Stem Cell Rep 5:448–459. doi:10.1016/j.stemcr.2015.08.001
2. Bikard D, Jiang W, Samai P et al (2013) Programmable repression and activation of bacterial gene expression using an engineered CRISPR-Cas system. Nucleic Acids Res 41:7429–7437. doi:10.1093/nar/gkt520
3. Black JB, Adler AF, Wang H-G et al (2016) Targeted epigenetic remodeling of endogenous loci by CRISPR/Cas9-based transcriptional activators directly converts fibroblasts to neuronal cells. Cell Stem Cell 19:406–414. doi:10.1016/j.stem.2016.07.001
4. Chakraborty S, Ji H, Kabadi AM et al (2014) A CRISPR/Cas9-based system for reprogramming cell lineage specification. Stem Cell Rep 3:940–947. doi:10.1016/j.stemcr.2014.09.013
5. Chavez A, Scheiman J, Vora S et al (2015) Highly efficient Cas9-mediated transcriptional programming. Nat Methods 12:326–328. doi:10.1038/nmeth.3312
6. Chavez A, Tuttle M, Pruitt BW et al (2016) Comparison of Cas9 activators in multiple species. Nat Methods 13:563–567. doi:10.1038/nmeth.3871
7. Chen B, Gilbert LA, Cimini BA et al (2013) Dynamic imaging of genomic loci in living human cells by an optimized CRISPR/Cas system. Cell 155:1479–1493. doi:10.1016/j.cell.2013.12.001
8. Cheng AW, Wang H, Yang H et al (2013) Multiplexed activation of endogenous genes by CRISPR-on, an RNA-guided transcriptional activator system. Cell Res 23:1163–1171. doi:10.1038/cr.2013.122
9. Cong L, Ran FA, Cox D et al (2013) Multiplex genome engineering using CRISPR/Cas systems. Science 339:819–823. doi:10.1126/science.1231143
10. Dominguez AA, Lim WA, Qi LS (2016) Beyond editing: repurposing CRISPR-Cas9 for precision genome regulation and interrogation. Nat Rev Mol Cell Biol 17:5–15. doi:10.1038/nrm.2015.2
11. Doudna JA, Charpentier E (2014) Genome editing. The new frontier of genome engineering with CRISPR-Cas9. Science 346:1258096. doi:10.1126/science.1258096
12. Gasiunas G, Barrangou R, Horvath P, Siksnys V (2012) Cas9-crRNA ribonucleoprotein complex mediates specific DNA cleavage for adaptive immunity in bacteria. Proc Natl Acad Sci U S A 109:E2579–E2586. doi:10.1073/pnas.1208507109
13. Gilbert LA, Larson MH, Morsut L et al (2013) CRISPR-mediated modular RNA-guided regulation of transcription in eukaryotes. Cell 154:442–451. doi:10.1016/j.cell.2013.06.044
14. Gilbert LA, Horlbeck MA, Adamson B et al (2014) Genome-scale CRISPR-mediated control of gene repression and activation. Cell 159:647–661. doi:10.1016/j.cell.2014.09.029
15. Hilton IB, D'Ippolito AM, Vockley CM et al (2015) Epigenome editing by a CRISPR-Cas9-based acetyltransferase activates genes from promoters and enhancers. Nat Biotechnol 33:510–517. doi:10.1038/nbt.3199

16. Jiang F, Doudna JA (2015) The structural biology of CRISPR-Cas systems. Curr Opin Struct Biol 30:100–111. doi:10.1016/j.sbi.2015.02.002
17. Jinek M, Chylinski K, Fonfara I et al (2012) A programmable dual-RNA-guided DNA endonuclease in adaptive bacterial immunity. Science 337:816–821. doi:10.1126/science.1225829
18. Kearns NA, Genga RMJ, Enuameh MS et al (2014) Cas9 effector-mediated regulation of transcription and differentiation in human pluripotent stem cells. Development 141:219–223. doi:10.1242/dev.103341
19. Kearns NA, Pham H, Tabak B et al (2015) Functional annotation of native enhancers with a Cas9-histone demethylase fusion. Nat Methods 12:401–403. doi:10.1038/nmeth.3325
20. Konermann S, Brigham MD, Trevino AE et al (2015) Genome-scale transcriptional activation by an engineered CRISPR-Cas9 complex. Nature 517:583–588. doi:10.1038/nature14136
21. Maeder ML, Linder SJ, Cascio VM et al (2013) CRISPR RNA-guided activation of endogenous human genes. Nat Methods 10:977–979. doi:10.1038/nmeth.2598
22. Mali P, Yang L, Esvelt KM et al (2013) RNA-guided human genome engineering via Cas9. Science 339:823–826. doi:10.1126/science.1232033
23. Perez-Pinera P, Kocak DD, Vockley CM et al (2013) RNA-guided gene activation by CRISPR–Cas9-based transcription factors. Nat Methods 10:973–976. doi:10.1038/nmeth.2600
24. Qi LS, Larson MH, Gilbert LA et al (2013) Repurposing CRISPR as an RNA-guided platform for sequence-specific control of gene expression. Cell 152:1173–1183. doi:10.1016/j.cell.2013.02.022
25. Sternberg SH, Redding S, Jinek M et al (2014) DNA interrogation by the CRISPR RNA-guided endonuclease Cas9. Nature 507:62–67. doi:10.1038/nature13011
26. Tanenbaum ME, Gilbert LA, Qi LS et al (2014) A protein-tagging system for signal amplification in gene expression and fluorescence imaging. Cell 159:635–646. doi:10.1016/j.cell.2014.09.039
27. Waddington CH (1957) The strategy ofthe genes. Allen, London
28. Xiong X, Chen M, Lim WA et al (2016) CRISPR/Cas9 for human genome engineering and disease research. Annu Rev Genomics Hum Genet 17:131–154. doi:10.1146/annurev-genom-083115-022258
29. Zalatan JG, Lee ME, Almeida R et al (2015) Engineering complex synthetic transcriptional programs with CRISPR RNA scaffolds. Cell 160:339–350. doi:10.1016/j.cell.2014.11.052

Part IV
Developing RNAa-Based Therapeutics

Chapter 11
RNA-Mediated Gene Activation: Identifying a Candidate RNA for Preclinical Development

David R. Corey

Abstract The ability to activate gene expression would provide new opportunities for drug development for diseases caused by inadequate or nonexistent expression of a therapeutic protein. Duplex RNAs that target gene promoters are one strategy for enhancing the expression of target genes. This chapter summarizes our understanding of mechanism behind gene activation by small RNAs that bind noncoding transcripts at gene promoters. We describe a path for choosing candidate genes for therapeutic development.

Keywords RNA • Gene activation • Noncoding RNA • Duplex RNA • Nucleic acid therapeutics

11.1 Introduction

The parent volume for this chapter is devoted to varied methods that share a common goal – activating gene expression with RNA. The objective of this contribution is to describe a method developed in my laboratory over the past decade involving duplex RNAs that act through nuclear RNAi to target noncoding transcripts that overlap transcription start sites. Our purpose is to provide the background necessary to make decisions regarding whether it is feasible to apply our approach to activate genes for therapy (Fig. 11.1).

11.2 Origin of Activating RNA Hypothesis

My laboratory never intended to investigate the ability of duplex RNAs to target gene promoters and control gene expression. Like most investigators performing research in mammalian cells, we had assumed that the activity of duplex RNAs and

D.R. Corey (✉)
Departments of Pharmacology & Biochemistry, University of Texas Southwestern Medical Center, 6001 Forest Park Road, Dallas, Texas 75390-9041, USA
e-mail: david.corey@utsouthwestern.edu

L.-C. Li (ed.), *RNA Activation*, Advances in Experimental Medicine and Biology 983, DOI 10.1007/978-981-10-4310-9_11

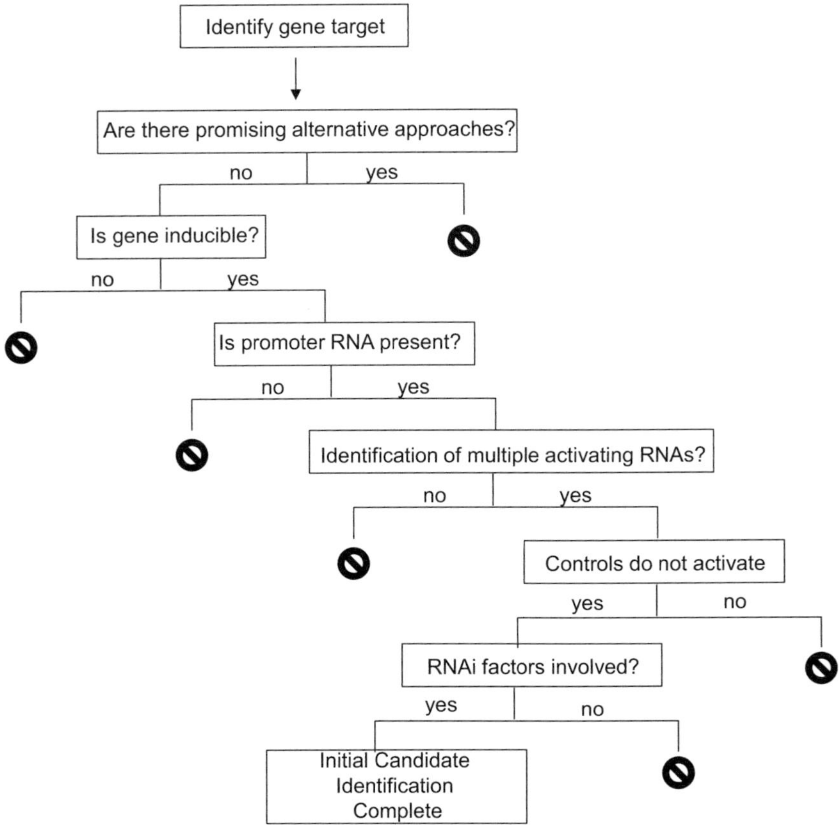

Fig. 11.1 Decision guide for identifying activating RNAs as candidates for therapeutic discovery

RNA interference (RNAi) [6] were restricted to inhibition of gene expression through targeting mRNA in the cytoplasm [35]. This attitude began to change in 2004 when two papers appeared reporting that RNAs complementary to human gene promoters could block expression of their target genes [16, 26]. While one of these papers was subsequently retracted [33], the door had been opened to reconsider the dogma limiting what duplex RNAs might achieve within human cells.

We had been examining the ability of peptide nucleic acids (PNAs) to inhibit gene expression by targeting gene promoters in human cells [9]. PNAs are an amide-linked nucleic acid mimic that contain an uncharged backbone capable of efficient invasion of duplex DNA. Significantly, our choice of sequence target was not informed by methylation status of the promoter or any other epigenetic criteria. Instead, we were motivated by work from David Sigman's laboratory showing that oligonucleotides could be used to target the open complex at transcription start sites [24]. This is significant – rather than target methylation sites that might be distant from the coding region of a gene, we focused on sites adjacent to the coding region. Our subsequent success may be due to adopting this "start site first" design rule.

It was a simple matter to model duplex RNAs after PNAs that had already been proven active, introduce them into cells, and examine gene expression. Our target gene was progesterone receptor (PR). PR had been chosen because of its biological significance and because it was known that its expression could be regulated by hormones. To our surprise (because we had not expected the experiment to yield positive results), we observed complete inhibition of PR expression [10]. This inhibition was robust and was reproduced routinely by multiple experiments over the next 6 years.

Because gene silencing by RNAs that were complementary to gene promoters ran counter to the dogma that RNAi requires complementarity to mRNA, we spent the next months challenging the finding. Control experiments were run with various mismatched RNAs. We tested numerous related promoter-targeted RNAs to identify additional active compounds. Nuclear run-on assays and RNA polymerase chromatin immunoprecipitation were performed to support the hypothesis that gene transcription was being regulated. 5′-RACE was used to fully define the transcription start sites and ensure that inhibition was not due to incorrect identification of the transcription start sites for PR mRNA.

As we tested additional duplex RNAs that were complementary to sequences throughout the promoter, we noticed a curious phenomenon. Some promoter-targeted RNAs not only did not inhibit PR expression but also reproducibly activated expression by 25–75%. Such relatively small effects could easily be explained by chance off-target interactions, but they led to the hypothesis that duplex RNAs might also have the potential to induce gene expression. Many transcription factors play a repressive role in one context and inducing one in another [14], encouraging further investigation.

To investigate this hypothesis, we switched from T47D breast cancer cells, where expression of PR is high, to MCF7 breast cancer cells where expression is much lower. When our panel of promoter-targeted RNAs was introduced into MCF7 cells, the RNA activated PR expression to levels equivalent to those observed in T47D cells [12].

11.3 Mechanism of Gene Activation

We have begun to understand the mechanism of promoter-directed RNAs. Argonaute (AGO) is a family of proteins that play a central role in RNAi [18, 23, 29]. There are four AGO proteins in human cells, AGO1–4. AGO2 is the "catalytic engine" responsible for driving recognition of duplex RNAs and, in some cases, subsequent cleavage of the target. While we initially reported that both AGO1 and AGO2 were involved in gene activation by promoter RNAs [11], more recent studies implicated AGO2 as the dominant factor [3]. GW182 paralogs TNRC6A/B/C, proteins also known to be involved in RNAi, were also required for RNA-mediated gene activation [22].

Involvement of RNAi factors suggested that RNA, rather than chromosomal DNA, might be the target for promoter RNAs. As we were performing our initial experiments in 2004–2006, an increasing body of work revealed that noncoding transcripts that overlap gene promoters were a common phenomenon [2, 15, 25].

These data suggested that we might be targeting a noncoding transcript rather than DNA. Supporting this suggestion, we identified a noncoding transcript at the PR promoter in both T47D and MCF7 cells [32]. Activation was blocked by removal of the transcript by additional of antisense oligonucleotides designed to promote cleavage by RNase H. RNA immunoprecipitation revealed that AGO2 was recruited to the noncoding transcript upon addition of activating or inhibitory duplex RNAs.

11.4 Nuclear RNAi

Despite multiple publications reporting the activity of RNAi factors inside mammalian cell nuclei [27, 31], we encountered significant resistance to the concept that cell nuclei contained RNAi factors. To clarify this issue, we took a multipronged approach to evaluating the localization of RNAi factors like AGO2 and TNRC6A [7]. Both microscopy and cell fractionation assays confirmed that RNAi factors were in mammalian cell nuclei. Activity assays suggested that loading occurs in the cytoplasm but that complexes loaded with AGO2 have the capacity to induce cleavage of target transcripts in nuclei.

Many of the major aspects of mechanism now appear to be in place [7, 14, 22]. Upon introduction into cells, small RNAs are loaded into AGO2 and pass into the nucleus. The GW182 paralogs TNRC6A/B/C associate with AGO2. Once inside the nucleus, the RNA–AGO2 complex can recognize noncoding transcripts at gene promoters to affect gene transcription. It is reasonable to hypothesize that duplex RNA can partner with RNAi factors to form complexes that have the potential to recognize most RNAs in cell nuclei.

11.5 Nuclear RNAi and Drug Discovery

11.5.1 Selection of Target Gene

How can gene activation by promoter RNAs be used as a basis for a drug discovery program? The first step is choosing an appropriate target. Target choice is the most critical and most difficult decision for any development program regardless of whether the program is based on small molecules, proteins, or nucleic acids. The development of duplex RNAs is no exception. Lessons from more established pharmacologic approaches still apply.

Table 11.1 Criteria for target selection

There should be a compelling need for gene activation to treat a disease
Direct administration of protein should not be a practical alternative
Small-molecule gene activators should be a practical alternative
There is reason to believe that activating RNAs will be advantageous relative to activating expression through gene therapy
Expression of target should be inducible
Duplex RNA must be deliverable to target tissue

Several criteria can be applied when selecting a suitable gene (Table 11.1). A primary requirement is that manipulating expression of a target gene should lead to expression of a beneficial protein product. Once a disease and gene target have been identified, the feasibility of competing approaches must be evaluated. Can administration of the protein directly achieve the same end? Is it possible to rationalize the design of small-molecule activators of protein expression? Because protein and small-molecule therapeutics are well-tested disease interventions, they may offer formidable competition.

The Ionis Pharmaceutical drug Kynamro offers a cautionary example [28]. Kynamro is an antisense oligonucleotide that has been approved by the Food and Drug Administration for treating familial hypercholesterolemia. Its development was a landmark for nucleic-acid therapeutics because clinical trials presented clear evidence that the drug successfully engaged its intended mRNA target upon systemic injection. As Kynamro was approved, a small molecule, Juxtapid, was also approved. Both drugs carry toxicity warnings and Kynamro has a lower price. Juxtapid, which can be administered as a pill, has come to dominate the market.

Gene therapy delivery of a vector or direct administration of a mRNA that can lead to expression are also competing approaches that must be considered. However, like activating RNAs, gene therapy and mRNA are both novel and will require more definitive evidence of efficacy in patients.

Our experience is that promising gene targets are inducible – they can be expressed at variable levels depending on the cellular environment [9, 10, 21, 22]. Variable expression is evidence that transcription can be changed. Genes with constant levels of expression may be less likely to be good candidates.

Upon systemic administration, duplex RNA can be most reliably delivered to the liver [13]. Delivery can be achieved by complexation with nanoparticles or by conjugation to molecules that facilitate receptor-mediated uptake. Our method uses duplex RNA, so the same preferences for tissue uptake are likely to apply. When choosing a target, there should be evidence that duplex RNA can be delivered to the tissue and affect gene expression.

11.5.2 Target Validation

Our proposed mechanism for RNA activation requires that RNA be expressed to overlap a gene promoter [14]. Transcriptome databases can be used to predict whether a promoter transcript is present. However, quantitative PCR using multiple primer sets complementary to the promoter region is straightforward. Therefore, while information from databases can be a useful guide, there is no substitute for experimental evidence supporting the transcript's existence in cell types relevant to the disease of interest. Transcripts can be in either the sense or antisense direction relative to mRNA transcription, and PCR primers should be able to detect both orientations.

If a promoter transcript is detected, it should be further characterized. 5′-RACE can be used do define transcript length and gain insight into transcript heterogeneity among transcription start sites. qRT-PCR or droplet digital PCR (ddPCR) can be used, in conjunction with careful standardization, to estimate the number of promoter RNA molecules per cell [4]. To act in *cis* relative to a promoter, the promoter RNA must be expressed at a few copies per cell. If the transcript is expressed at only a few copies per hundred or thousand cells, that scarcity would argue against being a biologically relevant target.

Confirming the existence of a target transcript and the location of transcription start sites should be experimentally straightforward. With this information in hand, the next step is to test the central hypothesis that duplex RNA can activate gene expression. Our laboratory has had success targeting regions near the most upstream transcription start site. We recommend designing at least ten duplex RNAs overlapping sequences from +1 to –100. These RNAs should be transfected into cells and expression monitored.

If one or more duplex RNAs appear to yield gene activation, the next step is to test the hypothesis that activation is either artefactual or due to an off-target effect. When RNA is introduced into cells, gene expression inside cells can be influenced in ways that are impossible to predict. These effects will vary depending on the sequence of the duplex RNA, complicating analysis and obscuring simple comparisons.

For many genes, the most convenient method for measuring expression will be quantitative RT-PCR (qRT-PCR). qPCR, however, is subject to artefactual results if not done with care [1]. For example, qRT-PCR requires measurement of the expression for a cellular gene as a standardization tool. If a duplex RNA alters the expression of this gene, and if this altered expressed is not noted, the result will be the appearance of a change in the expression of a target gene. If expression of the gene standard goes down, the duplex RNA would appear to be an activator of the target gene.

While sometimes less convenient than qRT-PCR, Western analysis also offers advantages. Because altered protein production is the desired goal, visualizing protein expression is the most direct method for evaluating initial success. Western analysis also avoids potential artefacts from amplification steps and offers a less error-prone approach to monitoring gene expression

11.5.3 Dealing with Off-Target Effects

Off-target effects are a longstanding and ubiquitous problem confronting designed nucleic acids inside cells [5, 20]. Every sequence has the potential to produce unexpected effects on gene expression. Unfortunately, in some cases these effects will lead to increased expression of the target gene through a mechanism that is unrelated to recognition at the target gene. Not all changes in gene expression were created equal. An off-target effect responsible for gene activation is unlikely to be specific or potent enough to form a reliable basis for drug development.

Discriminating between on- and off-target effects is relatively straightforward. As soon as a sequence is identified as a potential activating RNA, multiple scrambled and mismatched RNA duplexes should be designed. Scramble duplexes are designed by moving blocks of bases relative to the parent sequence. Mismatched duplexes are mismatched at multiple positions relative to the RNA target sequence, including within the putative seed sequence.

Our proposed mechanism does not demand stringent target sequence usage and we expect that, if activation is possible, more than one sequence should be active. Therefore, once one sequence is identified, other sequences should be tested to discover additional activating RNAs. These activating RNAs may be more potent and each has its own potential for development. Discovery of multiple activating RNAs also bolsters the case for an on-target mechanism. Data revealing that several complementary RNAs activate gene expression while multiple non-complementary RNAs do not would support a mechanism of action that requires recognition of the target gene.

11.5.4 Testing a Nuclear RNAi Mechanism

Our proposed mechanism for RNA-mediated gene activation involves RNAi factors and a central role for AGO2. After identifying duplex RNAs that demonstrate on-target activation, it is useful to confirm the mechanism of action. One strategy for achieving this is to silence expression of AGO2 in an initial transfection, and then add the activating RNA [22]. Reduced expression of AGO2 should reverse gene activation if AGO2 is involved in the mechanism.

Another mechanistic tool is RNA immunoprecipitation (RIP) [22]. Our mechanism predicts that the complex of small RNA and AGO2 should bind to a promoter transcript. RIP allows experimenters to monitor this recruitment. The experiment has the potential to further implicate the promoter transcript in the mechanism for gene activation and support that the mechanism involves RNAi factors.

Identifying an activating RNA is challenging, but it does not require great effort to come to a go/no go decision. If the target gene is important, the relatively small effort may be worth the risk. By systematically approaching the discovery of activating RNAs, it should be straightforward to identify unpromising targets in a timely fashion. For example, failure to identify a promoter transcript makes the target unsuited for RNA activation, at least by the mechanism that we have characterized. Likewise, failure to identify an activating RNA will also indicate that a project is unlikely to succeed.

11.6 First Steps Beyond Lead Identification

Once an activating RNA is identified, drug discovery can move forward. The duplex RNAs used to activate gene expression are identical in overall design to duplex RNAs used commonly to silence gene expression in the cytoplasm. As a result, challenges confronting biodistribution, chemical modification, and cellular uptake will be similar and lessons learned from duplex RNAs developed previously can be applied [13]. We have previously tested promoter-targeted RNAs that were chemically modified to improve properties in vivo and found that gene activation was compatible with standard chemical alterations [34].

Another challenge has been the transition from activating RNAs in human cell lines to animal models. Promoter sequences are often less conserved than sequences that code for proteins. Rodent models may also be likely to have different regulatory demands and may not have the same (or any) role for noncoding RNA.

11.7 Other Options for Nuclear RNAi

This chapter has focused on duplex RNAs that target gene promoters and activate expression. These are not the only opportunities presented by nuclear RNAi. Duplex RNAs can also influence alternative splicing [19]. For gene silencing, there may be some cases where a promoter-targeted duplex RNA has advantages relative to single-stranded antisense oligonucleotides or duplex RNAs that act in the cytoplasm.

Because RNAi factors and miRNAs are present in cell nuclei [7], it is reasonable to predict that miRNAs might be involved in regulating transcription or splicing. Single stranded antagonists of miRNAs (antimiRs) may be able to modulate the activity of these miRNAs and indirectly activate gene expression or alter splicing. miRNA mimics may be able to phenocopy natural miRNAs and push expression in a therapeutic direction.

11.8 Case Study: Activating Expression of Frataxin

Our laboratory's work developing RNA activators of frataxin expression is a case study for applying nuclear RNAi to early stage drug discovery. Friedreich's Ataxia (FA) is a severe neurodegenerative disorder with no curative treatments [30]. FA is caused by an expansion of a CUG trinucleotide repeat that leads to inadequate expression of frataxin. This mutation is within an intron rather than a promoter or coding region. The proposed mechanism involves R-loop formation by the expanded intron, which blocks the chromosomal locus, induces histone modifications, and reduces frataxin expression to a fraction of wild-type levels [8].

We reasoned that a duplex RNA might be able to bind to the expanded repeat [17]. By binding RNA: AGO2 complexes to the repeat, R-loop formation would be prevented and frataxin expression restored. One experimental challenge is that, unlike our prior experience with cyclooxygenase 2 (Masui et al. 2013) or progesterone receptor [12], but like our experience with LDL receptor [21], the difference between full and reduced expression of frataxin is only threefold (versus 20–30 fold for COX2 or PR). Smaller fold changes require more care to detect and verify.

Our initial experiments using qRT-PCR suggested that activation of frataxin expression was occurring. This activation could be achieved by several different duplex RNAs. Duplex RNAs that were mismatched or scrambled did not activate. Activation was consistently reproduced. RNA immunoprecipitation revealed recruitment of AGO2 to the target intronic transcript. Taken together, these results suggested that RNA-mediated gene activation could occur at the frataxin locus and encouraged subsequent experiments to identify compounds that would be potent enough to justify testing in animals.

11.9 Summary

RNA-mediated gene activation is a new frontier for drug discovery. While controversial in the past, in is now becoming clear that nuclear RNAi and gene activation are real phenomenon and present opportunities for the well-prepared investigator. Selecting appropriate gene targets is a critical first step. Well-controlled experiments and a skeptical, questioning attitude are essential. A methodical approach will allow a go/no go decision to be made quickly. Once a promising duplex RNA is identified, its development will be aided by the many lessons learned from the development of duplex RNAs that act in cell cytoplasm.

Acknowledgements This work was supported by the National Institutes of Health (GM118103 to DRC) and the Robert Welch Foundation (I-1244 to DRC). DRC holds the Rusty Kelley Professorship in Medical Science.

References

1. Bustin SA (2010) Whey the need for qPCR publication guidelines? The case for MIQE. Methods 40:217–226
2. Carninci P, Kasukawa T, Katayama S, Gough J, Frith MC, Maeda N, Oyama R, Ravasi T, Lenhard B, Wells C et al (2005) The transcriptional landscape of the mammalian genome. Science 309:1559–1563
3. Chu Y, Yue X, Younger ST, Janowski BA, Corey DR (2010) Involvement of argonaute proteins in gene silencing and activation by RNAs complementary to a noncoding transcript at the progesterone receptor promoter. Nucleic Acids Res 38:7736–7748
4. Dodd DW, Gagnon KT, Corey DR (2013) Digital quantitation of potential therapeutic target RNAs. Nucleic Acid Ther 23:188–194
5. Editorial (2003) Whither RNAi? Nat Cell Biol 5:489–490
6. Elbashir SM, Harborth J, Lendeckel W, Yalcin A, Weber K, Tuschl T (2001) Duplexes of 21-nucleotide RNAs mediate RNA interference in cultured mammalian cells. Nature 411:494–498
7. Gagnon KT, Li L, Chu Y, Janowski BA, Corey DR (2014) RNAi factors are present and active in human cell nuclei. Cell Rep 6:211–221
8. Groh M, Lufino MMP, Wade-Martins R, Gromak N (2014) R-loops associated with triplet repeat expansions promote gene silencing in Friedreich Ataxia and Fragile X syndrome. PLoS Genet 10:e1004318
9. Janowski BA, Kaihatsu K, Huffman KE, Schwartz JC, Ram R, Hardy D, Mendelson CR, Corey DR (2005a) Inhibiting transcription of chromosomal DNA with antigene peptide nucleic acids. Nat Chem Biol 1:210–215
10. Janowski BA, Huffman KE, Schwartz JC, Ram R, Hardy D, Shames DS, Minna JD, Corey DR (2005b) Inhibiting gene expression at transcription start sites in chromosomal DNA with antigene RNAs. Nat Chem Biol **1**:216–222
11. Janowski BA, Huffman KE, Schwartz JC, Ram R, Nordsell R, Shames DS, Minna JD, Corey DR (2006) Involvement of AGO1 and AGO2 in mammalian transcriptional silencing. Nat Struct Mol Biol 13:787–792
12. Janowski BA, Younger ST, Hardy DB, Ram R, Huffman KE, Corey DR (2007) Activating gene expression in mammalian cells with promoter-targeted duplex RNAs. Nat Chem Biol 3:166–173
13. Juliano RL (2016) The delivery of therapeutic oligonucleotides. Nucleic Acids Res 44:6518–6548
14. Kalantari R, Chiang C-M, Corey DR (2016) Regulation of mammalian transcription and splicing by nuclear RNAi. Nucleic Acids Res 44:524–537
15. Kapranov P, Cheng J, Dike S, Nix DA, Duttagupta R, Willingham AT, Stadler PF, Hertel J, Hackermuller J, Hofacker IL et al (2007) RNA maps reveal new RNA classes and a possible function for pervasive transcription. Science 316:1484–1488
16. Kawasaki H, Taira K (2004) Induction of DNA methylation and gene silencing by short interfering RNAs in human cells. Nature 431:211–217
17. Li L, Matsui M, Corey DR (2016) Activation of frataxin expression by duplex RNAs. Nat Commun 7:10606
18. Liu J, Carmell MA, Rivas FV, Marsden CG, Thomson JM, Song JJ, Hammond SM, Joshua-Tor L, Hannon GJ (2004) Argonaute2 is the catalytic engine of mammalian RNAi. Science 305:1437–1441
19. Liu J, Hu J, Corey DR (2012) Expanding the action of duplex RNAs into the nucleus: redirecting alternative splicing. Nucleic Acids Res 40:1240–1250
20. Matsui M, Corey DR (2017) Noncoding RNAs as drug targets. Nat Rev Drug Discov 16 (3):167–179

21. Matsui M, Sakurai F, Elbashir S, Foster DJ, Manoharan M, Corey DR (2010) Activation of LDL receptor expression by small RNAs complementary to a noncoding transcript that overlaps the LDLR promoter. Chem Biol 17:1344–1355
22. Matsui M, Chu Y, Zhang H, Gagnon KT, Shaikh S, Kuchimanchi S, Manoharan M, Corey DR, Janowski BA (2013) Promoter RNA links transcriptional regulation of inflammatory pathway genes. Nucleic Acids Res 41:10086–10109
23. Meister G, Landthaler M, Patkaniowska A, Dorsett Y, Teng G, Tuschl T (2004) Human Argonaute2 mediates RNA cleavage targeted by miRNAs and siRNAs. Mol Cell 15:185–197
24. Milne L, Xu Y, Perrin DM, Sigman DS (2000) An approach to gene-specific transcirption inhibition using oligonucleotides complementary to the template strand of the open complex. Proc Natl Acad Sci U S A 97:3136–3141
25. Morris KV, Mattick JS (2014) The rise of regulatory RNA. Nat Rev Genet 15:423–437
26. Morris KV, Chan SW, Jacobsen SE, Looney DJ (2004) Small interfering RNA-induced transcriptional gene silencing in human cells. Science 305:1289–1292
27. Ohrt T, Mutze J, Staroske W, Weinmann L, Hock J, Crell K, Meister G, Schwille P (2008) Fluorescence correlation spectroscopy and fluorescence cross-correlation spectroscopy reveal the cytoplasmic origination of loaded nuclear RISC in vivo in human cells. Nucleic Acids Res 36:6439–6449
28. Patel RS, Scopelliti EM, Savelloni J (2015) Therapeutic management of familial hypercholesterolemia: current and emerging drug therapies. Pharmacotherapy 35:1189–1203
29. Rand TA, Ginalski K, Grishin NV, Wang X (2004) Biochemical identification of Argonaute 2 as the sole protein required for RNA-induced silencing complex activity. Proc Natl Acad Sci U S A 101:14385–14389
30. Richardson TE, Kelly HN, Yu AE, Simpkins JW (2013) Therapeutic strategies in Friedreich's Ataxia. Brain Res 1514:91–97
31. Robb GB, Brown KM, Khurana J, Rana TM (2005) Specific and potent RNAi in the nucleus of human cells. Nat Struct Mol Biol 12:133–137
32. Schwartz JC, Younger ST, Nguyen NB, Hardy DB, Monia BP, Corey DR, Janowski BA (2008) Antisense transcripts are targets for activating small RNAs. Nat Struct Mol Biol 15:842–848
33. Taira K (2004) Induction of DNA methylation and gene silencing by short interfering RNAs in human cells. Nature 441:1176
34. Watts JK, Yu D, Charisse K, Montaillier C, Potier P, Manoharan M, Corey DR (2010) Effect of chemical modifications on modulation of gene expression by duplex antigene RNAs that are complementary to non-coding transcripts at gene promoters. Nucleic Acids Res 38:5242–5259
35. Zeng Y, Cullen BR (2002) RNA interference in human cells is restricted to the cytoplasm. RNA 8:855–860

Chapter 12
Treatment of Pancreatic Cancer by Aptamer Conjugated C/EBPα-saRNA

Sorah Yoon and John J. Rossi

Abstract Pancreatic cancer is estimated to become the second-leading cause of cancer-related mortality by 2020. While the death rates of most other cancers continue to decline recently, the death rates of pancreatic cancer are still increasing, with less than 5% of patients achieving 5-year survival. Despite great efforts to improve treatment with combinational therapies in pancreatic cancer patients, limited progress has been made. V-Ki-ras2 Kirsten rat sarcoma viral oncogene homolog (KRAS) has been depicted as a therapeutic target in pancreatic cancer for many years. However, the clinical outcome of KRAS-directed therapies has not been successful, suggesting that KRAS is an undruggable target. For the new druggable target, epigenetically silenced transcriptional factor C/EBPα (CCAAT/enhancer-binding protein α), upregulator of a strong inhibitor of cell proliferation (p21), is upregulated by small activating RNA (saRNA) in pancreatic cancer. For the cell type-specific delivery, pancreatic cancer-specific 2′-Fluoropyrimidine RNA-aptamers (2′F-RNAs) are conjugated with C/EBPα-saRNA *via* sticky bridge sequences. The conjugates of aptamer–C/EBPα-saRNA upregulate the expression of C/EBPα *in vitro* and inhibit the tumor growth *in vivo*. It suggests that aptamer-mediated targeted delivery of therapeutic C/EBPα-saRNA might be the effective therapeutics under the current therapeutic modality failure in pancreatic cancer.

Keywords Pancreatic cancer • Aptamer • C/EBPα-saRNA • Anticancer agents

S. Yoon
Department of Molecular and Cellular Biology, Beckman Research Institute of City of Hope, 1500 E Duarte Rd, Duarte, CA, USA
e-mail: syoon@coh.org

J.J. Rossi (✉)
Department of Molecular and Cellular Biology, Beckman Research Institute of City of Hope, 1500 E Duarte Rd, Duarte, CA, USA

Irell and Manella Graduate School of Biological Sciences, Beckman Research Institute of City of Hope, 1500 E Duarte Rd, Duarte, CA, USA
e-mail: jrossi@coh.org

L.-C. Li (ed.), *RNA Activation*, Advances in Experimental Medicine and Biology 983, DOI 10.1007/978-981-10-4310-9_12

12.1 Introduction

Pancreatic cancer is estimated to become the second-leading cause of cancer-related mortality by 2020 [22]. The death rates of pancreatic cancer are increasing, with less than 5% of patients achieving 5-year survival, while the death rates of most other cancers continue to decline recently [22]. The majority of patients present with either local or systemic recurrence within 2 years following resection and postoperative adjuvant chemotherapy [1, 7, 27]. Despite great efforts to improve treatment of patients with pancreatic cancer, limited progress has been made [18, 23]. Gemcitabine, as a single agent administered postoperatively, remains the current standard of care, although numerous new drugs are entering phase III trials. However, gemcitabine only improves the 1-year survival rate from 16 to 19%. Combinations of gemcitabine with other chemotherapeutic drugs or biological agents for unresectable pancreatic cancer, or adjuvant treatment following resection, have resulted in limited improvement: the 5-year survival of patients with pancreatic cancer remains less than 5% [1, 7, 8, 13, 16, 17, 27]. This limitation of conventional treatment is most likely due to the chemoresistance in pancreatic cancer [6, 30].

More than 90% of cases of pancreatic intraepithelial neoplasia of all grades have mutations to the V-Ki-ras2 Kirsten rat sarcoma viral oncogene homolog (KRAS) [12]. It is generally accepted that this has the critical role in pancreatic cancer progression. Therefore, for many years, KRAS has been recognized as an important therapeutic target in pancreatic cancer. However, despite the great efforts to develop KRAS-targeted therapeutics, the clinical outcomes have not been successful, and KRAS is now assumed to be an undruggable target [5]. To overcome current limitation of therapeutics and explore a new druggable target, epigenetically silenced transcriptional factor C/EBPα (CCAAT/enhancer-binding protein α), upregulator of a strong inhibitor of cell proliferation (p21), is upregulated by small activating RNA (saRNA) for pancreatic cancer treatment [32].

12.2 Small Activating RNAs (saRNAs)

Ever since two research papers (in 2006 and 2007) reported that short RNAs to specific promoter regions of certain genes upregulate expression at the transcriptional level without altering the genome in human cells, RNA activation (RNAa or saRNA) has been recognized as a novel strategy for nonintegrative gene activation in mammalian cells by short 21-mer nucleotide duplexes [10, 14]. Targeted increase in transcription is known to occur through an Argonaute-2-mediated mechanism and recognition of key promoter regions by the saRNAs to the target genes [10, 14]. Gene activation by saRNA shows a seed-region-dependent manner of the 5′ region of the antisense strand [15].

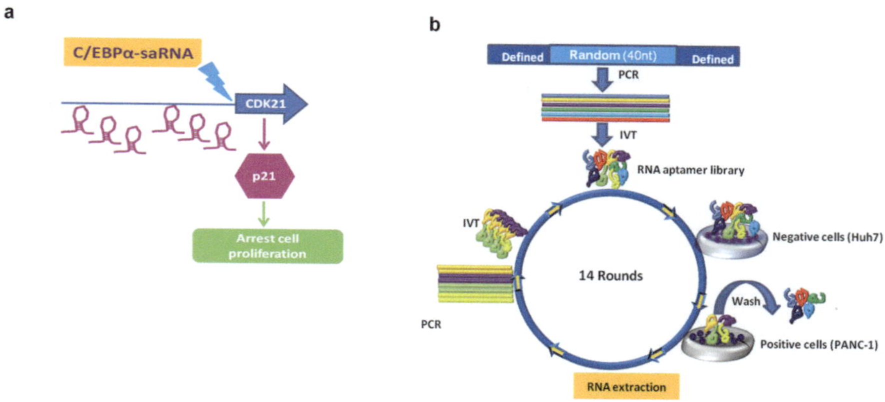

Fig. 12.1 C/EBPα–saRNA and Naïve whole cell based SELEX (**a**) Schematics of small activating RNA (saRNA) targeting C/EBPα to increase the expression of p21 for antitumor effects. (**b**) Schematic live-cell SELEX procedures; 2′F-modified RNA aptamer library is synthesized throughout PCR and *in vitro* transcription. After removal of nonspecific binding to the negative cells, the supernatant is incubated on the positive cells for positive selection. The RNA aptamer selection is repeated for 14 rounds of SELEX

SaRNAs targeting C/EBPα *in vivo* have previously shown to induce a potent antitumor effect in hepatocellular carcinoma through positive regulation of C/EBPα and its downstream targets including cyclin-dependent kinase inhibitor 1 (p21) [20]. More evidence of antitumor effects through activation of targets has been observed; E-cadherin in breast and lung carcinoma [11]. It suggests that strategic choice of gene activation by saRNAs can be exploited as novel therapeutics in cancer.

In pancreatic cancer, loss of heterozygosity of the KDM6B gene, which encodes a histone demethylase for trimethylated histone H3 lysine 27, has been reported. The loss of KDM6B enhances the aggressiveness and metastasis of pancreatic cancer through the downregulation of C/EBPα [31]. In KDM6B knock-out pancreatic cancer cells, forced expression of C/EBPα rescues the tumor cell sphere formation, suggesting that C/EBPα plays an important role in pancreatic cancer progression [31]. Since it is now clear that reduction of C/EBPα is recognized in pancreatic intraepithelial neoplasms, C/EBPα-saRNA is developed to increase expression level of C/EBPα and its downstream targets including cyclin-dependent kinase inhibitor 1 (p21) (Fig. 12.1a and Table 12.1).

12.3 Aptamer Selection for Targeting Modalities

With the aim of using aptamers as delivery vehicles for saRNA therapeutics, RNA aptamers that specifically recognize pancreatic cancer cells are selected. The small, structured single-stranded RNA aptamers are powerful tools for targeting cell-

Table 12.1 Sequence information of P19, P1, P19-stick, P1-stick, sense, and antisense strand of C/EBPα-saRNA

Name	Sequences
P19	GGGAGACAAGAAUAAACGCUCAAUGGC**GAAUGCCC**GCCUAAUAGGGCGUUAUGACUUGUUGAGUUCGACAGGA GGCUCACAACAGGC
P1	GGGAGACAAGAAUAAACGCUCAAUGCGCU**GAAUGCCC**AGCCGUGAAAGCGUCGAUUUCCAUCCUUCGACAGGAG GCUCACAACAGGC
P19-stick	5′-GGGAGACAAGAAUAAACGCUCAAUGGCGAAUGCCCGCCUAAUAGGGC GUUATGACUUGUUGAGUUCGACAGGAGGCUCACAACAGGCooooooo**GUACAUUCUAG AUACGC**-3′
P1-stick	5′-GGGAGACAAGAATAAACGCUCAAUGCGCUGAAUGCCCAGCCGUGAAAG CGUCGAUUUCCAUCCUUCGACAGGA GGCUCACAACAGGCooooooo**GUACAUUCUAGAUACGC**-3′
AW1 Sense	CGGUCAUUGUCACUGGUCAUU
AW1 Antisense	UGACCAGUGACAAUGACCGUU

Bold letters are stick sequences to anneal C/EBPα-saRNA or scramble RNA
o: C3 carbon linker

surface motifs and show great promise for clinical therapy. The tertiary structure of these nucleotide-based molecules allows for specific epitope recognition and cell internalization, making them the chemical equivalent of antibodies to deliver a therapeutic "payload" to the target cells. Aptamers can be selected to recognize a wide variety of targets, from small molecules to proteins and nucleic acids, and have been used in cultured cells and ex *vivo* organ cultures [2, 3, 9, 26, 28, 29]. RNA aptamers hold their three-dimensional structures by means of a well-defined set of complementary nucleic acid sequences which allow the chain to fold back into their natural conformation following denaturation. Therefore, aptamers are able to maintain their structural conformation even when exposed to physiologically harsh reducing conditions [19]. As a potential molecular vehicle for therapeutic delivery, RNA aptamers offer significant advantages, such as better structural stability, lower toxicity, and lower immunogenicity, over antibodies [19].

For targeting modalities, pancreatic cancer cell-specific aptamers are basically performed as described by using the ***S***ystematic ***E***volution of ***L***igands by ***EX***ponential enrichment (**SELEX**) methodology [4, 25], with few modifications for this study [33]. The DNA library containing 40 nt of random sequences is synthesized by Integrated DNA Technologies (IDT). The random region is flanked by constant regions for the amplification. The DNA random library is amplified by PCR and converted to RNA library using 2′-Fluoro modified pyrimidines throughout *in vitro* transcription. The human pancreatic adenocarcinoma cells, PANC-1, are used as target cells for the aptamer selection. To remove irrelevant binding, the hepatocellular carcinoma cells, Huh7, are used for the counterselection step. RNAs that bound to target cells is recovered, amplified by RT-PCR and *in vitro* transcription, and used in the following selection rounds. The schematic naïve cell SELEX is depicted in Fig. 12.1b. After 14 rounds of SELEX, two different groups of aptamers, named P19 and P1 are selected (Table 12.1).

12.4 Pancreatic Cancer Specificity of Aptamers

P19 and P1 hold multistem loops and structural similarity, including a common motif: **GAAUGCCC** (Table 12.1). A minimum-energy structural analysis of the selected aptamers is carried out using NUPACK software. As depicted, the calculated secondary structures of the RNA aptamers, P19 and P1, contain several stem-loop regions (Fig. 12.2a).

For the binding assays of aptamers, flow cytometry is assessed. The PANC-1 cells detached using a nonenzymatic cell dissociation solution are incubated with Cy3-labeled aptamers for 30 min at 37 °C. Cells are washed with binding buffer and immediately analyzed by flow cytometry. 4′6′-diamidino-2-phenylindole (DAPI) (1 μg/ml) is used to exclude dead cells. Flow cytometry analyses of the individual enriched aptamer clones, compared to the initial RNA library, show the enriched cell surface binding to the PANC-1 cells (Fig. 12.2b).

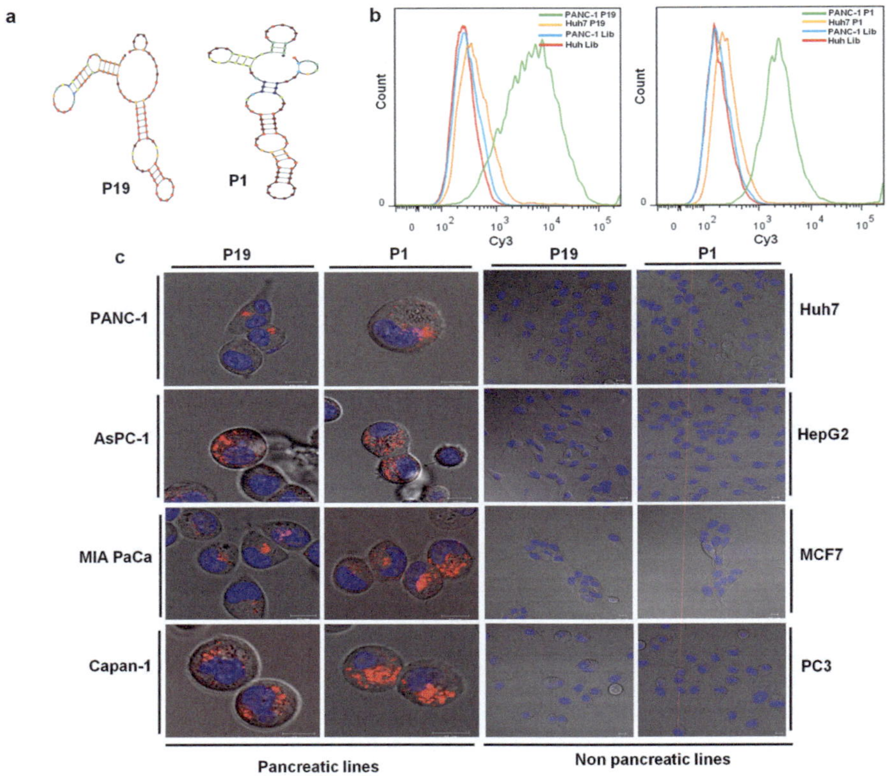

Fig. 12.2 **Aptamer secondary structures and cancer cell-specific internalization** (**a**) Secondary structures of P19 and P1 predicted using NUPACK software. (**b**) Binding efficiency of P19 and P1 aptamers by flow cytometry in PANC-1 and control Huh7 cells; Cy3-labeled P19 and P1 aptamer are used. (**c**) Internalization of Cy3-labeled P19 and P1 aptamer (100 nM) in the pancreatic cell lines (PANC-1, AsPC-1, MIA PaCa, and Capan-1) and nonpancreatic lines (Huh7, HepG2, MCF7 and PC3) by confocal microscopy; red: Cy3 labeled RNA, blue: Hoechst 33342. Scale bar: 10 μm. (**d**) Internalization of Cy3-labeled aptamers (100 nM) in normal primary pancreatic cells by confocal microscopy; blue: Hoechst 33342. Scale bar: 10 μm. (**e**) Flow cytometry analysis of normal T and B cells with Cy3-labeled P19 and P1 aptamers. (**f** and **g**) The dissociation constant (K_D) measured by flow cytometry using increasing concentrations of Cy3-labeled aptamers (from 15.6 to 500 nM); Mean fluorescence intensity (MFI) is measured and calculated using a one-site binding model for nonlinear regression

To verify the specificity and internalization of the enriched aptamers to pancreatic cancer cells, live-cell confocal imaging is employed. For the assay, a panel of four different pancreatic cancer cell lines (PANC-1, AsPC-1, MIA PaCa, and Capan-1) and nonpancreatic cancer lines (Huh7, HepG2, MCF7, and PC3) are seeded in 35-mm glass-bottom dishes and grown in medium for 24 h. The Cy3-labeled aptamers are added to the cells at 100 nM and incubated for 1 h. After washing the cells, the images are taken by 2-photon confocal microscope with

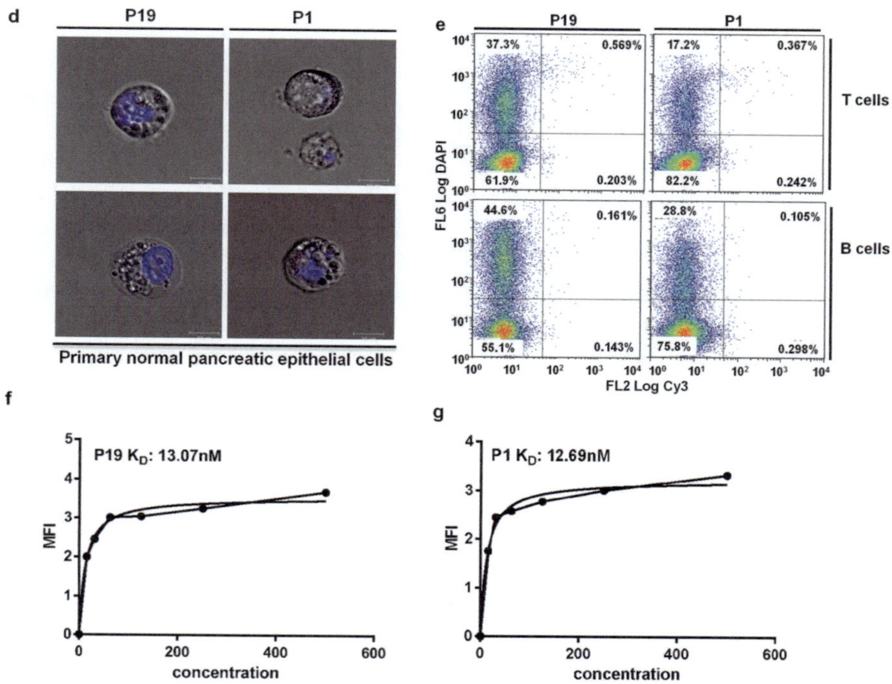

Fig. 12.2 (continued)

a C-Apo 40x/1.2NA water immersion objective. Confocal images confirm the uptake of P19 and P1 with punctuate staining in the cytoplasm on all of pancreatic cancer cells (Fig. 12.2c, *left panel*), but not on nonpancreatic cancer lines (Fig. 12.2c, *right panel*). The primary pancreatic epithelial cells dissociated from normal human pancreatic tissue are incubated with the Cy3-labeled aptamers, to determine whether P19 or P1 are ubiquitously internalized irrespective of the status of the pancreatic cells. No uptake is observed in normal primary pancreatic cells (Fig. 12.2d). No binding is also observed when normal T cells and B cells are incubated with the Cy3-labeled aptamers (Fig. 12.2e). It has demonstrated that P19 and P1 hold the pancreatic cancer specificity.

To determine the apparent dissociation constant (K_D) of aptamers to PANC-1 cells, the mean fluorescence intensity (MFI) is calculated for each concentration and for the unselected library controls. The values for the controls is considered to be background fluorescence and is subtracted from the values for the aptamers, as described previously [21]. The dissociation constants are calculated using a one-site binding model. The nonlinear curve regression is performed using Graph Pad Prism (GraphPad Software, La Jolla, California, USA). The measured binding affinities of P19 and P1 aptamers are 13.07 nM and 12.69 nM, respectively (Figs. 12.2f and g).

12.5 Aptamer–saRNA Conjugates

2′-Fluoropyrimidine RNA (2′F-RNA) pancreatic cancer-specific aptamers are used as a targeting modality to deliver C/EBPα-saRNA into pancreatic cancer cells for activation of C/EBPα expression (Fig. 12.3a). To make the conjugates of P19-C/EBPα-saRNA or P1- C/EBPα-saRNA, P1-STICK, P19-STICK, sense-STICK, and antisense RNAs are chemically synthesized in the Synthetic and Biopolymer Chemistry Core at the City of Hope. The P1-STICK and P19-STICK RNAs are refolded in binding buffer, heated to 95 °C for 3 min, and then slowly cooled to 37 °C. The incubation is continued at 37 °C for 10 min. The sense-STICK and antisense strand are annealed to the complementary strand using the same molar amounts to form the STICK-C/EBPα RNAs or scrambled RNAs. The same amount of the refolded P1- and P19-STICK is added and incubated at 37 °C for 10 min in binding buffer to form the P1- and P19- STICK-C/EBPα RNAs or the P1- and P19 STICK-scrambled RNAs (Fig. 12.3b).

To determine the internalization of the P19-saRNA conjugates and P1-saRNA conjugates, PANC-1 are seeded in 35-mm glass-bottom dishes before 24 h. The Cy3-labeled conjugates are added to the cells at 100 nM and incubated for 1 h. After washing the cells, the images are taken using a Zeiss 2-photon confocal microscope system with a C-Apo 40x/1.2NA water immersion objective. The P19 and P1 conjugates show the endocytic uptake in pancreatic cancer cells (Fig. 12.3c). This finding clearly demonstrates that the efficient uptake of aptamer linked with saRNA is comparable to lipid-mediated transient transfection methods.

12.6 *In Vitro* Assay of Aptamer–saRNA Conjugates

To investigate gene activation *in vitro,* the P19 and P1 conjugated C/EBPα-saRNA or scrambled RNAs are added to PANC-1 cells in culture media in the absence of any transfection reagent. For analyzing gene activation, PANC-1 cells are seeded into 24-well plates at a density of 1×10^5 cells per well. P19 and P1, conjugated with C/EBPα-saRNAs or scrambled saRNAs, are added directly to the cells, in duplicate, at a final concentration of 80 nM, for RNA and protein extraction. The treatment is repeated 24 h later and the cells are harvested at the 72 h time point. The total RNA is extracted for reverse transcription and target cDNA amplification by real-time PCR. Cells treated with the conjugated C/EBPα-saRNA aptamers show significantly higher levels of C/EBPα mRNA and its downstream target, p21, when compared to a scrambled-saRNA. The P19-C/EBPα-saRNA constructs induce a fivefold increase in C/EBPα transcript ($p = 0.029$) and a 50-fold increase in p21 transcript level (($p = 0.03$), two-tailed T-test with Welch's correction at 95% CI) (Fig. 12.4a and b). P1-C/EBPα-saRNA constructs induce a 1.3-fold increase in C/EBPα transcript ($p = 0.01$) and a 1.4-fold increase in p21 transcript level (($p = 0.026$), two-tailed T-test with Welch's correction at 95% CI) (Fig. 12.4a and b).

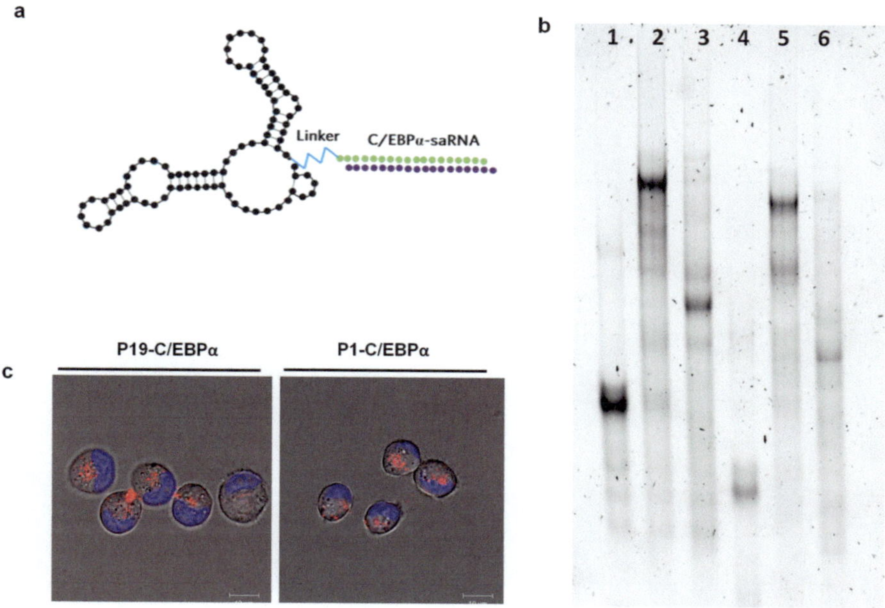

Fig. 12.3 The aptamer–saRNA conjugates and internalization in PANC-1 cells (a) Schematics of the aptamer–saRNA conjugates; C/EBPα-saRNAs are conjugated with aptamer *via* linker sequences. (**b**) Construction of aptamer conjugates with saRNA; *1*: C/EBPα-saRNA, *2*: P1-C/EBPα saRNA, *3*: P19-C/EBPα saRNA, *4*: Scramble RNAs (Scr), *5*: P1-Scr, *6*: P19-Scr. 12% Native PAGE gel. (**c**) Internalization of Cy3-labeled P19- or P1-conjugated C/EBPα-saRNAs in PANC-1 cells; *red*: Cy3-labeled RNA, *blue*: Hoechst 33342, Scale bar: 10 μm

Since C/EBPα is known to stabilize the cyclin-dependent kinase inhibitor, p21 to elevate expression levels and block cyclin-dependent kinases for cell cycle arrest [24]; The WST-1 cell proliferation assay is performed on PANC-1 cells treated with the conjugated aptamers for 96 h. Cells are cultured into 96-well plate at a density of 2.5×10^5 cells per well as three independent replicates. The WST-1 reagent is added and incubated for duration of 1 h with spectrophotometry readings at 420 and 620 nm taken every 15 min. Over 80% reduction in cell proliferation is observed following treatment with either P19-C/EBPα-saRNA (Fig. 12.4c, *left panel*) or P1-C/EBPα-saRNA (Fig. 12.4c, *right panel*); ($p = 0.011$, paired T-test at 5% CI). A strong antiproliferative effect through mediation of p21 has been shown in cell proliferation assay.

For protein expression assay, PANC-1 cells are seeded into 24-well plates at a density of 1×10^5 cells per well. P19 and P1, conjugated with C/EBPα-saRNAs or scrambled saRNAs, are added directly to the cells, in duplicate, at a final concentration of 80 nM, for RNA and protein extraction. The treatment is repeated 24 h later and the cells are harvested at the 72h time point. The total protein is extracted using a conventional RIPA buffer. The total protein content is then quantitated using a Bio-Rad Bradford Assay. The total protein extracts are separated by

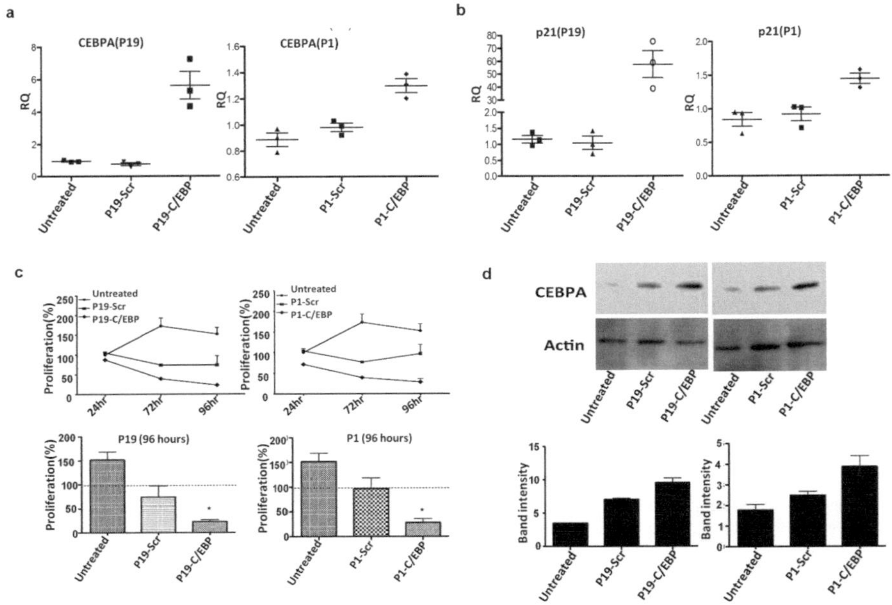

Fig. 12.4 Biological effects of conjugated aptamers in PANC-1 cells (**a**) Relative transcript expression (qPCR) for C/EBPα mRNA. (**b**) Quantification of p21 mRNA by real-time PCR. (**c**) A WST-1 cell proliferation assay in PANC-1 cells treated with P19- or P1-conjugated C/EBP-α-saRNA or scrambled saRNA aptamers. (**d**) Western blot analysis in PANC-1 cells treated with P19- or P1-conjugated C/EBPα-saRNA or scrambled saRNA aptamers; Band intensity from three representative blots is analyzed (*lower panel*)

SDS-PAGE and transferred onto PVDF membranes, then probed with antibodies against C/EBPα or actin. The proteins of interest are detected with an HRP-conjugated secondary antibody and visualized with LI-COR Western Sure ECL substrate. A Western blot analysis of CEBPA protein extracted from the treated PANC-1 cells demonstrates three times higher band intensity in cells treated with P19-C/EBPα-saRNA when compared to P1-C/EBPα-saRNA (Fig. 12.4d, *top* and *bottom panel*). At least a twofold increase in CEBPA signal from cells treated with scramble conjugated aptamer is observed (Fig. 12.4d, *top* and *bottom panel*).

Linking a duplex C/EBPα-saRNA molecule to pancreatic cancer-specific P19 and P1 aptamers demonstrates a significant increase in the C/EBPα transcript and protein levels. An interesting point in the *in vitro* study is that the P19-C/EBP-α-saRNA has induced higher expression of the C/EBPα transcript than the P1-C/EBPα-saRNA. However, protein expression demonstrates a similar increase of C/EBPα. The authors suggest the mechanism that autologous regulation or another transcriptional factor might be involved.

12.7 *In Vivo* Assay of Aptamer–saRNA Conjugates

To investigate the antitumor effects of C/EBPα-saRNA *in vivo*, the traceable tumor animal models are established. For traceable xenografted animal model, the firefly luciferase-expressed cells are constructed. To construct the firefly luciferase-expressed cells, the firefly-luciferase fragment is inserted into a pcDNA-3.1(+) backbone, which also encodes ampicillin resistance for selection in bacteria and the neomycin resistance gene for selection in mammalian cells. The recombinant constructs are purified using the plasmid midi kit. PANC-1 and AsPC-1 cells are transfected with the recombinant constructs for 24 h. The following day, the culture medium is replaced with standard medium containing 1.2 mg/mL G418 (Merck, Germany) for stable clone selection. Two weeks after selection, a single stable cell is picked and maintained in medium containing 1.2 mg/mL G418. Luciferase expression is assessed using the Luciferase Assay System.

To establish *in vivo* animal model for antitumor effects, the subcutaneous implantation is performed by injecting 25 μl of a monocellular suspension containing 10^6 PANC-1 or AsPC-1 cells expressing luciferase and a 25 μl growth factor-reduced matrigel matrix under the dorsal skin of 6-week-old female nonobese diabetic/severe combined immunodeficient (NOD/SCID) mice. When tumors are developed to about 1×1 cm in approximately 3 weeks after inoculation, mice are distributed in each group. Each subgroup of mice is injected with 100 pmol, 250 pmol or 1 nmol aptamer–STICK–saRNA systemically *via* tail vein four times per week for 3 weeks and sacrificed 1 week after the last injection.

Tumor growth is monitored by evaluating bioluminescence using an IVIS 200 live-animal imaging system. The tumors are evaluated before the first injection and 1 week after the last injection. Prior to the *in vivo* imaging, the mice are anesthetized using isoflurane. A solution of 150 μg/kg D-luciferin (Biosynth, USA) is then injected by the intraperitoneal route. The mice are imaged and bioluminescent signals are analyzed using the Living Image. Tumor size is measured with a ruler and calculated by the formula $0.52 \times \text{length} \times \text{width} \times \text{width}$.

In group of the 100 and 250 pmol of P19-C/EBPα-saRNA and P1-C/EBP-α-saRNA conjugates systemically into human pancreatic cancer xenografts, a significant reduction in tumor growth (Fig. 12.5a), photon counts (Fig. 12.5b left and c), and bioluminescent signal changes (Figs. 12.5b right and c) are observed in the treated groups, when compared to scramble control-treated and untreated animals. This observation confirms again the known antiproliferative effects of C/EBPα in PDAC. When compared to the antitumor effects of C/EBPα with gemcitabine over a period of 4 weeks, P19-C/EBPα-saRNA demonstrated 30% more efficient antitumor response when compared to gemcitabine by 4 weeks (Fig. 12.5d). The same study is performed in xenograft animal using gemcitabine-resistant human pancreatic adenocarcinoma cells (AsPC-1). The significant antitumor effect is observed only in the P19-C/EBPα-saRNA treated group. Gemcitabine shows no response (Fig. 12.5e). No evidence of blood toxicity in response to P19-C/EBPα-saRNA is seen as indicated by an analysis of hemoglobin, white blood cells, neutrophil, and platelet counts (Fig. 12.5f). It is the first study to prove an antitumor response in a gemcitabine-resistant pancreatic tumor-xenograft animal model. P19-C/EBPα-saRNA induced a 40% decrease in tumor growth with no evidence of toxicity to the host.

Taken together, the xenograft models when treated with the aptamer-saRNA conjugates demonstrate a significant reduction in tumor burden. Targeted delivery

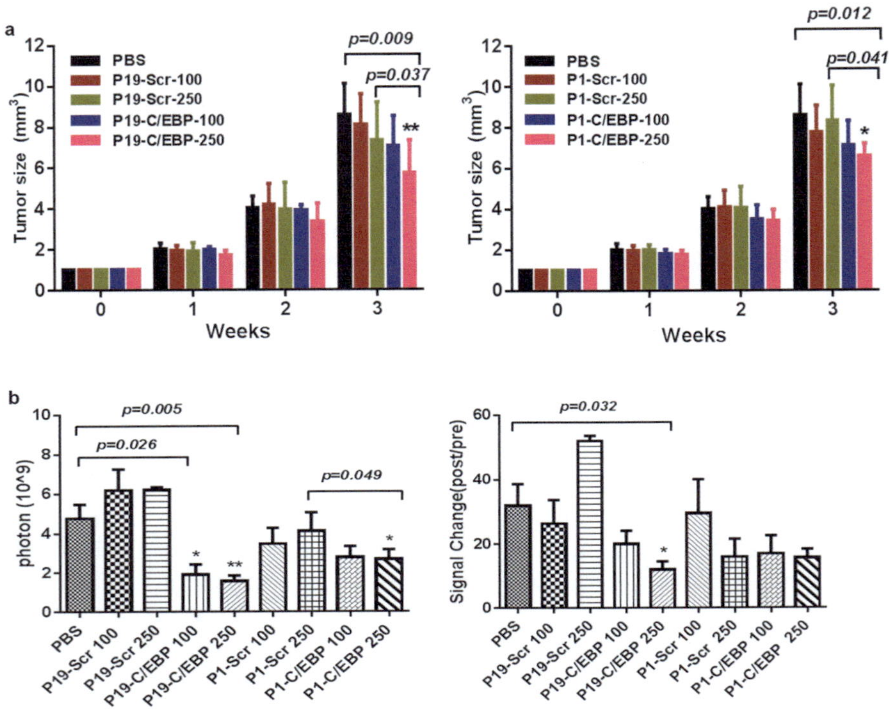

Fig. 12.5 ***In vivo*** **effects of P19- and P1-C/EBPα–saRNA** (**a**) Monitoring of tumor growth by measuring tumor size in PANC-1 engrafted mice; P19- and P1-conjugated C/EBPα-saRNA are injected in via tail vein injection at 100 and 250 pmol. Tumor size is calculated by the formula 0.52 × length × width × width. Data are presented as the mean ± SD. (**b**) Monitoring of tumor growth evaluating bioluminescence before the first injection and 1 week after the last injection; the photons (*left*) and bioluminescent signal changes (*right*) are quantified. (**c**) Representative bioluminescent images of the xenografts in each treatment group. (**d**) Comparison of the antitumor effect of C/EBPα-saRNA with gemcitabine. P19-conjugated C/EBPα-saRNA is injected in PANC-1-engrafted mice by tail vein injection at 1 nmol. Gemcitabine is injected by intraperitoneal injection at 3 mg on day 5 and 7 for 4 weeks. Data are presented as the mean ± SD. (**e**) The antitumor effects of P19-conjugated C/EBPα-saRNA in gemcitabine-resistant AsPC-1 cells. P19 conjugated C/EBPα-saRNA is injected in AsPC-1-engrafted mice by tail vein injection at 1 nmol. Data are presented as the mean ± SD. Tumor growth is monitored by evaluating bioluminescence before the first injection and 1 week after the last injection. The percentage of the cells that have metastasized to the ascites is measured. (**f**) The cytotoxicity of P19-conjugated C/EBPα-saRNA; to assess cytotoxicity following gemcitabine treatment, the blood parameter for hemoglobin, white blood cell count, platelets, and neutrophils are measured. Data are presented as the mean ± SD

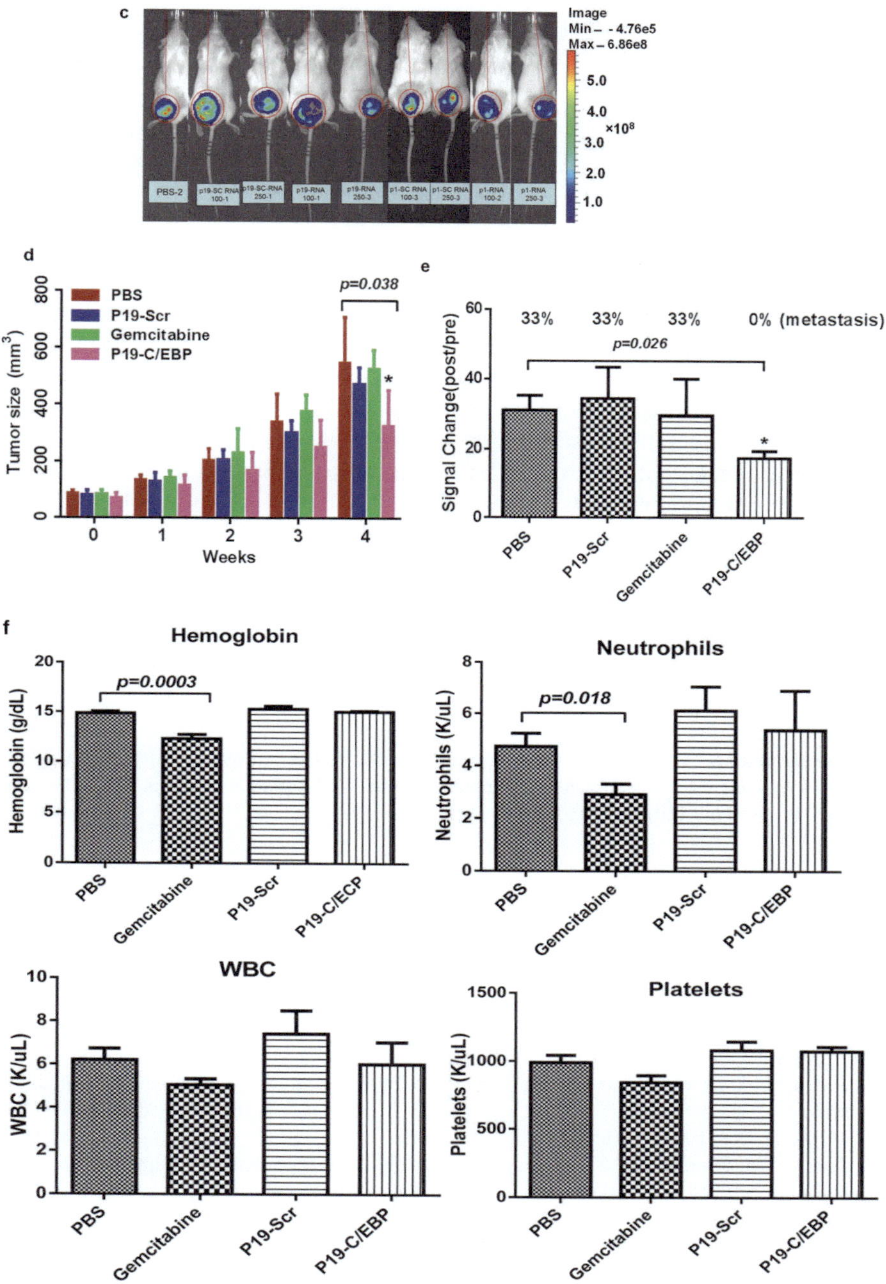

Fig. 12.5 (continued)

of C/EBPα through pancreatic cancer specific aptamer might be an effective therapeutics for pancreatic cancer in a clinically relevant way.

12.8 Summary

For the pancreatic cancer-targeted delivery, 2′F-RNA combinatorial library is employed to isolate pancreatic cancer-specific aptamer (P19 and P1) through a whole cell-based SELEX. Both P19 and P1 aptamers are linked to C/EBPα-saRNAs as a novel strategy to reactivate its epigenetically silenced target gene. The antitumor effects of P19- or P1-C/EBPα-saRNA are assessed *in vitro* and *in vivo* using human pancreatic adenocarcinoma (PANC-1) and gemcitabine-resistant AsPC-1 cell xenograft mouse models. The saRNA-linked aptamers demonstrate an efficient uptake within PANC-1 cells, which is comparable to lipid-mediated transient transfection methods. A strong antiproliferative effect through mediation of p21 is observed *in vitro*. The xenograft models when treated with the aptamer–saRNA constructs also demonstrate a significant reduction in tumor burden. It suggests that C/EBPα might be a novel druggable target for pancreatic cancer under current therapeutic modality failure. This study demonstrates a novel approach for site-directed targeting of gene activation using nucleotide-based molecules with cancer-specific aptamers.

Acknowledgments This chapter is based on the published paper "Targeted Delivery of C/EBP-α-saRNA by Pancreatic Ductal Adenocarcinoma-specific RNA Aptamers Inhibits Tumor Growth *In Vivo*" in *Mol Ther* (2016); 24 (6), 1106–1116.

References

1. Alexakis N, Halloran C, Raraty M, Ghaneh P, Sutton R, Neoptolemos JP (2004) Current standards of surgery for pancreatic cancer. Br J Surg 91(11):1410–1427. doi:10.1002/bjs.4794
2. Blank M, Weinschenk T, Priemer M, Schluesener H (2001) Systematic evolution of a DNA aptamer binding to rat brain tumor microvessels. Selective targeting of endothelial regulatory protein pigpen. J Biol Chem 276(19):16464–16468. doi:10.1074/jbc. M100347200M100347200 [pii]
3. Daniels DA, Chen H, Hicke BJ, Swiderek KM, Gold L (2003) A tenascin-C aptamer identified by tumor cell SELEX: systematic evolution of ligands by exponential enrichment. Proc Natl Acad Sci U S A 100(26):15416–15421. doi:10.1073/pnas.2136683100
4. Ellington AD, Szostak JW (1990) In vitro selection of RNA molecules that bind specific ligands. Nature 346(6287):818–822. doi:10.1038/346818a0
5. Eser S, Schnieke A, Schneider G, Saur D (2014) Oncogenic KRAS signalling in pancreatic cancer. Br J Cancer. doi:10.1038/bjc.2014.215
6. Fulda S (2009) Apoptosis pathways and their therapeutic exploitation in pancreatic cancer. J Cell Mol Med 13(7):1221–1227. doi:JCMM748 [pii]10.1111/j.1582–4934.2009.00748.x
7. Ghaneh P, Costello E, Neoptolemos JP (2007) Biology and management of pancreatic cancer. Gut 56(8):1134–1152. doi:10.1136/gut.2006.103333

8. Guidelines for the Management of Patients with Pancreatic Cancer Periampullary and Ampullary Carcinomas (2005) Gut 54 Suppl 5:v1–16. doi:10.1136/gut.2004.057059
9. Hicke BJ, Marion C, Chang YF, Gould T, Lynott CK, Parma D, Schmidt PG, Warren S (2001) Tenascin-C aptamers are generated using tumor cells and purified protein. J Biol Chem 276 (52):48644–48654. doi:10.1074/jbc.M104651200M104651200 [pii]
10. Janowski BA, Younger ST, Hardy DB, Ram R, Huffman KE, Corey DR (2007) Activating gene expression in mammalian cells with promoter-targeted duplex RNAs. Nat Chem Biol 3 (3):166–173. doi:10.1038/nchembio860
11. Junxia W, Ping G, Yuan H, Lijun Z, Jihong R, Fang L, Min L, Xi W, Ting H, Ke D, Huizhong Z (2010) Double strand RNA-guided endogeneous E-cadherin up-regulation induces the apoptosis and inhibits proliferation of breast carcinoma cells in vitro and in vivo. Cancer Sci 101(8):1790–1796. doi:10.1111/j.1349-7006.2010.01594.x
12. Kanda M, Matthaei H, Wu J, Hong SM, Yu J, Borges M, Hruban RH, Maitra A, Kinzler K, Vogelstein B, Goggins M (2012) Presence of somatic mutations in most early-stage pancreatic intraepithelial neoplasia. Gastroenterology 142(4):730–733. doi:10.1053/j.gastro.2011.12.042e739
13. Klinkenbijl JH, Jeekel J, Sahmoud T, van Pel R, Couvreur ML, Veenhof CH, Arnaud JP, Gonzalez DG, de Wit LT, Hennipman A, Wils J (1999) Adjuvant radiotherapy and 5-fluorouracil after curative resection of cancer of the pancreas and periampullary region: phase III trial of the EORTC gastrointestinal tract cancer cooperative group. Ann Surg 230 (6):776–782. discussion 782–774
14. Li LC, Okino ST, Zhao H, Pookot D, Place RF, Urakami S, Enokida H, Dahiya R (2006) Small dsRNAs induce transcriptional activation in human cells. Proc Natl Acad Sci U S A 103 (46):17337–17342. doi:10.1073/pnas.0607015103
15. Meng X, Jiang Q, Chang N, Wang X, Liu C, Xiong J, Cao H, Liang Z (2016) Small activating RNA binds to the genomic target site in a seed-region-dependent manner. Nucleic Acids Res 44(5):2274–2282. doi:10.1093/nar/gkw076
16. Neoptolemos JP, Stocken DD, Friess H, Bassi C, Dunn JA, Hickey H, Beger H, Fernandez-Cruz L, Dervenis C, Lacaine F, Falconi M, Pederzoli P, Pap A, Spooner D, Kerr DJ, Buchler MW (2004) A randomized trial of chemoradiotherapy and chemotherapy after resection of pancreatic cancer. N Engl J Med 350(12):1200–1210. doi:10.1056/NEJMoa032295
17. Oettle H, Post S, Neuhaus P, Gellert K, Langrehr J, Ridwelski K, Schramm H, Fahlke J, Zuelke C, Burkart C, Gutberlet K, Kettner E, Schmalenberg H, Weigang-Koehler K, Bechstein WO, Niedergethmann M, Schmidt-Wolf I, Roll L, Doerken B, Riess H (2007) Adjuvant chemotherapy with gemcitabine vs observation in patients undergoing curative-intent resection of pancreatic cancer: a randomized controlled trial. JAMA 297(3):267–277. doi:10.1001/jama.297.3.267
18. Pancreatic cancer in the UK (2011) Lancet 378(9796):1050. doi:10.1016/S0140-6736(11)61465-7
19. Que-Gewirth NS, Sullenger BA (2007) Gene therapy progress and prospects: RNA aptamers. Gene Ther 14(4):283–291. doi:10.1038/sj.gt.3302900
20. Reebye V, Saetrom P, Mintz PJ, Huang KW, Swiderski P, Peng L, Liu C, Liu X, Lindkaer-Jensen S, Zacharoulis D, Kostomitsopoulos N, Kasahara N, Nicholls JP, Jiao LR, Pai M, Spalding DR, Mizandari M, Chikovani T, Emara MM, Haoudi A, Tomalia DA, Rossi JJ, Habib NA (2013) Novel RNA oligonucleotide improves liver function and inhibits liver carcinogenesis in vivo. Hepatology. doi:10.1002/hep.26669
21. Sefah K, Shangguan D, Xiong X, O'Donoghue MB, Tan W (2010) Development of DNA aptamers using cell-SELEX. Nat Protoc 5(6):1169–1185. doi:10.1038/nprot.2010.66
22. Siegel RL, Miller KD, Jemal A (2016) Cancer statistics, 2016. CA Cancer J Clin 66(1):7–30. doi:10.3322/caac.21332
23. Stathis A, Moore MJ (2010) Advanced pancreatic carcinoma: current treatment and future challenges. Nat Rev Clin Oncol 7(3):163–172. doi:10.1038/nrclinonc.2009.236

24. Timchenko NA, Wilde M, Nakanishi M, Smith JR, Darlington GJ (1996) CCAAT/enhancer-binding protein alpha (C/EBP alpha) inhibits cell proliferation through the p21 (WAF-1/CIP-1/SDI-1) protein. Genes Dev 10(7):804–815
25. Tuerk C (1997) Using the SELEX combinatorial chemistry process to find high affinity nucleic acid ligands to target molecules. Methods Mol Biol 67:219–230
26. Ulrich H, Magdesian MH, Alves MJ, Colli W (2002) In vitro selection of RNA aptamers that bind to cell adhesion receptors of *Trypanosoma cruzi* and inhibit cell invasion. J Biol Chem 277(23):20756–20762. doi:10.1074/jbc.M111859200M111859200 [pii]
27. Vincent A, Herman J, Schulick R, Hruban RH, Goggins M (2011) Pancreatic cancer. Lancet 378(9791):607–620. doi:10.1016/S0140-6736(10)62307-0
28. Wang J, Jiang H, Liu F (2000) In vitro selection of novel RNA ligands that bind human cytomegalovirus and block viral infection. RNA 6(4):571–583
29. Wilson DS, Szostak JW (1999) In vitro selection of functional nucleic acids. Annu Rev Biochem 68:611–647. doi:10.1146/annurev.biochem.68.1.611
30. Wong HH, Lemoine NR (2009) Pancreatic cancer: molecular pathogenesis and new therapeutic targets. Nat Rev Gastroenterol Hepatol 6(7):412–422. doi:nrgastro.2009.89 [pii]10.1038/nrgastro.2009.89
31. Yamamoto K, Tateishi K, Kudo Y, Sato T, Yamamoto S, Miyabayashi K, Matsusaka K, Asaoka Y, Ijichi H, Hirata Y, Otsuka M, Nakai Y, Isayama H, Ikenoue T, Kurokawa M, Fukayama M, Kokudo N, Omata M, Koike K (2014) Loss of histone demethylase KDM6B enhances aggressiveness of pancreatic cancer through downregulation of C/EBPalpha. Carcinogenesis. doi:10.1093/carcin/bgu136
32. Yoon S, Huang KW, Reebye V, Mintz P, Tien YW, Lai HS, Saetrom P, Reccia I, Swiderski P, Armstrong B, Jozwiak A, Spalding D, Jiao L, Habib N, Rossi JJ (2016) Targeted delivery of C/EBPalpha -saRNA by pancreatic ductal adenocarcinoma (PDAC)-specific RNA aptamers inhibits tumor growth in vivo. Mol Ther. doi:10.1038/mt.2016.60
33. Yoon S, Lee G, Han D, Song JY, Kang KS, Lee YS (2010) Neutralization of infectivity of porcine circovirus type 2 (PCV2) by capsid-binding 2'F-RNA aptamers. Antivir Res 88 (1):19–24. doi:10.1016/j.antiviral.2010.06.012

Chapter 13
Treatment of Liver Cancer by C/EBPA saRNA

Xiaoyang Zhao, Jon Voutila, Stephanos Ghobrial, Nagy A. Habib, and Vikash Reebye

Abstract The prognosis for hepatocellular carcinoma (HCC) remains poor and has not improved in over two decades. Most patients with advanced HCC who are not eligible for surgery have limited treatment options due to poor liver function or large, unresectable tumors. Although sorafenib is the standard-of-care treatment for these patients, only a small number respond. For the remaining, the outlook remains bleak. A better approach to target "undruggable" molecular pathways that reverse HCC is therefore urgently needed. Small activating RNAs (saRNAs) may provide a novel strategy to activate expression of genes that become dysregulated in chronic disease. The transcription factor CCAAT/enhancer-binding protein alpha (C/EBPα), a critical regulator of hepatocyte function, is suppressed in many advanced liver diseases. By using an saRNA to activate C/EBPα, we can exploit the cell's own transcription machinery to enhance gene expression without relying on exogenous vectors that have been the backbone of gene therapy. saRNAs do not integrate into the host genome and can be modified to avoid immune stimulation. In preclinical models of liver disease, treatment with C/EBPα saRNA has shown reduction in tumor volume and improvement in serum markers of essential liver function such as albumin, bilirubin, aspartate aminotransferase (AST), and alanine transaminase (ALT). This saRNA that activates C/EBPα for advanced HCC is the first saRNA therapy to have entered a human clinical trial. The hope is that this new tool will help break the dismal 20-year trend and provide a more positive prognosis for patients with severe liver disease.

Keywords Oligonucleotide therapy • Small activating RNA • Hepatocellular carcinoma • CEBPA transcription factor • Liver disease

---

X. Zhao • S. Ghobrial • N.A. Habib (✉) • V. Reebye
Department of Surgery, Hammersmith Hospital, Imperial College London, London, UK
e-mail: nagy.habib@imperial.ac.uk

J. Voutila
MiNA Therapeutics Ltd., London, UK

L.-C. Li (ed.), *RNA Activation*, Advances in Experimental Medicine and Biology 983, DOI 10.1007/978-981-10-4310-9_13

13.1 Introduction

13.1.1 Liver Cancer

Liver cancer is the second leading cause of cancer-related deaths and the sixth most common malignancy worldwide [2, 22]. Hepatocellular carcinoma (HCC), which accounts for approximately 80% of the total worldwide liver cancer burden, represents the predominant histological subtype [14]. HCC is most frequently caused by liver cirrhosis linked epidemiologically to major risk factors and prevalence of predisposing conditions, including most commonly hepatitis B and C virus infection, chronic alcohol consumption, and nonalcoholic fatty liver disease. HCC population analysis by region has shown alcoholic diseases and hepatitis C are the main causes in Western countries, while hepatitis C is strongly associated in Asia and sub-Saharan African countries [12]. Surgical resection is considered the primary option for HCC treatment offering the prospect of cure and long-term survival when compared with other strategies [7]. However, due to poor liver function, main vessel tumor invasion, and extrahepatic metastases, the overall resectable rate for HCC is low (10–25%), and the general prognosis is poor with a 3–5% overall survival rate, which has not improved in over two decades [1, 13, 18].

Molecular targeted therapies against novel key signaling pathways are gaining momentum, especially for advanced HCC. To date, only sorafenib, a multikinase inhibitor targeting Raf-1, B-Raf, VEGFRs, and PDGFR-β, has shown significant improvement in treating advanced HCC [4]. Although sorafenib is currently considered the standard-of-care treatment for advanced HCC, it can only prolong median life expectancy by 1 year [3]. Thus, there remains an unmet need for novel therapeutic drugs for HCC.

Reactivation of genes that are suppressed in diseased states using nonviral methods has gained immense footing in the form of small activating RNAs (saRNAs). With a detailed understanding of the sequences within the promoter and enhancer regions of target genes; 21mer oligonucleotide sequences are generated using a platform technology. The identified gene target sense and antisense strands, when synthesised and duplexed, are able to exploit the innate Argonaute-2 dependent pathway of mammalian cells to initiate a transcriptionally active complex at the seed region of the target gene for de novo transcription of messenger RNA. These saRNAs provide a clinically safe strategy for enhancing potentially undruggable molecular targets to prevent or even reverse disease progression.

Here, we showed that it was possible to synthetically design an saRNA duplex to activate a vital liver-enriched transcription factor, CEBPα, known to be suppressed in liver disease. Reactivation of CEBPα expression in a liver cancer/cirrhotic animal model showed significant improvement of liver function and reduction in tumor burden [16]. Such was the impact caused by enhancing CEBPα expression in diseased liver models, the saRNA duplex has been further developed into a clinical candidate (MTL-CEBPA) with launch of a multi-centre Phase I clinical trial (OUTREACH) for patients with advanced HCC [5].

13.1.2 CCAAT/Enhancer-Binding Protein Alpha (C/EBPα)

The CCAAT/enhancer-binding protein (C/EBP) family of transcription factors are involved in a wide variety of cellular processes, including cell growth, metabolism, differentiation, and the immune response. The first discovered member of this family was named C/EBPα, and it is most highly expressed in the liver, adipose tissue, and the myeloid lineage [15]. C/EBPα has been recognized as a general tumor suppressor due to its well-characterized antimitotic functions through inhibition of the p21, E2F, and CDK2/4 pathways [19]. In the liver, C/EBPα regulates genes involved in mature hepatocyte differentiation and maintenance of metabolic balance and body weight homeostasis [10, 21]. Since patients with HCC frequently have poor liver function [8, 20], this makes C/EBPα an attractive target for gene upregulation in advanced liver cancer. This is supported by a C/EBPα knock-in mouse model showing reduced susceptibility to HCC [23] as well as a rat HCC model showing reduced C/EBPα expression in tumor tissues [6]. A retrospective analysis of human HCC samples has confirmed that C/EBPα protein is downregulated in HCC and is associated with poor survival [25]. Furthermore, other studies of C/EBPα function demonstrate that it is critical for hepatic regulation of glucose and fat homeostasis [9] in addition to having antifibrotic properties [11, 24]. This places C/EBPα as a unique target for the amelioration of many liver diseases including fibrosis, cirrhosis, nonalcoholic fatty liver disease, steatohepatitis, and HCC.

13.2 C/EBPα saRNA in the Treatment of HCC

13.2.1 C/EBPα saRNA in a Rat HCC Model

The first published report using C/EBPα saRNA identified an oligonucleotide sequence that increased C/EBPα mRNA twofold after transfection in the human HCC cell line, HepG2 [16]. The upregulation of C/EBPα mRNA also caused a twofold increase in albumin expression, consistent with C/EBPα's role in liver function. This saRNA was then tested *in vivo* in a rat diethylnitrosamine (DEN)-induced spontaneous HCC model. Intravenous delivery was achieved with the use of polyamidoamine (PAMAM) dendrimer [27]. Tail vein injection of C/EBPα saRNA–dendrimer complex into DEN-treated rats showed a significant reduction in tumor burden compared to treatment with a control oligonucleotide. This was accompanied by significant increases in serum albumin and significant decreases in serum bilirubin, aspartate aminotransferase (AST), and alanine aminotransferase (ALT). Quantitative reverse transcription polymerase chain reaction (RT-PCR) analysis of liver tissue from treated animals confirmed upregulation of C/EBPα and albumin mRNA as well as critical hepatocyte markers such as HNF1α and HNF4α. Taken together, these results confirm the previously published reports of

C/EBPα's function as a tumor suppressor as well as a regulator of hepatocyte function. This exciting study provided the basis to further develop C/EBPα saRNA for clinical use.

13.2.2 Preclinical Development of C/EBPα saRNA for Treatment of HCC

Since publication, the C/EBPα saRNA reported in Reebye et al. has been sequence-optimized and modified to prevent immune stimulation for clinical use [26]. This clinical saRNA candidate, referred to as CEBPA-51, retains the activity of the original C/EBPα saRNA oligonucleotide. Upon transfection into human HCC cell lines, C/EBPα mRNA is upregulated as well as albumin and cytochrome P450 enzyme expression (unpublished data). CEBPA-51 treatment also shows significant inhibition of cell proliferation in these transformed tumorigenic cell lines. At the publication time of this chapter, CEBPA-51 has been encapsulated in Marina Biotech's liposomal carrier molecules for *in vivo* use as MTL-CEBPA. These amphoteric liposomes have anionic and cationic groups that provide serum stability and high tropism to the liver with efficient cellular uptake. The SMARTICLES, moreover, provide a pH-triggered endosomal escape for intracellular delivery of the double-stranded saRNA payload. MTL-CEBPA has shown an 80% decrease in tumor burden in the rat DEN-induced spontaneous HCC model (unpublished data) and over 70% increase in survival of a chronic liver fibrosis model [17]. MTL-CEBPA entered into Phase I clinical trials in early 2016 for patients with advanced liver cancer, representing the first-in-human study of activating oligonucleotide therapy [5].

Most patients have limited treatment options for advanced HCC due to poor liver function or large, unresectable tumors making them ineligible for surgical intervention. C/EBPα is an attractive target for these patients because it has proven anticancer properties as well as being critical for hepatocyte function. Data from human HCC patients showing downregulation of C/EBPα in advanced disease mirror the progressive decrease in liver function. This provides a unique opportunity for an saRNA therapy to reactivate gene expression in tissues that lose gene regulation during disease progression. saRNA targeting C/EBPα has now been shown to reduce tumor burden as well as improve liver function, providing the basis for advancement into the clinic. We hope that by improving liver function, MTL-CEBPA might downgrade patients previously not eligible for surgical intervention to become eligible, or even facilitate liver regeneration in patients with a wide spectrum of chronic liver disease. Upregulation of transcription factors with saRNA is poised to be a game-changer for unmet needs in many undruggable diseases that are caused by suppression of key genes.

References

1. Belghiti J, Panis Y, Farges O, Benhamou JP, Fekete F (1991) Intrahepatic recurrence after resection of hepatocellular carcinoma complicating cirrhosis. Ann Surg 214(2):114–117
2. Bray F, Jemal A, Grey N, Ferlay J, Forman D (2012) Global cancer transitions according to the Human Development Index (2008–2030): a population-based study. Lancet Oncol 13 (8):790–801. doi:10.1016/S1470-2045(12)70211-5
3. Choi KJ, Baik IH, Ye SK, Lee YH (2015) Molecular targeted therapy for hepatocellular carcinoma: present status and future directions. Biol Pharm Bull 38(7):986–991. doi:10. 1248/bpb.b15-00231
4. Chuma M, Terashita K, Sakamoto N (2015) New molecularly targeted therapies against advanced hepatocellular carcinoma: from molecular pathogenesis to clinical trials and future directions. Hepatol Res 45(10):E1–E11. doi:10.1111/hepr.12459
5. ClinicalTrials.gov (2016) First-in-human safety and tolerability study of MTL-CEBPA in patients with advanced liver cancer. https://ClinicalTrials.gov/show/NCT02716012
6. Flodby P, Liao DZ, Blanck A, Xanthopoulos KG, Hallstrom IP (1995) Expression of the liver-enriched transcription factors C/EBP alpha, C/EBP beta, HNF-1, and HNF-4 in preneoplastic nodules and hepatocellular carcinoma in rat liver. Mol Carcinog 12(2):103–109
7. Johnson PJ (2005) Non-surgical treatment of hepatocellular carcinoma. HPB (Oxford) 7 (1):50–55. doi:10.1080/13651820410024076
8. Johnson PJ, Berhane S, Kagebayashi C, Satomura S, Teng M, Reeves HL, O'Beirne J, Fox R, Skowronska A, Palmer D, Yeo W, Mo F, Lai P, Inarrairaegui M, Chan SL, Sangro B, Miksad R, Tada T, Kumada T, Toyoda H (2015) Assessment of liver function in patients with hepatocellular carcinoma: a new evidence-based approach-the ALBI grade. J Clin Oncol 33(6):550–558. doi:10.1200/JCO.2014.57.9151
9. Matsusue K, Gavrilova O, Lambert G, Brewer HB Jr, Ward JM, Inoue Y, LeRoith D, Gonzalez FJ (2004) Hepatic CCAAT/enhancer binding protein alpha mediates induction of lipogenesis and regulation of glucose homeostasis in leptin-deficient mice. Mol Endocrinol 18 (11):2751–2764. doi:10.1210/me.2004-0213
10. McKnight SL (2001) McBindall—a better name for CCAAT/enhancer binding proteins? Cell 107(3):259–261
11. Mei S, Wang X, Zhang J, Qian J, Ji J (2007) In vivo transfection of C/EBP-alpha gene could ameliorate CCL(4)-induced hepatic fibrosis in mice. Hepatol Res 37(7):531–539. doi:10.1111/ j.1872-034X.2007.00074.x
12. Mokdad AA, Lopez AD, Shahraz S, Lozano R, Mokdad AH, Stanaway J, Murray CJ, Naghavi M (2014) Liver cirrhosis mortality in 187 countries between 1980 and 2010: a systematic analysis. BMC Med 12:145. doi:10.1186/s12916-014-0145-y
13. Nagasue N, Uchida M, Makino Y, Takemoto Y, Yamanoi A, Hayashi T, Chang YC, Kohno H, Nakamura T, Yukaya H (1993) Incidence and factors associated with intrahepatic recurrence following resection of hepatocellular carcinoma. Gastroenterology 105(2):488–494
14. Perz JF, Armstrong GL, Farrington LA, Hutin YJ, Bell BP (2006) The contributions of hepatitis B virus and hepatitis C virus infections to cirrhosis and primary liver cancer worldwide. J Hepatol 45(4):529–538. doi:10.1016/j.jhep.2006.05.013
15. Ramji DP, Foka P (2002) CCAAT/enhancer-binding proteins: structure, function and regulation. Biochem J 365(Pt 3):561–575. doi:10.1042/BJ20020508
16. Reebye V, Saetrom P, Mintz PJ, Huang KW, Swiderski P, Peng L, Liu C, Liu X, Lindkaer-Jensen S, Zacharoulis D, Kostomitsopoulos N, Kasahara N, Nicholls JP, Jiao LR, Pai M, Spalding DR, Mizandari M, Chikovani T, Emara MM, Haoudi A, Tomalia DA, Rossi JJ, Habib NA (2014) Novel RNA oligonucleotide improves liver function and inhibits liver carcinogenesis in vivo. Hepatology 59(1):216–227. doi:10.1002/hep.26669
17. Reebye V, Voutila J, Huang KW, Muragundla A, Jayaprakash A, Vadnal P, Huber H, Habib R, Saetrom P, Rossi J, Habib N (2015) Systemic administration of a novel development

candidate, MTL-CEBPA, up-regulates the liver-enriched transcription factor C/EBP-alpha and reverses CCl4-induced liver failure in vivo. Hepatology 62:269a–270a
18. Schmidt S, Follmann M, Malek N, Manns MP, Greten TF (2011) Critical appraisal of clinical practice guidelines for diagnosis and treatment of hepatocellular carcinoma. J Gastroenterol Hepatol 26(12):1779–1786. doi:10.1111/j.1440-1746.2011.06891.x
19. Schuster MB, Porse BT (2006) C/EBPalpha: a tumour suppressor in multiple tissues? Biochim Biophys Acta 1766(1):88–103. doi:10.1016/j.bbcan.2006.02.003
20. Simonetti RG, Camma C, Fiorello F, Politi F, D'Amico G, Pagliaro L (1991) Hepatocellular carcinoma. A worldwide problem and the major risk factors. Dig Dis Sci 36(7):962–972
21. Soriano HE, Kang DC, Finegold MJ, Hicks MJ, Wang ND, Harrison W, Darlington GJ (1998) Lack of C/EBP alpha gene expression results in increased DNA synthesis and an increased frequency of immortalization of freshly isolated mice [correction of rat] hepatocytes. Hepatology 27(2):392–401. doi:10.1002/hep.510270212
22. Stewart BW, Wild C, International Agency for Research on Cancer, World Health Organization (2014) World cancer report 2014. International Agency for Research on Cancer. WHO Press, Lyon/Geneva
23. Tan EH, Hooi SC, Laban M, Wong E, Ponniah S, Wee A, Wang ND (2005) CCAAT/enhancer binding protein alpha knock-in mice exhibit early liver glycogen storage and reduced susceptibility to hepatocellular carcinoma. Cancer Res 65(22):10330–10337. doi:10.1158/0008-5472.CAN-04-4486
24. Tao LL, Cheng YY, Ding D, Mei S, Xu JW, Yu J, Ou-Yang Q, Deng L, Chen Q, Li QQ, Xu ZD, Liu XP (2012) C/EBP-alpha ameliorates CCl(4)-induced liver fibrosis in mice through promoting apoptosis of hepatic stellate cells with little apoptotic effect on hepatocytes in vitro and in vivo. Apoptosis 17(5):492–502. doi:10.1007/s10495-012-0700-y
25. Tseng HH, Hwang YH, Yeh KT, Chang JG, Chen YL, Yu HS (2009) Reduced expression of C/EBP alpha protein in hepatocellular carcinoma is associated with advanced tumor stage and shortened patient survival. J Cancer Res Clin Oncol 135(2):241–247. doi:10.1007/s00432-008-0448-5
26. Voutila J, Reebye V, Roberts T, Blakey D, Habib R, Huber H, Saetrom P, Rossi JJ, Habib NA (2016) Development and mechanism of a small activating RNA targeting CEBPA, a novel therapeutic in clinical trials for patients with liver cancer. Poster presented at the The Oligo meeting, Montreal, Canada
27. Zhou J, Neff CP, Liu X, Zhang J, Li H, Smith DD, Swiderski P, Aboellail T, Huang Y, Du Q, Liang Z, Peng L, Akkina R, Rossi JJ (2011) Systemic administration of combinatorial dsiRNAs via nanoparticles efficiently suppresses HIV-1 infection in humanized mice. Mol Ther 19(12):2228–2238. doi:10.1038/mt.2011.207

Chapter 14
Enhancing Angiogenesis in Mice by VEGF-Targeting Small Activating RNAs

Tiia A. Turunen, Seppo Ylä-Herttuala, and Mikko P. Turunen

Abstract The prevalence of cardiovascular diseases is steadily increasing, and it is the leading cause of death worldwide. Therefore, new treatments, such as gene therapy are needed. During the last decade, the role of small noncoding RNAs (ncRNAs) in the regulation of gene expression at the transcriptional level has been shown. Promoter-targeted small RNAs recruit histone-modifying enzymes and can either repress or induce target gene expression. As an example, we have targeted mouse VEGF-A promoter with small hairpin RNAs (shRNAs) and identified two shRNAs which either repressed or induced VEGF-A expression on messenger RNA and protein level in vitro, depending on the targeted location. The changes in expression levels correlate with changes in the levels of epigenetic markers, such as histone modifications associated with repressed or active state of chromatin. In ischemic mouse hindlimbs, upregulation of VEGF-A expression increased vascularity and blood flow. When VEGF-A was upregulated in mouse myocardial infarction model, the blood vessel formation in the risk zone was observed and infarct size was significantly decreased already 2 weeks after treatment. We suggest that epigenetic upregulation of VEGF-A by ncRNAs can be transferred to clinical use for the treatment of ischemic diseases in the near future.

Keywords VEGF-A • Angiogenesis • Lentiviral vectors • Gene therapy • Animal models

14.1 Introduction

14.1.1 Cardiovascular Disease

Cardiovascular diseases (CVD) are the leading cause of death worldwide. An estimated 17.5 million people died from CVD in 2012, representing 31% of all global deaths according to the WHO. Clinically, the outcome of the disease is often

T.A. Turunen • S. Ylä-Herttuala • M.P. Turunen (✉)
A.I.Virtanen Institute for Molecular Sciences, University of Eastern Finland, Kuopio, Finland
e-mail: mikko.turunen@uef.fi

L.-C. Li (ed.), *RNA Activation*, Advances in Experimental Medicine and Biology 983, DOI 10.1007/978-981-10-4310-9_14

myocardial infarction, stroke, or peripheral vascular disease, all of which are influenced by atherosclerosis where arteries are narrowed by plaques and therefore the blood flow to the tissue is reduced [38]. This restricted blood flow results in problems in the transport of nutrients and oxygen to the tissue and ultimately leads to the formation of ischemic areas. Plaque formation is a slow process that develops throughout the lifespan. Plaque rupture leads to blood clots that block blood vessels either in the heart, initiating myocardial infarction, or elsewhere in the circulatory system, such as in the brain or in the legs.

Currently, the most common operative treatments for CVD are bypass surgery that aims to replace the occluded blood vessels and thus restore the blood flow in the ischemic area, and the insertion of drug-eluting stents, that in the occluded vessel release drugs aiming to prevent reocclusion and clot formation [29]. However, these treatments are not suitable for all patients and therefore additional strategies, such as gene therapy approaches, are being developed. Basic hypothesis is that delivery of angiogenic genes, such as Vascular Endothelial Growth Factors (VEGFs), into the damaged area would induce angiogenesis and thus help to restore blood flow to ischemic areas.

14.1.2 Vascular Endothelial Growth Factors

Vascular endothelial growth factor family of proteins are key mediators of angiogenesis, occuring by sprouting, bridging, intussusception, and/or enlargement of capillaries from the preexisting ones. The first identified vascular endothelial growth factor (VEGF-A) and the master regulator of angiogenesis and vascular permeability was cloned in 1989 [19]. Four other members in the human VEGF family have also been identified: VEGF-B, VEGF-C, VEGF-D, and placental growth factor (PlGF) [1, 15, 21, 27]. The downstream signals of VEGFs in vascular endothelium are mediated by three tyrosine kinase-signaling receptors (VEGF receptor [VEGFR]-1, -2, and -3) [33].

VEGF-A genes in human and mouse consist of eight exons separated by seven introns, resulting in a coding region of 14 kb [36, 40]. The murine gene includes the proximal promoter of around 1.2 kb and long untranslated regions (UTRs). The 5′ end of the gene contains important regulatory elements for VEGF-A activity and 1.3 kb deletion of that region demolishes the gene activity. Like other growth-related genes, the promoter lacks consensus TATA or CCAAT sequences, but the RNA Polymerase II transcription is instead initiated through the binding of sequence-specific transcription factors. The promoter has multiple characterized transcription factor binding sites, such as Sp1, AP1, AP2, Egr-1, STAT-3, and NF-κB [28, 36]. VEGF-A gene expression has been shown to be induced in hypoxia, ischemic conditions, and inflammatory signals. The promoter contains a binding site for the hypoxia-inducible factor 1 (HIF-1) on the hypoxia response element [7].

VEGFs have been used in vascular medicine for their vasculoprotective and angiogenic properties [43]. Various studies have shown that gene transfers of VEGF-A, -C, and -D prevent restenosis after arterial injury in several animal models [3, 11, 18, 34]. However, even though these animal studies were promising, no beneficial clinical effect of VEGF-A gene transfer on restenosis was found [10] which suggests that the most suitable target for gene therapy with VEGFs might be the treatment of myocardial or peripheral ischemia.

14.2 Inducing Angiogenesis in Mice

14.2.1 Small Hairpin RNAs Can Regulate VEGF-A Via Epigenetic Mechanisms

In mammalian cells, transcriptional silencing of gene activity by small ncRNAs was first published in 2004, regulating the expression of human eukaryotic elongation factor 1 α promoter [23]. Soon after, the transcriptional gene activation was shown in human cells on E-cadherin, p21, and VEGF-A [20]. These studies used siRNAs or dsRNAs, respectively, of 21 nt in length as the effector molecules, delivered by feline immunodeficiency virus transduction or by transfection into the target cells. In our studies, we have targeted the murine VEGF-A promoter with shRNAs [13, 41, 42]. Four different shRNAs were designed with 19 nt complementary to different locations on the promoter, avoiding all known transcription factor-binding sites to maintain the endogenous transcriptional regulation of the gene and to ensure that the effect we observed was not due to purely inhibition of transcription factor binding (Fig.14.1). The shRNAs were delivered into cells using lentiviral vector, coexpressing GFP to monitor the transduction efficiency. Lentiviral delivery involves integration of the shRNA sequences under the U6-promoter into the target cell genome, from which they are then expressed and processed by the cellular machinery into effective mature small RNAs. In C166 mouse endothelial cell line, one of these shRNAs, targeting a location at 856 bp upstream from the TSS (shRNA-856), clearly downregulated VEGF-A expression as measured in mRNA and protein levels [41]. On the other hand, when targeting a location at 451 bp upstream from the TSS (shRNA-451), there was a significant upregulation of VEGF-A mRNA and protein. The two other shRNAs (targeting at 599 upstream or 335 downstream from the TSS) did not show any effect on VEGF-A levels. Thus, we showed that the expression of VEGF-A gene can be either induced or repressed in a target sequence-dependent manner. Importantly, we have also shown that this transcriptional regulation affects the expression of all of the VEGF-A isoforms [42].

In the transcriptional regulation of VEGF-A expression, the effect is observed as changes in the levels of specific histone modifications both at the shRNA target sites on the promoter and on the TSS. In untreated C166 cells, the VEGF-A

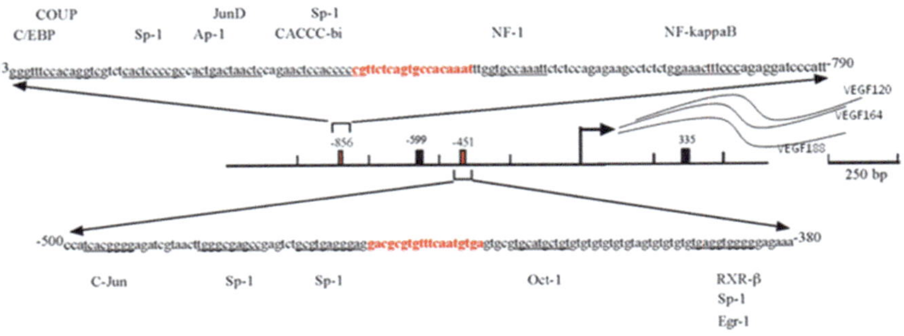

Fig. 14.1 Design of shRNAs targeting the VEGF-A promoter; the number in the names refers to the first nucleotides in the shRNA sequences relative to the TSS. The figure also shows the known and predicted binding sites of TFs on VEGF-A promoter around the shRNA-856 and shRNA-451 target sites (in *red*)

promoter contains high levels of histone modifications associated with active gene promoters, such as H3K4 methylation and H3K9 acetylation, and is therefore accessible for regulatory factors to bind their response elements [41]. As expected, the VEGF-A-downregulating shRNA-856 decreased these modifications on the promoter. On the other hand, the activating shRNA-451 increased the H3K4 methylation especially on the TSS of the gene. The other two shRNAs did not show any effect on histone modifications on the promoter or TSS. Methylation of DNA was not changed upon treatments with either of the shRNAs (shRNA-856 or shRNA-451) [42]. When treating another mouse endothelial cell line, MS1 cells, with the shRNAs, the transcriptional regulation was not observed [41]. There were also no changes in histone modification levels on VEGF-A promoter in these cells. MS1 cells express a lower basal level of VEGF-A although the histone modification pattern on the promoter seems similar as observed in C166 with chromatin immunoprecipitation. When treated with DNA methyl transferase inhibitor, the MS1 cells became responsive to the shRNA-856 and downregulation of VEGF-A expression [42]. Interestingly, the same inhibitor treatment in C166 cells abolished the effect observed with both shRNAs, suggesting that the preexisting epigenetic status and level of DNA methylation may play a role in explaining the differences between these cell lines. Studies by others have similarly observed the disappearance of the TGS effect when treating the cells first with DNA methyl transferase inhibitor or HDAC inhibitor [23]. On the other hand, in many studies DNA methylation levels do not seem to change upon transcriptional gene silencing or activation [20, 30, 39]. It may be that the TGS/TGA effect does not change the DNA methylation status but mainly affects histone modification levels, but the preexisting cell-type-specific epigenetic status with the combination of specific DNA methylation and histone modification state defines the response of the cells to the TGS/TGA stimulus. This could potentially be utilized in future experiments to target specific cell type or to limit the effect to only cells with specific basal level of expression of the gene of interest.

Corresponding to the epigenetic changes observed in VEGF-A promoter, many histone-modifying enzymes were recruited to the targeted loci in response to shRNAs. The binding of histone demethylase LSD-1 that specifically demethylates H3K4 and thus acts as a transcriptional corepressor [35] was increased in cells treated with shRNA-856 that downregulates VEGF-A expression [41]. Binding of histone acetyl transferases SRC-1 and CBP [26, 37] was decreased in TGS of mVEGF-A but increased upon TGA stimuli with shRNA-451. Factors participating in RNAi have been detected and shown to function in cell nuclei [8] and we also observed recruitment of Ago2 to the VEGF-A promoter in response to TGS and TGA [41]. Ago2 has been associated with TGS and TGA in other studies as well [12, 24, 31]. It may be that Ago2 is the key effector for the transcriptional gene regulation, whereas the other Argonaute proteins do not have such a strong involvement [6]. However, some studies have shown that Ago1 participates especially in the regulation of TGS [14, 16]. The mechanisms of TGS and TGA are not yet clearly constructed, so the pathway to silence the gene expression may differ from the activation of gene transcription, although similar components have been observed in both.

14.2.2 Off-Target Effects of shRNAs Mediating Transcriptional Regulation

The off-target effects of the TGS and TGA of VEGF-A gene were studied using microarray [13]. The transcriptional silencing of VEGF-A seems to induce more widespread effect on gene expression pattern than the TGA; this can be due to the effects of downregulating VEGF-A in the endothelial cells, where VEGF-A is an important growth factor and TGS with shRNA-856 leads to morphological changes in cell culture, such as growth arrest and reduced adherence. Therefore, endothelial cells have a more profound response to the TGS stimuli with more changes in the expression of other genes. In the microarray experiment, the expression changes induced by the shRNAs seem to correlate well with changes to VEGF-A regulation. In silico screening for putative shRNA targets revealed that there were only small changes in the expression levels of those putative target genes.

14.2.3 In Vivo Gene Therapy Using VEGF-A-Activating shRNAs

The experiments by us and others utilizing TGS and TGA in vitro were promising, and therefore we have assessed the in vivo therapeutic use of these small ncRNAs in mouse disease models. Upregulation of VEGF-A expression is desired, for example, in ischemic conditions, where injury has restricted the blood flow, and therefore

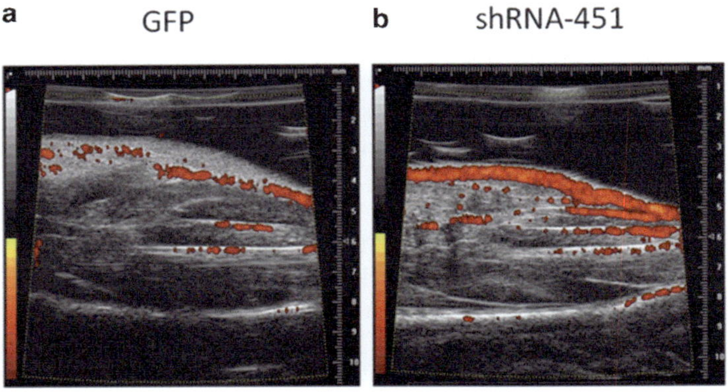

Fig. 14.2 saRNA-induced angiogenesis in mice; ultrasound imaging of the muscle vasculature at 7 d after treatment of hindlimb ischemia with GFP control (a) or shRNA-451 (b); *red color* shows blood flow in the muscle

the transfer of nutrients and oxygen to the tissue is disturbed. In the first in vivo experiment, ischemia was introduced to hindlimbs in mice and treated with shRNA-451 that activates VEGF-A gene expression [41]. After seven days, the muscles were imaged with ultrasound and the mice then sacrificed and the tissues collected. Ultrasound imaging showed a clear increase in the vasculature and improvement in the blood flow when compared to the control with GFP expression only (Fig. 14.2a–b). The shRNA-856 that downregulates VEGF-A expression showed no difference in the vasculature in the ultrasound. As securing the blood flow is essential for the tissue, the cells have different regulatory pathways to maintain the essential vasculature within the tissue, so other complementary mechanisms play a role than just the downregulation of VEGF-A expression by shRNA.

The muscle tissues were analyzed for VEGF-A expression at protein and mRNA level and for the epigenetic status of the promoter after the treatments. Similar effects as observed earlier in cell culture were seen in muscle. VEGF-A was regulated both at the protein and mRNA level with both upregulating and downregulating shRNAs, and the histone-modification changes correlated with the transcriptional regulation. Thus, both TGA and TGS were shown to function efficiently in in vivo model as well, as an "epigenetherapy" treatment.

Epigenetherapy was studied further in another mouse model, where myocardial infarction was induced in the hearts by occlusion of left main descending coronary artery [42]. The lentivirus encoding for the GFP control alone or the shRNA-451 with GFP was injected into heart tissue at the time of ligation of the artery. The shRNA treatment induced VEGF-A protein in the tissue compared to the GFP control, as monitored with ELISA. Magnetic resonance imaging of the mice revealed that 14 d after the treatment, shRNA treated mice had significantly smaller infarct size than the control group. Histology confirmed the reduction in the infarct size (Fig. 14.3) and also showed blood vessel formation in the border zone, indicating that the treatment induces sprouting of new vessels in the infarcted heart.

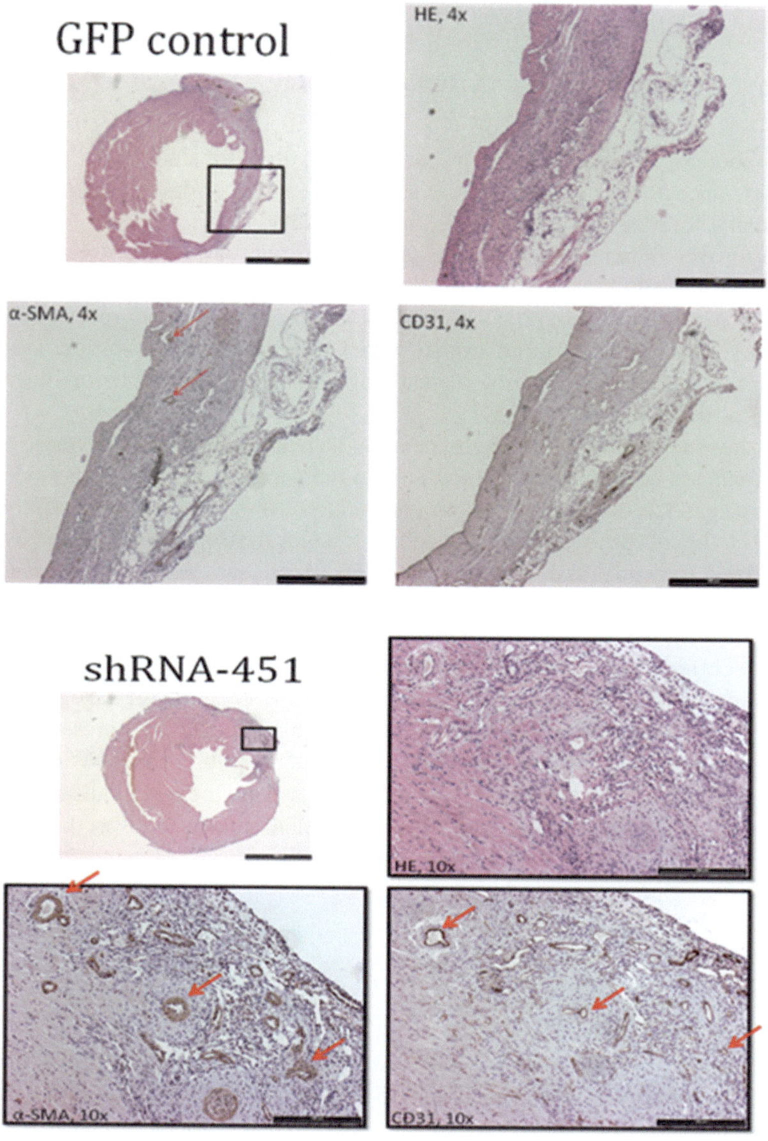

Fig. 14.3 VEGF-A upregulation by shRNA results in a therapeutic effect in myocardial infarction model as observed with histological analysis. *Red arrows* point to the formation of arteriols at the border zone

14.3 Future Perspectives for Therapeutics

14.3.1 Moving to Large Animal Models

Small laboratory animals, such as mice, are widely used for studying cardiovascular diseases since they are easy to use and not very expensive. For example, the myocardial infarction model [9] used in our studies is performed in 1–2 min and therefore over hundred mice can be operated in one day. However, the clinical relevance of results obtained from these studies is an open question since they do not fully resemble human physiological and pathophysiological processes. Therefore, the development of large animal models of myocardial infarction and using them for testing the therapeutic efficiency of saRNAs is required before beginning of clinical trials in humans.

We have developed a chronic myocardial ischemia model in pigs where bottleneck stents are used to restrict blood flow in the coronary arteries. The model will cause rest and stress ischemia and mimics human stress-induced myocardial ischemia [32]. Local gene transfer to ischemic myocardium will be done using a specialized mapping and injection catheter which can both identify ischemic regions in myocardium and then deliver therapeutic constructs to the ischemic area [17]. With the catheter-mediated local intramyocardial gene transfer, potential off-target effects will be significantly reduced and treatment effect can be targeted only to the compromised myocardium. This model will be used for the evaluation of the efficacy and safety of shRNA vectors upregulating VEGF expression in pigs [25]. Since size of the heart, catheter systems, and available non-invasive imaging tools used in clinical medicine are all applicable to pigs, the model allows building up a preclinical safety and efficacy dossier which can be used as the basis for regulatory guidelines for human applications.

14.3.2 Future Perspectives

The translational development of saRNAs for clinical use suffers from one major limitation: the promoter sequences and noncoding RNAs are not conserved between rodents, pigs, and humans. Therefore, the development and therapeutic validation for saRNAs always starts from the beginning when moving into a new species. When the mechanism of action of saRNAs is finally completely understood, hopefully in the near future, the development of large animal models for testing saRNAs will progress .In our earlier work, we have delivered our saRNA in vivo using lentiviral vectors which integrate the transgene or ncRNA to the target cell chromatin and thus maintain the expression level stability, as we have shown in mouse brains [22]. A very recent study showed that lentiviral Equine Infectious Anemia Virus (EIAV) vector expressing endostatin and angiostatin used for the treatment of macular degeneration was well tolerated in human clinical trial and maintained its activity for several years [5]. However, considering the treatment of myocardial infarction in humans by lentivirus-mediated continuous production of ncRNAs in the heart might raise several safety concerns. For example, the natural action of other important endogenous ncRNAs, such as miRNAs, could be tampered by saturation of important effector proteins involved in RNAi. Therefore, it is important to develop new effective, but transient, in vivo delivery strategies for noncoding RNAs. One possibility is to use exosomes as future delivery vehicles since they have recently been identified as important endogenous carriers of various types of RNAs and have been shown to mediate RNAi in vivo [2, 4].

14.4 Conclusions

Little over a decade after the discovery of gene regulation at the transcriptional level by noncoding RNAs, clinical trials based on promoter-targeted noncoding RNAs have not yet been initiated. During these years, many mechanistic details considering the biology of these RNAs have been unraveled, and when the therapeutic efficiency has been proved in large animal models, the first human clinical trials using saRNAs for the treatment of myocardial infarction or ischemia will be justified.

References

1. Achen MG, Jeltsch M, Kukk E, Mäkinen T, Vitali A, Wilks AF, Alitalo K, Stacker SA (1998) Vascular endothelial growth factor D (VEGF-D) is a ligand for the tyrosine kinases VEGF receptor 2 (Flk1) and VEGF receptor 3 (Flt4). Proc Natl Acad Sci U S A 95(2):548–553
2. Alvarez-Erviti L, Seow Y, Yin H, Betts C, Lakhal S, Wood MJA (2011) Delivery of siRNA to the mouse brain by systemic injection of targeted exosomes. Nat Biotechnol 29(4):3–4
3. Asahara T, Bauters C, Zheng LP, Takeshita S, Bunting S, Ferrara N, Symes JF, Isner JM (1995) Synergistic effect of vascular endothelial growth factor and basic fibroblast growth factor on angiogenesis in vivo. Circulation 92:II365–II371
4. Bryniarski K, Ptak W, Martin E, Nazimek K, Szczepanik M, Sanak M, Askenase PW (2015) Free extracellular miRNA functionally targets cells by transfecting exosomes from their companion cells. PLoS One 10(4):e0122991
5. Campochiaro PA, Lauer AK, Sohn EH, Mir TA, Naylor S, Anderton MC, Kelleher M, Harrop R, Ellis S, Mitrophanous KA (2016) Lentiviral vector gene transfer of endostatin/angiostatin for macular degeneration (GEM) study. Hum Gene Ther 28:99–111
6. Chu Y, Yue X, Younger ST, Janowski BA, Corey DR (2010) Involvement of argonaute proteins in gene silencing and activation by RNAs complementary to a noncoding transcript at the progesterone receptor promoter. Nucleic Acids Res 38(21):7736–7748
7. Forsythe JA, Jiang BH, Iyer NV, Agani F, Leung SW, Koos RD, Semenza GL (1996) Activation of vascular endothelial growth factor gene transcription by hypoxia-inducible factor 1. Mol Cell Biol 16(9):4604–4613
8. Gagnon KT, Li L, Chu Y, Janowski BA, Corey DR (2014) RNAi factors are present and active in human cell nuclei. Cell Rep 6:211–221

9. Gao E, Lei YH, Shang X, Huang ZM, Zuo L, Boucher M, Fan Q, Chuprun JK, Ma XL, Koch WJ (2010) A novel and efficient model of coronary artery ligation and myocardial infarction in the mouse. Circ Res 107(12):1445–1453
10. Hedman M, Hartikainen J, Syvänne M, Stjernvall J, Hedman A, Kivelä A, Vanninen E, Mussalo H, Kauppila E, Simula S, Närvänen O, Rantala A, Peuhkurinen K, Nieminen M, Laakso M, Ylä-Herttuala S (2003) Safety and feasibility of catheter-based local intracoronary vascular endothelial growth factor gene transfer in the prevention of postangioplasty and in-stent restenosis and in the treatment of chronic myocardial ischemia: phase II results of the Kuopio. Circulation 107(21):2677–2683
11. Hiltunen MO, Laitinen M, Turunen MP, Jeltsch M, Hartikainen J, Rissanen TT, Laukkanen J, Niemi M, Kossila M, Häkkinen TP, Kivelä A, Enholm B, Mansukoski H, Turunen AM, Alitalo K, Ylä-Herttuala S (2000) Intravascular adenovirus-mediated VEGF-C gene transfer reduces neointima formation in balloon-denuded rabbit aorta. Circulation 102(18):2262–2268
12. Hu J, Chen Z, Xia D, Wu J, Xu H, Ye Z (2012) Promoter-associated small double-stranded RNA interacts with heterogeneous nuclear ribonucleoprotein A2/B1 to induce transcriptional activation. Biochem J 447(3):407–416
13. Husso T, Turunen MP, Yla-Herttuala S (2015) Regulation of VEGF-A expression in endothelial cells by transcriptional gene activation or transcriptional gene silencing: analysis of genome wide transcriptional response. Gene Technol 4(3):e243
14. Janowski BA, Huffman KE, Schwartz JC, Ram R, Nordsell R, Shames DS, Minna JD, Corey DR (2006) Involvement of AGO1 and AGO2 in mammalian transcriptional silencing. Nat Struct Mol Biol 13(9):787–792
15. Joukov V, Pajusola K, Kaipainen A, Chilov D, Lahtinen I, Kukk E, Saksela O, Kalkkinen N, Alitalo K (1996) A novel vascular endothelial growth factor, VEGF-C, is a ligand for the Flt4 (VEGFR-3) and KDR (VEGFR-2) receptor tyrosine kinases. EMBO J 15(2):290–298
16. Kim DH, Villeneuve LM, Morris KV, Rossi JJ (2006) Argonaute-1 directs siRNA-mediated transcriptional gene silencing in human cells. Nat Struct Mol Biol 13(9):793–797
17. Lähteenvuo JE, Lähteenvuo MT, Kivelä A, Rosenlew C, Falkevall A, Klar J, Heikura T, Rissanen TT, Vähäkangas E, Korpisalo P, Enholm B, Carmeliet P, Alitalo K, Eriksson U, Ylä-Herttuala S (2009) Vascular endothelial growth factor-B induces myocardium-specific angiogenesis and arteriogenesis via vascular endothelial growth factor receptor-1- and neuropilin receptor-1-dependent mechanisms. Circulation 119(6):845–856
18. Laitinen M, Zachary I, Breier G, Pakkanen T, Hakkinen T, Luoma J, Abedi H, Risau W, Soma M, Laakso M, Martin JF, Yla-Herttuala S (1997) VEGF gene transfer reduces intimal thickening via increased production of nitric oxide in carotid arteries. Hum Gene Ther 8 (15):1737–1744
19. Leung D, Cachianes G, Kuang W, Goeddel D, Ferrara N (1989) Vascular endothelial growth factor is a secreted angiogenic mitogen. Science 246(4935):1306–1309
20. Li L-C, Okino ST, Zhao H, Pookot D, Place RF, Urakami S, Enokida H, Dahiya R (2006) Small dsRNAs induce transcriptional activation in human cells. Proc Natl Acad Sci U S A 103 (46):17337–17342
21. Maglione D, Guerriero V, Viglietto G, Delli-Bovi P, Persico MG (1991) Isolation of a human placenta cDNA coding for a protein related to the vascular permeability factor. Proc Natl Acad Sci U S A 88(20):9267–9271
22. Mäkinen PI, Koponen JK, Kärkkäinen AM, Malm TM, Pulkkinen KH, Koistinaho J, Turunen MP, Ylä-Herttuala S (2006) Stable RNA interference: comparison of U6 and H1 promoters in endothelial cells and in mouse brain. J Gene Med 8(4):433–441
23. Morris KV, Chan SW-L, Jacobsen SE, Looney DJ (2004) Small interfering RNA-induced transcriptional gene silencing in human cells. Science 305(5688):1289–1292
24. Napoli S, Pastori C, Magistri M, Carbone GM, Catapano CV (2009) Promoter-specific transcriptional interference and C-MYC gene silencing by siRNAs in human cells. EMBO J 28(12):1708–1719

25. Nurro J, Halonen PJ, Kuivanen A, Tarkia M, Saraste A, Honkonen K, Lähteenvuo J, Rissanen TT, Knuuti J, Ylä-Herttuala S (2016) AdVEGF-B $_{186}$ and AdVEGF-D $^{\Delta N \Delta C}$ induce angiogenesis and increase perfusion in porcine myocardium. Heart 102:1716–1720
26. Ogryzko VV, Schiltz RL, Russanova V, Howard BH, Nakatani Y (1996) The transcriptional coactivators p300 and CBP are histone acetyltransferases. Cell 87(5):953–959
27. Olofsson B, Pajusola K, von Euler G, Chilov D, Alitalo K, Eriksson U (1996) Genomic organization of the mouse and human genes for vascular endothelial growth factor B (VEGF-B) and characterization of a second splice isoform. J Biol Chem 271(32):19310–19317
28. Pagès G, Pouysségur J (2005) Transcriptional regulation of the vascular endothelial growth factor gene – a concert of activating factors. Cardiovasc Res 65(3):564–573
29. Park S-J, Park D-W (2012) New era of left main coronary artery treatment strategy. Cardiovasc Interv Ther 27(1):1–7
30. Park CW, Chen Z, Kren BT, Steer CJ (2004) Double-stranded siRNA targeted to the huntingtin gene does not induce DNA methylation. Biochem Biophys Res Commun 323(1):275–280
31. Portnoy V, Lin SHS, Li KH, Burlingame A, Hu Z-H, Li H, Li L-C (2016) saRNA-guided Ago2 targets the RITA complex to promoters to stimulate transcription. Cell Res 26(3):320–335
32. Rissanen TT, Nurro J, Halonen PJ, Tarkia M, Saraste A, Rannankari M, Honkonen K, Pietila M, Leppanen O, Kuivanen A, Knuuti J, Ylä-Herttuala S (2013) The bottleneck stent model for chronic myocardial ischemia and heart failure in pigs. Am J Physiol Heart Circ Physiol 305(9):H1297–H1308
33. Roy H, Bhardwaj S, Ylä-Herttuala S (2006) Biology of vascular endothelial growth factors. FEBS Lett 580(12):2879–2887
34. Rutanen J, Turunen A-M, Teittinen M, Rissanen TT, Heikura T, Koponen JK, Gruchala M, Inkala M, Jauhiainen S, Hiltunen MO, Turunen MP, Stacker SA, Achen MG, Ylä-Herttuala S (2005) Gene transfer using the mature form of VEGF-D reduces neointimal thickening through nitric oxide-dependent mechanism. Gene Ther 12(12):980–987
35. Shi Y, Lan F, Matson C, Mulligan P, Whetstine JR, Cole PA, Casero RA, Shi Y (2004) Histone demethylation mediated by the nuclear amine oxidase homolog LSD1. Cell 119(7):941–953
36. Shima DT, Kuroki M, Deutsch U, Ng YS, Adamis AP, D'Amore PA (1996) The mouse gene for vascular endothelial growth factor. J Biol Chem 271(7):3877–3883
37. Spencer TE, Jenster G, Burcin MM, Allis CD, Zhou J, Mizzen CA, McKenna NJ, Onate SA, Tsai SY, Tsai MJ, O'Malley BW (1997) Steroid receptor coactivator-1 is a histone acetyltransferase. Nature 389(6647):194–198
38. Tabas I, García-Cardeña G, Owens GK (2015) Recent insights into the cellular biology of atherosclerosis. J Cell Biol 209(1):13–22
39. Ting AH, Schuebel KE, Herman JG, Baylin SB (2005) Short double-stranded RNA induces transcriptional gene silencing in human cancer cells in the absence of DNA methylation. Nat Genet 37(8):906–910
40. Tischer E, Mitchell R, Hartman T, Silva M, Gospodarowicz D, Fiddes J, Abraham J (1991) The human gene for vascular endothelial growth factor. J Biol Chem 266(18):11947–11954
41. Turunen MP, Lehtola T, Heinonen SE, Assefa GS, Korpisalo P, Girnary R, Glass CK, Väisänen S, Ylä-Herttuala S (2009) Efficient regulation of VEGF expression by promoter-targeted lentiviral shRNAs based on epigenetic mechanism a novel example of epigenetherapy. Circ Res 105(6):604–609
42. Turunen MP, Husso T, Musthafa H, Laidinen S, Dragneva G, Laham-Karam N, Honkanen S, Paakinaho A, Laakkonen JP, Gao E, Vihinen-Ranta M, Liimatainen T, Ylä-Herttuala S (2014) Epigenetic upregulation of endogenous VEGF-A reduces myocardial infarct size in mice. PLoS One 9(2): e89979
43. Ylä-Herttuala S, Bridges C, Katz M, Korpisalo P (2017) Angiogenic gene therapy in cardiovascular diseases: dream or vision? Eur Heart J 0:1-7

Chapter 15
Suppression of Prostate Cancer Metastasis by *DPYSL3*-Targeted saRNA

Benyi Li and Changlin Li

Abstract Metastasis is the sole cause of cancer death and there is no curable means in clinic. Cellular protein CRMP4 (*DPYSL3* gene) was previously defined as a metastasis suppressor in human prostate cancers since its expression is dramatically reduced in lymphatic metastatic diseases and *DPYSL3* overexpression in prostate cancer cells significantly suppressed cancer cell migration and invasion. To develop a CRMP4-based antimetastasis therapeutic approach, the small activating RNA (saRNA) technique was utilized to enhance CRMP4 expression in prostate cancer cells. A total of 14 saRNAs were synthesized and screened in multiple prostate cancer cell lines. Two saRNAs targeting the isoform-2 promoter region were determined to have significant activating effect on *DPYSL3* gene expression at the mRNA and protein levels. These saRNA also largely reduced prostate cancer cell migration and invasion in vitro and in vivo. Most significantly, PSMA aptamer-mediated prostate cancer cell homing of these saRNAs blocked distal metastasis in an orthotopic nude mouse model. In conclusion, our data demonstrated that saRNA-based *DPYSL3* gene enhancement is capable of suppressing tumor metastasis in prostate cancer, which provides a potential therapeutic approach for cancer management.

Keywords Prostate cancer • DPYSL3 • CRMP4 • Small activating RNA • RNAa

15.1 Metastasis Is a Key Obstacle in Prostate Cancer Management

In the US, prostate cancer remains the most diagnosed cancer in men and the second leading cause after lung cancer for cancer-related death [40]. Prostate cancer deaths are always due to systemic metastases, while systemic metastasis often occurs

B. Li (✉)
Department of Urology, University of Kansas Medical Center, Kansas City, KS 66160, USA
e-mail: bli@kumc.edu

C. Li
Institute of Precision Medicine, Jining Medical University, Jining 272000, China

L.-C. Li (ed.), *RNA Activation*, Advances in Experimental Medicine and Biology 983, DOI 10.1007/978-981-10-4310-9_15

following local therapy failure in high-risk patients [11]. High-risk prostate cancers are defined as PSA >20 ng/ml, biopsy Gleason score 8–10, or clinical stage >T2-3a [2]. Prostate cancers often migrate to extraprostatic tissue and local lymph nodes before distant organ metastasis. It was reported that up to 12% of first-diagnosed patients are lymph node positive (stage N1M0) [6] and that about 79% of biochemical relapse patients (PSA rising after local therapy) had at least one positive lymph node [27]. Although patients with locally advanced high-risk prostate cancers are recommended to receive extended pelvic lymph node dissection of involved lymph nodes, and/or whole pelvic nodal radiation therapy, early biochemical relapse is not uncommon in these patients [7, 17], indicating residual lymph nodes with micrometastatic diseases. The ten-year progression-free survival probabilities were 79% for organ-confined disease, but only 12% for disease with lymph node metastases [35]. Most significantly, radial distance of extraprostatic tumor extension is significantly associated with prostate specific antigen (PSA) recurrence in patients with locally advanced disease [43], suggesting that local invasion/extension is one of the key issues in managing prostate cancers. In fact, current imaging technology has its limit in identifying all lymph nodes with micrometastasis, but a single lymph node metastasis increased the risk of cancer-specific death almost fourfold and patients with two or more lymph node involvement had another twofold increase of risk compared to patients with only one positive lymph node [3–5]. Therefore, suppressing metastasis is key to achieving cancer-free survival for locally advanced high-risk patients and to reducing or delaying systemic metastasis.

15.2 *DPYSL3* (CRMP4) Is a Metastasis Suppressor in Prostate Cancer

Dihydropyrimidinase-like 3 (*DPYSL3*), also termed as Collapsin Response Mediator Protein-4 (CRMP4), belongs to the CRMP family of cytosolic phosphoproteins, which are involved in semaphorin/collapsin-induced cellular events [36]. So far, five members of the CRMP gene family (CRMP1-5) have been cloned and they have 50–70% sequence homology [14, 15, 19, 33, 36]. Recent studies indicated that CRMP1 is an invasion suppressor in human glioma and lung cancer cells, and its reduced expression correlates with poor clinical outcomes in non-small-cell lung cancer [29, 38, 39]. CRMP-2 and CRMP-5 were found to be extensively expressed in colorectal cancers and high-grade lung neuroendocrine carcinomas, respectively [28, 48].

In searching for metastasis-associated proteins, we found recently published literature stating that CRMP4 expression was dramatically reduced in metastatic prostate cancers compared to benign prostate tissue or localized prostate cancers [16]. This finding is consistent with cDNA microarray data as shown in Oncomine database (www.oncomine.com) that CRMP4 mRNA expression is significantly downregulated in metastatic prostate cancers compared to primary cancers and is

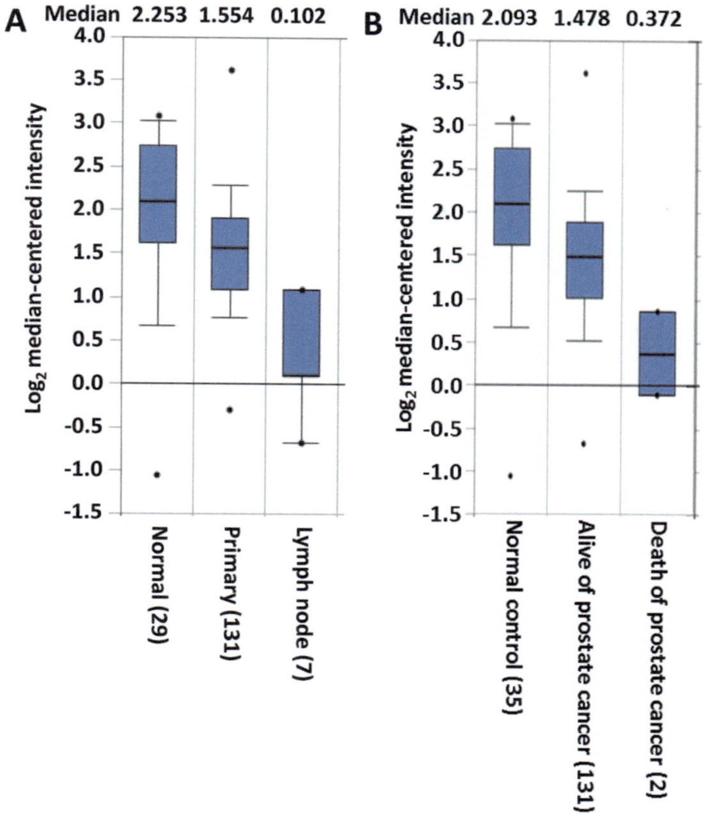

Fig. 15.1 ***DPYSL3* expression in prostate cancers.** Data mining was performed on the Oncomine™ database with Taylor Prostate 3 dataset [44]

associated with patient survival [44] (Fig. 15.1). Meanwhile, we demonstrated that overexpression of CRMP4 protein in prostate cancer cells strongly suppressed cell invasion in vitro and xenograft tumor metastasis in vivo [16, 24]. These data clearly indicate that CRMP4 is a tumor metastasis suppressor in prostate cancers.

15.3 Small Activating RNA Enhances *DPYSL3* Gene Expression

It has been demonstrated that small double-stranded RNAs targeting gene promoter region could induce prolonged gene activation at the transcriptional level [8, 23, 26, 31, 47]. This phenomenon is termed as RNA activation (RNAa) and is evolutionarily conservative [20, 32]. The small RNAs are referred as small activating RNAs (saRNA). Mechanistic studies revealed that RNAa machinery in cells is related to

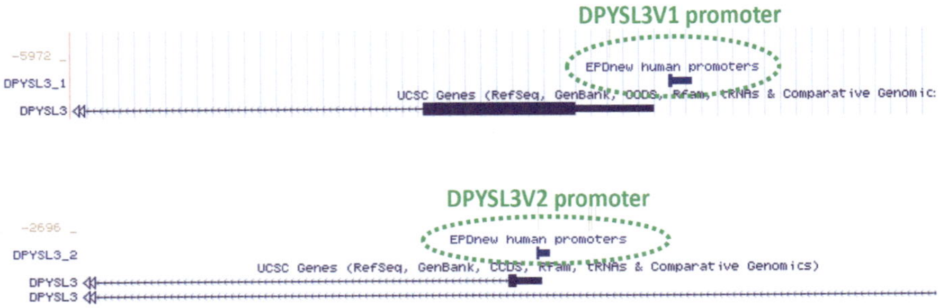

Fig. 15.2 ***DPYSL3 v1–v2* promoter location**. Genomic information was extracted from Genome Browser on UCSC website (https://genome.ucsc.edu)

RNA interferencing (RNAi), but with fundamental differences [20, 21, 37]. Also, saRNAs induce epigenetic changes on the target promoter, resulting in alterations of gene expression [22, 45]. Therefore, it is feasible that saRNAs with optimal properties can be used as therapeutic agents to induce de novo activation of silenced tumor metastasis suppressor genes, like *DPYSL3*.

Human *DPYSL3* gene has two transcriptional variants due to distinct promoter usage, as illustrated in Fig. 15.2. These two isoforms of *DPYSL3* gene encode two proteins that differ in their N-terminal amino acid sequence of exon 1 region. The isoform-1 has 2055 nt in cDNA nucleotide sequence while isoform-2 is 1713 nt. These isoforms are translated to proteins of CRMP4b (*DPYSL3v1*, 684 aa, 75 KD) and CRMP4a (*DPYSL3v2*, 570 aa, 64 KD). We examined the expression profiles of these two isoforms in human prostate cancers and prostate cancer cell lines. In the online database Oncomine, 9 out of 14 published datasets showed a significant reduction of *DPYSL3* gene expression in malignant tissues compared to the benign tissues [24]. Further analysis of one published dataset revealed that *DPYSL3* expression was largely reduced in lymph node metastasis (Fig. 15.1a), which was associated with cancer death (Fig. 15.1b).

To understand if *DPYSL3* isoforms are differently expressed in prostate cancer tissues, we conducted a real-time PCR analysis of prostate tissues obtained from radical prostatectomy. Quantitative data revealed that *DPYSL3v2* transcript was the dominant one with a remarkably higher level than *DPYSL3v1* transcript. However, *DPYSL3v2* levels were significantly lower in malignant tissues compared to that in case-matched surrounding benign tissues [24]. These results were consistent with our previous report [16].

To enhance *DPYSL3* gene expression, multiple saRNA molecules were designed to target *DPYSL3v2* promoter. The saRNA targeting sites are shown in Fig. 15.3. Four out of 14 saRNAs targeting the *DPYSL3v2* promoter significantly enhanced gene expression, especially, two out (saV2-5 and saV2-9) of these four saRNAs exerted similar enhancing effect in all cell lines [24]. Their sense sequences are as follows: V2-5 5′-AGGCAGATGCAAAGAGGAA-3′ and V2-9 5′-GCAGCATTCATGTTCTTTC-3′. A luciferase reporter construct driven by a 1250-bp segment (–1275/–25 from the transcription start site) of DPYSL3v2

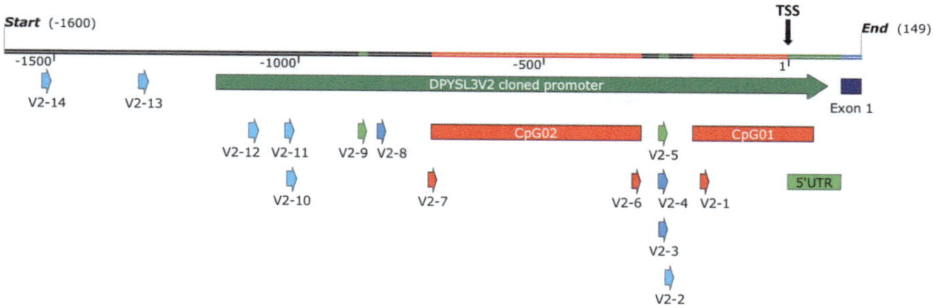

Fig. 15.3 Locations of the saRNAs on *DPYSL3v2* promoter region. Small activating RNAs were designed as described previously [20]. Their locations on the –2000/+100 promoter related to the TSS site are illustrated. Two CpG islands were proposed using an online MethPrimer program. (http://www.urogene.org/methprimer/)

promoter [24] was created to verify the saRNA-induced *DPYSL3* promoter activation. The lentiviral luciferase reporter (DPYSL3v2p-LUC) was indeed significantly activated when tested in PC-3 cells, of which the saV2-9 saRNA displayed the most potent effect compared to saV2-5. This effect was further confirmed by deleting the targeting segments of saV2-5 (v2-5p-LUC) and saV2-9 (v2-9l-LUC) saRNAs on the promoter individually.

We tested the antimigration activity of the saRNAs in PC-3 cells [41, 42, 46]. The V2-5 and v2-9 saRNAs were transfected into the cells. The Boyden chamber assay coupled with a basement membrane matrix preparation, *Matrigel*, as the matrix barrier, was used as the migration assay [16]. The saV2-9 saRNA significantly reduced cell migration, while cotransfection of *DPYSL3* siRNAs abolished saV2-9-induced suppressing effect on cell migration [24], demonstrating the significance of *DPYSL3v2* gene upregulation as a critical event in suppression of cell migration.

15.4 PSMA Aptamer-Mediated Prostate Cancer-Specific *DPYSL3* Gene Expression in Vivo

Successful delivery of therapeutic agents efficiently and specifically to the target tissue is desirable for treatment of human cancers. Nucleic acid aptamers are RNA or DNA molecules that can fold into unique conformational structure and interact with protein targets due to their complementary surface charge and shape [50]. Aptamers are short, single-strand oligonucleotides without immunogenicity or toxicity [13]. The sequence-defined aptamers can be synthesized in large batches and can be conjugated onto other particles such as polymer-based nanocarriers to achieve targeted cell-specific delivery [12].

Prostate-specific membrane antigen (PSMA) is a well-known tumor antigen [18]. It is primarily expressed on the surface of prostate cancer epithelial cells and also is highly expressed in metastatic prostate cancer cells and the microvasculature of most studied tumors. This restricted expression pattern warranted it as a promising target for the diagnosis, detection, and management of prostate cancer. As such, PSMA is currently being used for molecular imaging, cancer vaccine development, and targeted drug delivery in prostate cancers [1]. Particularly, an RNA-based aptamer targeting human PSMA has been developed, which interacts specifically with the PSMA extracellular domain and is being widely used in targeted drug delivery and molecular imaging [1, 25]. We have successfully used PSMA aptamer for prostate-specific delivery of therapeutic nanoparticles to prostate cancer cells [9, 49, 51]. Most significantly, recent reports defined two truncated PSMA aptamers with only 39–41mer in length but retaining the PSMA-binding function [10, 34]. These truncated PSMA aptamers make the chemical synthesis much easier to carry out and binding less nonspecific [30].

In this report, we utilized the second-generation aptamer (A10-3.2) to synthesize aptamer-saRNA chimera for in vivo delivery (Fig. 15.4). PSMA-dependent cellular intake was confirmed with a fluorescent-labeled APT-saV2-9 conjugate in PSMA-positive prostate cancer C4-2 [24]. We then generated orthotopic xenografts in nude mice with C4-2 cells, which were stably infected with lentiviral CMV–LUC reporter to express luciferase for in vivo imaging. Once the orthotopic tumor was established, intraperitoneal injection of the A10-3.2/V2-9 saRNA was performed at 1.0 nM per treatment for ten consecutive days. In comparison to the A10-3.2–V2-9saRNA chimera treatment group, animals in the scramble control saRNA group displayed significantly more metastatic lessons at regional lymph nodes and lungs [24]. These data strongly suggest that PSMA-mediated *DPYSL3* saRNAs have a potent inhibitory effect on prostate cancer metastasis.

15.5 RhoA Signal Pathway Is Involved in *DPYSL3*-Mediated Metastasis Suppression

In the study, we analyzed the involvement of RhoA signal pathways in *DPYSL3* (CRMP4 protein)-mediated metastasis suppression. For this purpose, we evaluated the expression levels of RhoA signal pathway proteins in prostate cancer cell lines (Fig. 15.5a). Our data revealed that most of the cell lines tested showed varying levels of RhoA signaling proteins except a lower level of Rac1 in LAPC-4, LNCaP, and C4-2 cells. We determined the CRMP4 interaction with RhoA family proteins in PC-3 cells. As shown in Fig. 15.5b, CRMP4a (*DPYSL3b*) overexpression in PC-3 cells interacted with only RhoA protein but not with other members including Rho C, cdc42, and Rac1. These data suggest that CRMP4a might be acting on RhoA to suppress prostate cancer migration or invasion.

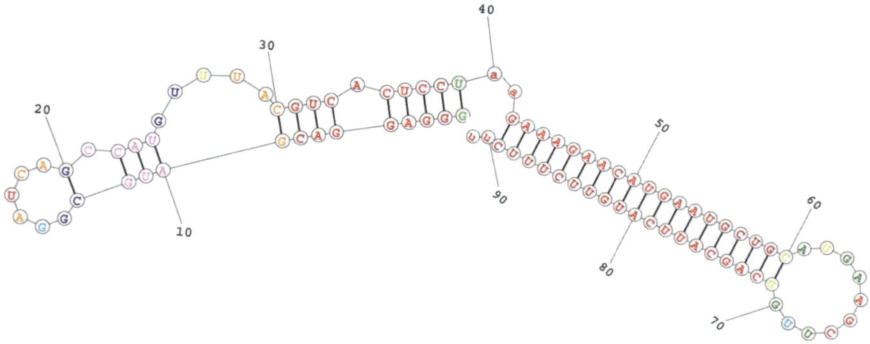

Fig. 15.4 A10-3.2 linked to V2-9 saRNA. A chimera of the truncated PSMA aptamer A10-3.2 with the V2-9 saRNA was shown and the folding structure was predicted with the online tool at http://rna.urmc.rochester.edu/RNAstructureWeb/

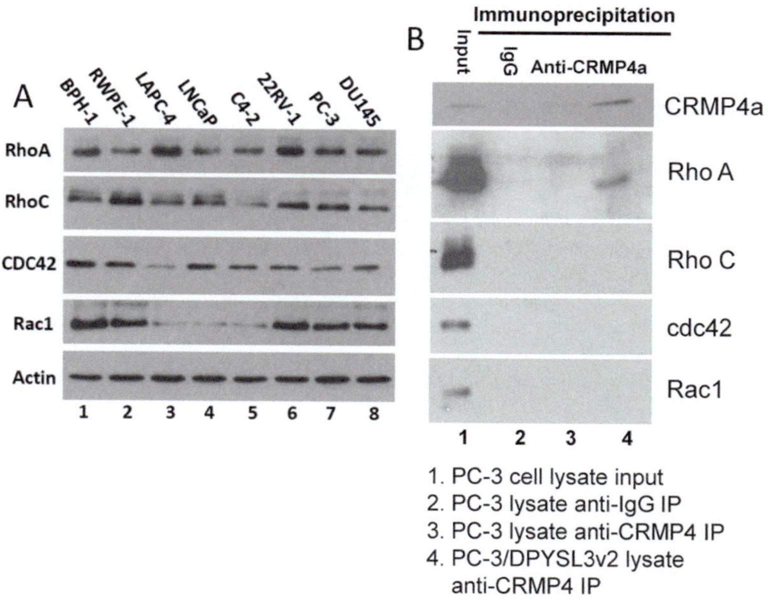

Fig. 15.5 CRMP4 interacts with RhoA in PC3 cells. (**a**) The expression levels of RhoA family proteins were analyzed in multiple prostate-derived benign and malignant cell lines by Western blot assays. (**b**) Coimmunoprecipitation assay was conducted with cellular lysates from PC-3 cells with or without *DPYSL3v2* gene overexpression

15.6 Conclusion

It has been shown that the saRNA approach is a powerful means to enhance endogenous gene expression in vivo, such as tumor suppressor genes that are often downregulated due to transcriptional repression. Our data have demonstrated here that targeting the *DPYSL3* gene promoter with saRNAs that have optimized sequences enhanced *DPYSL3* gene expression in prostate cancer cells, leading to reduced cancer cell migration in vitro and distal metastasis in vivo. This study suggested that the positive saRNAs are potential novel therapeutic agents in prostate cancer management.

References

1. Bander NH (2006) Technology insight: monoclonal antibody imaging of prostate cancer. Nat Clin Pract Urol 3:216–225
2. Bastian PJ, Boorjian SA, Bossi A, Briganti A, Heidenreich A, Freedland SJ, Montorsi F, Roach M 3rd, Schroder F, Van Poppel H, Stief CG, Stephenson AJ, Zelefsky MJ (2012) High-risk prostate cancer: from definition to contemporary management. Eur Urol 61:1096–1106
3. Boorjian SA, Thompson RH, Siddiqui S, Bagniewski S, Bergstralh EJ, Karnes RJ, Frank I, Blute ML (2007) Long-term outcome after radical prostatectomy for patients with lymph node positive prostate cancer in the prostate specific antigen era. J Urol 178:864–870. discussion 870-1
4. Briganti A, Blute ML, Eastham JH, Graefen M, Heidenreich A, Karnes JR, Montorsi F, Studer UE (2009a) Pelvic lymph node dissection in prostate cancer. Eur Urol 55:1251–1265
5. Briganti A, Karnes JR, Da Pozzo LF, Cozzarini C, Gallina A, Suardi N, Bianchi M, Freschi M, Doglioni C, Fazio F, Rigatti P, Montorsi F, Blute ML (2009b) Two positive nodes represent a significant cut-off value for cancer-specific survival in patients with node positive prostate cancer. A new proposal based on a two-institution experience on 703 consecutive N+ patients treated with radical prostatectomy, extended pelvic lymph node dissection and adjuvant therapy. Eur Urol 55:261–270
6. Cai T, Nesi G, Tinacci G, Giubilei G, Gavazzi A, Mondaini N, Zini E, Bartoletti R (2011) Clinical importance of lymph node density in predicting outcome of prostate cancer patients. J Surg Res 167:267–272
7. Castellucci P, Ceci F, Graziani T, Schiavina R, Brunocilla E, Mazzarotto R, Pettinato C, Celli M, Lodi F, Fanti S (2014) Early biochemical relapse after radical prostatectomy: which prostate cancer patients may benefit from a restaging 11C-Choline PET/CT scan before salvage radiation therapy? J Nucl Med 55:1424–1429
8. Chen Z, Place RF, Jia ZJ, Pookot D, Dahiya R, Li LC (2008) Antitumor effect of dsRNA-induced p21(WAF1/CIP1) gene activation in human bladder cancer cells. Mol Cancer Ther 7:698–703
9. Chen R, Zhao Y, Huang Y, Yang Q, Zeng X, Jiang W, Liu J, Thrasher JB, Forrest ML, Li B (2015) Nanomicellar TGX221 blocks xenograft tumor growth of prostate cancer in nude mice. Prostate 75:593–602
10. Dassie JP, Liu XY, Thomas GS, Whitaker RM, Thiel KW, Stockdale KR, Meyerholz DK, Mccaffrey AP, Mcnamara JO 2nd, Giangrande PH (2009) Systemic administration of optimized aptamer-siRNA chimeras promotes regression of PSMA-expressing tumors. Nat Biotechnol 27:839–849
11. Datta K, Muders M, Zhang H, Tindall DJ (2010) Mechanism of lymph node metastasis in prostate cancer. Future Oncol 6:823–836
12. Farokhzad OC, Cheng J, Teply BA, Sherifi I, Jon S, Kantoff PW, Richie JP, Langer R (2006a) Targeted nanoparticle-aptamer bioconjugates for cancer chemotherapy in vivo. Proc Natl Acad Sci U S A 103:6315–6320
13. Farokhzad OC, Karp JM, Langer R (2006b) Nanoparticle-aptamer bioconjugates for cancer targeting. Expert Opin Drug Deliv 3:311–324

14. Fukada M, Watakabe I, Yuasa-Kawada J, Kawachi H, Kuroiwa A, Matsuda Y, Noda M (2000) Molecular characterization of CRMP5, a novel member of the collapsin response mediator protein family. J Biol Chem 275:37957–37965
15. Gaetano C, Matsuo T, Thiele CJ (1997) Identification and characterization of a retinoic acid-regulated human homologue of the unc-33-like phosphoprotein gene (hUlip) from neuroblastoma cells. J Biol Chem 272:12195–12201
16. Gao X, Pang J, Li LY, Liu WP, Di JM, Sun QP, Fang YQ, Liu XP, Pu XY, He D, Li MT, Su ZL, Li BY (2010) Expression profiling identifies new function of collapsin response mediator protein 4 as a metastasis-suppressor in prostate cancer. Oncogene 29:4555–4566
17. Gao L, Yang L, Lv X, Bu S, Wan F, Qian S, Wei Q, Han P, Fan T (2014) A systematic review and meta-analysis of comparative studies on the efficacy of extended pelvic lymph node dissection in patients with clinically localized prostatic carcinoma. J Cancer Res Clin Oncol 140:243–256
18. Ghosh A, Heston WD (2004) Tumor target prostate-specific membrane antigen (PSMA) and its regulation in prostate cancer. J Cell Biochem 91:528–539
19. Goshima Y, Nakamura F, Strittmatter P, Strittmatter SM (1995) Collapsin-induced growth cone collapse mediated by an intracellular protein related to UNC-33. Nature 376:509–514
20. Huang V, Qin Y, Wang J, Wang X, Place RF, LIN G, Lue TF, Li LC (2010) RNAa is conserved in mammalian cells. PLoS One 5:e8848
21. Janowski BA, Younger ST, Hardy DB, Ram R, Huffman KE, COREY DR (2007) Activating gene expression in mammalian cells with promoter-targeted duplex RNAs. Nat Chem Biol 3:166–173
22. Li LC (2008) The multifaceted small RNAs. RNA Biol 5:61–64
23. Li LC, Okino ST, Zhao H, Pookot D, Place RF, Urakami S, Enokida H, Dahiya R (2006) Small dsRNAs induce transcriptional activation in human cells. Proc Natl Acad Sci U S A 103:17337–17342
24. Li C, Jiang W, HU Q, Li LC, Dong L, Chen R, Zhang Y, Tang Y, Thrasher JB, Liu CB, Li B (2016) Enhancing DPYSL3 gene expression via a promoter-targeted small activating RNA approach suppresses cancer cell motility and metastasis. Oncotarget 7:22893–22910
25. Lupold SE, Hicke BJ, Lin Y, Coffey DS (2002) Identification and characterization of nuclease-stabilized RNA molecules that bind human prostate cancer cells via the prostate-specific membrane antigen. Cancer Res 62:4029–4033
26. Mao Q, Li Y, Zheng X, Yang K, Shen H, Qin J, Bai Y, Kong D, Jia X, Xie L (2008) Upregulation of E-cadherin by small activating RNA inhibits cell invasion and migration in 5637 human bladder cancer cells. Biochem Biophys Res Commun 375:566–570
27. Meijer HJ, Van Lin EN, Debats OA, Witjes JA, Span PN, Kaanders JH, Barentsz JO (2012) High occurrence of aberrant lymph node spread on magnetic resonance lymphography in prostate cancer patients with a biochemical recurrence after radical prostatectomy. Int J Radiat Oncol Biol Phys 82:1405–1410
28. Meyronet D, Massoma P, Thivolet F, Chalabreysse L, Rogemond V, Schlama A, Honnorat J, Thomasset N (2008) Extensive expression of collapsin response mediator protein 5 (CRMP5) is a specific marker of high-grade lung neuroendocrine carcinoma. Am J Surg Pathol 32:1699–1708
29. Mukherjee J, Desouza LV, Micallef J, Karim Z, Croul S, Siu KW, Guha A (2009) Loss of collapsin response mediator Protein1, as detected by iTRAQ analysis, promotes invasion of human gliomas expressing mutant EGFRvIII. Cancer Res 69:8545–8554
30. Pan W, Clawson GA (2009) The shorter the better: reducing fixed primer regions of oligonucleotide libraries for aptamer selection. Molecules 14:1353–1369
31. Place RF, Li LC, pookot D, Noonan EJ, Dahiya R (2008) MicroRNA-373 induces expression of genes with complementary promoter sequences. Proc Natl Acad Sci U S A 105:1608–1613
32. Place RF, Noonan EJ, Foldes-Papp Z, Li LC (2010) Defining features and exploring chemical modifications to manipulate RNAa activity. Curr Pharm Biotechnol 11:518–526

33. Quach TT, Mosinger B Jr, Ricard D, Copeland NG, Gilbert DJ, Jenkins NA, Stankoff B, Honnorat J, Belin MF, Kolattukudy P (2000) Collapsin response mediator protein-3/unc-33-like protein-4 gene: organization, chromosomal mapping and expression in the developing mouse brain. Gene 242:175–182
34. Rockey WM, Hernandez FJ, Huang SY, Cao S, Howell CA, Thomas GS, Liu XY, Lapteva N, Spencer DM, Mcnamara JO, Zou X, Chen SJ, Giangrande PH (2011) Rational truncation of an RNA aptamer to prostate-specific membrane antigen using computational structural modeling. Nucleic Acid Ther 21:299–314
35. Roehl KA, Han M, Ramos CG, Antenor JA, Catalona WJ (2004) Cancer progression and survival rates following anatomical radical retropubic prostatectomy in 3,478 consecutive patients: long-term results. J Urol 172:910–914
36. Schmidt EF, Strittmatter SM (2007) The CRMP family of proteins and their role in Sema3A signaling. Adv Exp Med Biol 600:1–11
37. Schwartz JC, Younger ST, Nguyen NB, Hardy DB, Monia BP, Corey DR, Janowski BA (2008) Antisense transcripts are targets for activating small RNAs. Nat Struct Mol Biol 15:842–848
38. Shih JY, Yang SC, Hong TM, Yuan A, Chen JJ, YU CJ, Chang YL, Lee YC, Peck K, Wu CW, Yang PC (2001) Collapsin response mediator protein-1 and the invasion and metastasis of cancer cells. J Natl Cancer Inst 93:1392–1400
39. Shih JY, Lee YC, Yang SC, Hong TM, Huang CY, Yang PC (2003) Collapsin response mediator protein-1: a novel invasion-suppressor gene. Clin Exp Metastasis 20:69–76
40. Siegel RL, Miller KD, Jemal A (2015) Cancer statistics, 2015. CA Cancer J Clin 65:5–29
41. Sobel RE, Sadar MD (2005a) Cell lines used in prostate cancer research: a compendium of old and new lines—part 1. J Urol 173:342–359
42. Sobel RE, Sadar MD (2005b) Cell lines used in prostate cancer research: a compendium of old and new lines—part 2. J Urol 173:360–372
43. Sung MT, Lin H, Koch MO, Davidson DD, Cheng L (2007) Radial distance of extraprostatic extension measured by ocular micrometer is an independent predictor of prostate-specific antigen recurrence: a new proposal for the substaging of pT3a prostate cancer. Am J Surg Pathol 31:311–318
44. Taylor BS, Schultz N, Hieronymus H, Gopalan A, Xiao Y, Carver BS, Arora VK, Kaushik P, Cerami E, Reva B, Antipin Y, Mitsiades N, Landers T, Dolgalev I, Major JE, Wilson M, Socci ND, Lash AE, Heguy A, Eastham JA, Scher HI, Reuter VE, Scardino PT, Sander C, Sawyers CL, Gerald WL (2010) Integrative genomic profiling of human prostate cancer. Cancer Cell 18:11–22
45. Turunen MP, Lehtola T, Heinonen SE, Assefa GS, Korpisalo P, Girnary R, Glass CK, Vaisanen S, YLA-Herttuala S (2009) Efficient regulation of VEGF expression by promoter-targeted lentiviral shRNAs based on epigenetic mechanism: a novel example of epigenetherapy. Circ Res 105:604–609
46. Van Bokhoven A, Varella-Garcia M, Korch C, Johannes WU, Smith EE, Miller HL, Nordeen SK, Miller GJ, Lucia MS (2003) Molecular characterization of human prostate carcinoma cell lines. Prostate 57:205–225
47. Whitson JM, Noonan EJ, Pookot D, Place RF, Dahiya R (2009) Double stranded-RNA-mediated activation of *p21*-gene-induced apoptosis and cell cycle arrest in renal cell carcinoma. Int J Cancer 125:446–452
48. Wu CC, Chen HC, Chen SJ, Liu HP, Hsieh YY, Yu CJ, Tang R, Hsieh LL, Yu JS, Chang YS (2008) Identification of collapsin response mediator protein-2 as a potential marker of colorectal carcinoma by comparative analysis of cancer cell secretomes. Proteomics 8:316–332
49. Yang J, Xie SX, Huang Y, Ling M, Liu J, Ran Y, Wang Y, Thrasher JB, Berkland C, Li B (2012) Prostate-targeted biodegradable nanoparticles loaded with androgen receptor silencing constructs eradicate xenograft tumors in mice. Nanomedicine (London) 7:1297–1309
50. Zhang H, Zhou L, Zhu Z, Yang C (2016) Recent progress in aptamer-based functional probes for bioanalysis and biomedicine. Chemistry 22:9886–9900
51. Zhao Y, Duan S, Zeng X, Liu C, Davies NM, Li B, Forrest ML (2012) Prodrug strategy for PSMA-targeted delivery of TGX-221 to prostate cancer cells. Mol Pharm 9:1705–1716

Chapter 16
Development of Therapeutic dsP21-322 for Cancer Treatment

Moo Rim Kang, Gongcheng Li, Tiejun Pan, Jin-Chun Xing, and Long-Cheng Li

Abstract Small activating RNAs (saRNAs) are a class of artificially designed short duplex RNAs targeted at the promoter of a particular gene to upregulate its expression via a mechanism known as RNA activation (RNAa) and hold great promise for treating a wide variety of diseases including those undruggable by conventional therapies. The therapeutic benefits of saRNAs have been demonstrated in a number of preclinical studies carried out in different disease models including cancer. With many tumor suppressor genes (TSGs) downregulated due to either epigenetic mechanisms or haploinsufficiency resulting from deletion/mutation, cancer is an ideal disease space for saRNA therapeutics which can restore the expression of TSGs via epigenetic reprogramming. The $p21^{WAF1/CIP}$ gene is a TSG frequently downregulated in cancer and an saRNA for $p21^{WAF1/CIP}$ known as dsP21-322 has been identified to be a sequence-specific $p21^{WAF1/CIP}$ activator in a number of cancer types. In this chapter, we review preclinical development of medicinal dsP21-322 for cancer, especially prostate cancer and bladder cancer, and highlight its potential for further clinical development.

Keywords RNAa • saRNA • $p21^{WAF1/CIP}$ • Cancer • Chemical modification • Lipid nanoparticle

---

Moo Rim Kang and Gongcheng Li contributed equally to this work.

M.R. Kang
Ractigen Therapeutics, Nantong, China

G. Li • T. Pan
Department of Urology, Wuhan General Hospital, Guangzhou Command PLA, Wuhan, China

J.-C. Xing
Department of Urology, The First Affiliated Hospital of Xiamen University, Xiamen, China

L.-C. Li (✉)
Department of Urology, The First Affiliated Hospital of Xiamen University, Xiamen, China

Laboratory of Molecular Medicine, Peking Union Medical College Hospital, Chinese Academy of Medical Sciences, Beijing, China
e-mail: long@urogene.org

L.-C. Li (ed.), *RNA Activation*, Advances in Experimental Medicine and Biology 983, DOI 10.1007/978-981-10-4310-9_16

16.1 Introduction

Small RNA molecules, such as short interfering RNAs (siRNAs) and microRNAs (miRNAs), have been known to mainly suppress gene expression via the RNA interference (RNAi) mechanism by binding to mRNA and consequently leading to degradation of mRNA and/or inhibition of translation. However, increasing evidence reveals that small RNAs partnering with Argonaute (AGO) proteins could induce sequence-specific transcriptional gene activation by targeting gene promoters, a phenomenon known as RNA activation (RNAa) [16, 19, 21]. Such promoter-targeting and transcription-activating small RNAs are referred to as small activating RNA (saRNA) [19, 33]. Since the first report of RNAa phenomenon in 2006 by Li et al. [19], worldwide groups have published over 50 papers reporting RNAa of over 30 different genes of therapeutic interest, many of which are tumor suppressor genes (TSGs).

Cancer is the result of an imbalance between overactive oncogenes and inactive TSGs and represents a disease space well suited for RNAa-based treatment. A number of preclinical studies have already demonstrated, in different cancer types, antitumor efficacy of targeted activation of TSGs by saRNAs and one of them has edged its way into clinical trial.

The p53–p21 axis is a fundamental molecular pathway that regulates cell growth and death, and alterations in this pathway represent one of the major contributors to tumorigenesis [35]. Upon DNA damage, p53 arrests cell cycle progression and induces senescence or apoptosis by transcriptional activation of various target genes, thereby enforcing a fail-safe mechanism to counter uncontrolled proliferation [9]. $p21^{WAF1/CIP}$ (p21), a founding member of the Cip/Kip family of cyclin-dependent kinase (CDK) inhibitors (CDKI), acts as a critical downstream effector of the p53 tumor suppressor pathway [1] and exerts antiproliferative activity mainly by binding to and inhibiting the kinase activity of CDKs, resulting in suppressed phosphorylation of Rb protein by the cyclin/CDK complexes [13]. In addition to its function as a CDKI, p21 also interacts with the DNA replication factor PCNA, exerting additional control of cell cycle progression, DNA replication, and DNA repair [34]. Although p53 appears to be an ideal RNAa target for treating cancer, it is frequently affected by genetic aberrations in cancer such as mutation and deletion, leading to a protein product with abnormal function [27, 28]. In contrast, p21 is rarely affected by structural aberrations or mutations, but loss of p21 expression is frequently associated with the progression of cancer [6, 29]. Ectopic overexpression of p21 has been shown to inhibit tumor growth both in vitro and in vivo [11, 31, 39]. Thus, p21 can serve as a surrogate target for correcting the function of the p53-p21 axis.

In this chapter, we review studies from our group and the literature on in vitro and in vivo investigation of a p21 saRNA named dsP21-322 which can induce p21 expression in a battery of cancer cells and the step-wise development of medicinal dsP21-322 for the treatment of prostate cancer and bladder cancer.

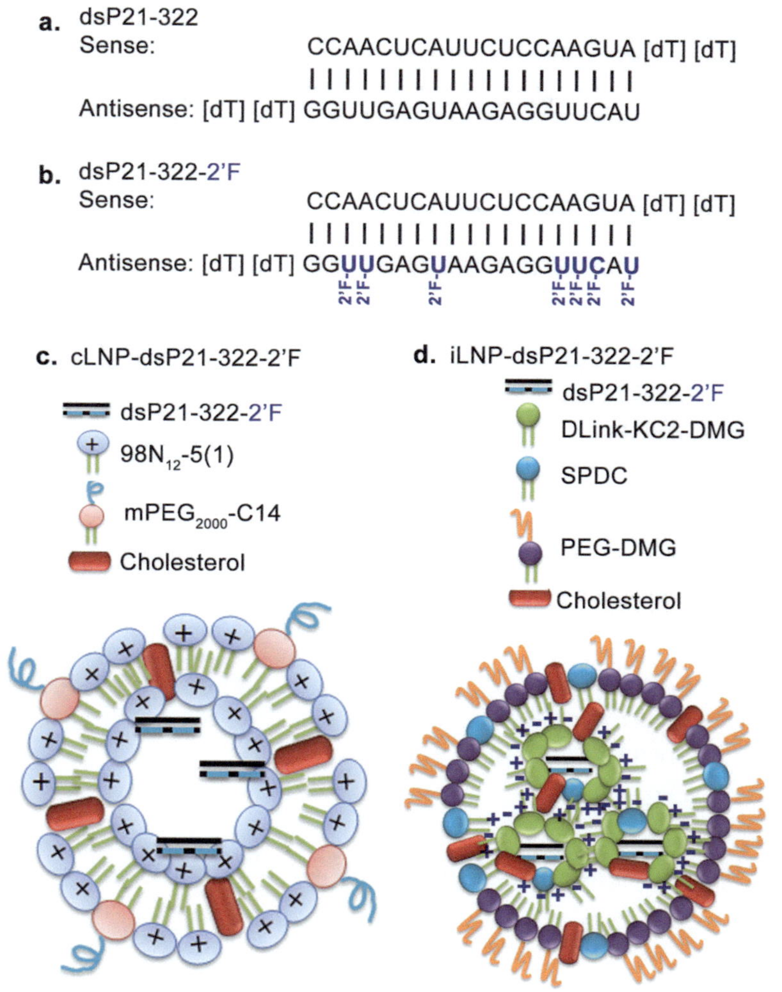

Fig. 16.1 Development of therapeutic dsP21-322. (**a**) Sequence composition of dsP21-322 (**b**) Chemical modification of dsP21-322; indicated is the antisense strands modified with 2′-fluoro-modifications in every cytidine and uridine. (**c**) and (**d**) Composition of cationic LNP-dsP21-322-2′F (**c**) and ionizable cationic LNP-dsP21-322-2′F (**d**)

16.2 Identification of saRNA dsP21-322

In an attempt to induce endogenous p21 expression, Li et al. picked a 19-nt saRNA target on the p21 promoter at the –322 location relative to p21 transcription start site (TSS) by following the rational siRNA design rules and synthesized the corresponding duplex saRNA (dsP21-322, Fig. 16.1a) [19]. Transfecting dsP21-322 into PC-3 (prostate cancer), HeLa (cervical cancer), and MCF-7 (breast cancer) cells elicited 12.5-, 10.1-, and 2.4-fold induction of p21 mRNA levels respectively [19]. The induction was further confirmed at the protein level by Western blot analysis. To identify a lead p21 saRNA for preclinical development, in a subsequent study by Place et al., eight more saRNAs were designed to target p21 promoter ranging from –208 to –466 positions, of which, three significantly activated at least threefold expression of p21 expression. However, none of them were superior to dsP21-322 in RNAa activity [23]. Therefore, dsP21-322 was chosen for subsequent development.

16.3 In Vitro Effects of dsP21-322 on Cancer Cells

Over a dozen studies have shown that dsP21-322 induces endogenous p21 expression at a magnitude ranging from 2 to 14 folds in a variety of cell lines, most of which are cancer cells (Table 16.1). dsP21-322-induced p21 upregulation in cancer cells often results in predicted functional and morphological changes characteristic of p21 functional consequences including an arrested cell morphology, inhibited cell growth and survival, perturbed cell cycle, and increased apoptosis in cancer cells [5, 17, 18, 23, 40]. Mechanistically, dsP21-322-induced G_0/G_1 cell cycle arrest by decreasing phosphorylated Rb levels leading to increased Rb-E2F complexes to inhibit the cell cycle [17, 23], and induced apoptosis by suppression of antiapoptotic protein Bcl-xL and activation of caspase-3 and PARP [40, 41]. Moreover, it was reported that dsP21-322 induced apoptosis via decreasing survivin expression following p21 activation in A498 renal cancer cells and SHG-44 glioma cells [10, 38]. Thus, dsP21-322 induces p21 expression in cancer cells, leading to the inhibition of cell proliferation, survival and cell cycle, and the induction of apoptotic cell death.

16.4 Chemical Modifications of dsP21-322

It is widely recognized that chemical modification of duplex RNAs (i.e., siRNAs) improves in many ways their pharmacologic properties including enhancing nuclease resistance and suppressing immunostimulatory potential and off-target effects [32]. Commonly used modifications include 2′-fluoro (2′F) or 2′-O-methyl (2′Ome) to the ribose ring [36]. To develop medicinal dsP21-322, both 2′F and 2′Ome modifications were applied to different positions of either one or both strands of dsP21-322.

16.4.1 2′F Modification of dsP21-322 Retains RNAa Activity

After several rounds of screening, 2′F modification to the pyrimidines in the antisense strand—previously identified as the guide strand [19]—of dsP21-322 (Fig. 16.1b) was found to retain its RNAa activity comparable to unmodified dsP21-322 [23], whereas other modifications including 2′F and 2′Ome modification of all pyrimidines of both strands resulted in partial or complete loss of RNAa activity, which is consistent with a previous report in which 2′F modification to both

Table 16.1 List of studies testing *in vitro* or/and in vivo antitumor effects of dsP21-322

Cell lines	Fold induction of p21 mRNA	Growth inhibition	Morphology change	Cell cycle arrest	Apoptosis		References
Prostate cancer (PC-3)	12.5	–	–	–	–		[19]
Cervical cancer (HeLa)	10.1	–	–	–	–		
Breast cancer (MCF-7)	2.4	–	–	–	–		
Monkey kidney (COS1)	3.7	–	–	–	–		[15]
Chimpanzee fibroblast (WES)	3.7	–	–	–	–		
Prostate cancer (PC-3)	Semi-quantitative RT-PCR	–	–	–	–		[22]
Cervical cancer (Hela)		–	–	–	–		
Renal cancer (A498)		–	–	–	–		
Prostate cancer (PC-3)	5.5	–	–	–	–		[14]
Prostate cancer (PC-3)	6.0	–	–	–	–		[40]
Bladder cancer (T24)	4.0	–	–	–	–		
Renal cancer (ACHN)	2.0	–	–	–	–		
Bone Osteosarcoma (U2-OS)	4.1	–	–	–	–		
Cervical cancer (HeLa)	5.0	–	–	–	–		
Liver cancer (Hep3B)	3.8	–	–	–	–		
Embryonic kidney (293T)	4.0	–	–	–	–		
Prostate cancer (PC-3)	9.3	–	–	–	–		[24]
Bladder cancer (T24)	2.0	84.4%	+	G_0/G_1	45.3%		[5]
Bladder cancer (J82)	3.0	65.1%	+	G_0/G_1	26.6%		
Bladder cancer (T24)	2.0	77.2%	–	G_0/G_1	50.80%	Bcl-xL ↓caspase-3 ↑PARP ↑	[41]

(continued)

Table 16.1 (continued)

Cell lines	Fold induction of p21 mRNA	Growth inhibition	Morphology change	Cell cycle arrest	Apoptosis		References
Renal cancer (A498)	5.0	75.0%	+	G_0/G_1	21%	survivin ↓	[38]
Lung cancer (A549)	RT-PCR	–	–	G_0/G_1	N/D		[37]
Osteosarcoma (Saos2)							
Liver cancer (BEL7402)	2.9	–	+	–		Bcl-xL ↓cl-xL%%12.	[40]
Liver cancer (SMMC-7721)	2.6	–	+	–	6.3%	↑	
Liver cancer (MHCC97L)	2.8	–	+	–		caspase-3 ↑	
Liver cancer (MHCC97H)	3.2	–	+	–		PARP ↑	
Liver cancer (MHCCLM3)	4.0	65.9%	+	–	12.3%		
Bladder cancer (KU-7)	5	–	+	G_0/G_1	25.70%	caspase-3 ↑	[17]
				(p-Rb↓)		PARP ↑	
Liver cancer (HepG2)	2.8	29.4%	+	G_0/G_1	36.6%	caspase-3 ↑	[18]
Liver cancer (Hep3B)	2.4	33.4%	+	G_2/M	24.6%	PARP ↑	
Prostate cancer (PC-3)	14.0	90.0%	+	G_0/G_1	–		[23]
Prostate cancer (LNCaP)	5.0	–	–	–	–		
Prostate cancer (DU-145)	4.0	–	–	–	–		
Glioma (SHG-44)	2.0	63.50%	–	G_0/G_1	41.90%	survivin ↓	[10]

strands of a saRNA could be detrimental to its RNAa activity [36]. The estimated EC_{50} on p21 activation of dsP21-322-2′F was about 1 nmol/L as measured at 72 h in PC-3 cells [23]. A similar result was obtained in KU-7 bladder cancer cells, with the EC_{50} for unmodified dsP21-322 and dsP21-322-2′F at 1.42 and 1.7 nmol/L, respectively [17]. Furthermore, dsP21-322-2′F caused similar effects on G_0/G_1 cell cycle arrest and induced apoptosis of both prostate [23] and bladder [17] cancer cells.

16.4.2 2′F Modification of dsP21-322 Reduces Immune Stimulation

Therapeutic dsRNAs can be recognized by toll-like receptors (TLRs) and may thus act as a potent activator of the innate immune system [8]. Indeed, unmodified duplex RNAs (e.g., dsP21-322) possessed strong immunostimulatory activity as assessed by measuring IFN-α and TNF-α released from cultured peripheral blood mononuclear cells (PBMCs) which have been treated with the duplex RNA for 24 h. However, cytokine stimulation by the 2′F modified dsP21-322 was reduced to a negligible level [23].

16.4.3 2′F Modification of dsP21-322 Increases Stability in Serum and Urine

Therapeutic dsRNAs without protection could be prematurely degraded by nuclease, either intracellularly or extracellularly. As RNA is vulnerable to enzymatic degradation, chemical modification improves duplex RNAs stability. 2′F modification of dsP21-322 significantly increased its nuclease resistance and extended its half-life in active mouse serum from ~6 to ~14 h [23]. In human urine, unmodified dsP21-322 was very unstable and completely degraded within 3 h; however, chemically modified dsP21-322-2′F significantly increased duplex stability extending its life to 24 h in urine [17]. Surprisingly, either unmodified or modified duplex RNAs showed no sign of degradation after 96 h of incubation in mouse urine [17]. These data indicate that chemical modification improves duplex stability in mouse serum and human urine; however, inherent differences in nuclease activity/composition between human and mouse urine may influence in vivo properties in either species.

16.5 Preparation of dsP21-322-2′F for Delivery

Clinical development saRNA therapeutics faces the same delivery-related challenges as do siRNA drugs, such as efficient cellular uptake and target site accumulation, rapid clearance from the circulation, susceptibility to degradation by nucleases, and undesired stimulation of the innate immune system. Currently, lipid nanoparticle (LNP) delivery systems represent the most advanced platforms for systemic delivery of siRNAs [4] and have been used to deliver siRNAs into a number of solid tumors in different organs especially the liver in animal studies including nonhuman primates [7].

16.5.1 Cationic LNP-Formulated dsP21-322-2′F for Intratumoral Delivery

Cationic liposomes are commonly used in nucleic acid delivery because of the electrostatic complexation of positively charged lipids with negatively charged dsRNAs. Lipidoids, a new class of lipid-like delivery molecules, are based on the conjugate addition of alkyl-acrylates or alkyl-acrylamides to the primary or secondary amines and show improved local and systemic delivery of RNA therapeutics [2, 3]. Based on lipidoids $98N_{12}-5(1)$, dsP21-322-2′F was formulated with $98N_{12}-5(1)$, cholesterol and $mPEG_{2000}$-C14 at the molar ratio of ~42:48:10 (known as LNP01 formulation) and the resultant particles had a mean size of ~50–60 nm and dsRNA entrapment efficiency >95% [23] (Fig. 16.1c). Compared to nonformulated dsP21-322-2′F, cationic LNP-formulated dsP21-322-2′F (cLNP-dsP21-322-2′F) was more stable in active mouse serum extending the half-life of dsP21-322-2′F from ~14 to ~38 h, and with potency for activating p21 gene expression and decreasing cell viability equivalent to that of the original dsP21-322 [23].

16.5.2 Ionizable Cationic LNP-Formulated dsP21-322-2′F Intravesical Delivery

The surface charge of liposomes is important for their interaction with cell membranes and for endosomal escape. Because cationic LNP may possess toxicity, its overall charge can be reduced by using ionizable cationic lipids [12] which almost neutralizes the charge of cLNP, resulting in improved endosomal escape and reduced toxicity [30]. The ionizable cationic lipid DLin-KC2-DMA (2,2-dilinoleyl-4-(2-dimethylaminoethyl)-[1,3]-dioxolane) has proven to be highly effective in systemically delivering siRNAs to silence a hepatic gene (Factor VII) [25]. DLin-KC2-DMA mixed with DSPC, cholesterol, and PEG–DMG at the molar ratio of ~50:10:38.5:1.5 (DLin-KC2-DMA:DSPC:cholesterol:PEG–DMG) resulted in formulation LNP09 which has been used to deliver siRNAs to the liver in several clinical trials [30]. By taking advantage of this new LNP formulation, dsP21-322-2′F was encapsulated in LNP09 (iLNP-dsP21-322-2′F) for local delivery to the bladder after confirming its RNAa activity in vitro and testing its stability in urine [17] (Fig. 16.1d). Surprisingly, iLNP-dsP21-322-2′F showed no sign of degradation even after 6 d in human urine at 37 °C whereas the unformulated dsP21-322-2′F could only survive for 24 h before it was completely degraded [17].

The inner surface of the bladder wall is lined with a layer of glycosaminoglycan (GAG) which provides protection to the underlying urothelium from harmful substances in the urine. The GAG layer could also represent a barrier to instilled dsRNAs. To assess whether LNP09-formulated dsP21-322 could penetrate the GAG layer to reach the urothelium and be internalized by urothelial cells, FITC-conjugated and LNP09-formulated dsP21-322-2′F was delivered to normal mouse bladders. After 9 h, fluorescent signal could be visualized in the cells lining the inner bladder wall as well as the interstitial spaces of the urothelium [17]. These findings indicate that LNP09 formulation, although optimized for liver delivery, is well suited for regional delivery of dsRNA into the bladder by stabilizing dsRNA in urine and infiltrating the GAG layer.

16.6 In Vivo Efficacy Studies of dsP21-322

16.6.1 In Vivo Antitumor Effects of cLNP-Formulated dsP21-322-2′F for Prostate Cancer

To evaluate cLNP-dsP21-322-2′F activity in vivo, xenograft models of human prostate cancer were established in the franks of nude mice by subcutaneously injecting prostate cancer cells (PC-3). When the resulted subcutaneous tumors became visible, they were treated by an intratumoral injection of cLNP-dsP21-322-2′F or cLNP-dsP21-322 at 5 mg/kg every 3 days for three times [23]. Control mice received injections of PBS or a control saRNA similarly modified and formulated (cLNP-dsCon-2′F). At 3 weeks following the initial treatment when all mice were sacrificed, significant reduction (~50%) in tumor volume and weight in the treatment groups were noted compared to control groups (PBS or cLNP-dsCon-2′F) [23]. Moreover, in the two treatment groups (unmodified cLNP-dsP21-322 and cLNP-dsP21-322-2′F), mice in the latter group had smaller tumors (compared a mean weight of 0.29–0.35 g) with less variation in tumor weight [23], suggesting the 2′F modification might have provided dsP21-322 improved pharmacologic properties probably by increasing its stability and suppressing its immune stimulation. To assess in vivo activation of p21, immunohistochemistry and immunoblot analysis of p21 protein were performed in tumor removed 3 days following the last dose. Upregulation of p21 protein was evident by both assays for tumors treated with cLNP-dsP21-322-2′F [23].

In conclusion, cLNP-dsP21-322-2′F treatment activates p21 expression in vivo and inhibits growth of prostate cancer tumors. Despite the positive results, administration of dsP21-322 in patients with prostate cancer especially for those with metastatic tumors is faced with a formidable challenge of delivering the saRNA specifically to prostate cancer cells in the body. To solve this issue, we recently conjugated a *DPYSL3* saRNA to an RNA aptamer which specifically recognized prostate-specific membrane antigen (PSMA). Systemic administration of this conjugate to nude mice bearing orthotopic xenografts of prostate cancer significantly suppressed distal metastasis [20]. The same conjugation strategy could also be applied to dsP21-322 for systemic delivery.

16.6.2 Preclinical Evaluation of iLNP-Formulated dsP21-322-2′F in Orthotopic Bladder Cancer Models

Encouraged by the results from treating xenograft prostate cancer with cLNP-dsP21-322-2′F, we wished to test further its antitumor efficacy in a clinically relevant model. The bladder is an easily accessible hollow organ ideal for local drug delivery. Intravesical administration of saRNA can have the benefit of bringing high concentrations of drugs into direct contact with diseased tissue without the side effects associated with systemic administration. Most importantly, a clinically unmet need for the management of bladder cancer is the eradication of residual tumor cells and prevention of recurrence after transurethral resection (TUR) of superficial tumors which accounts for 70% of initially diagnosed bladder cancer. Despite the fact that routine intravesical administrations of chemical and/or immunological agents (e.g., mitomycin, bacillus Calmette-Guérin, etc.) follows TUR, tumors recur in 50% of the treated patients with increased grades and stages [26]. Therefore, RNAa therapy could offer a solution to this unmet medical need via its novel mechanism of action.

To assess the antitumor efficacy of iLNP-dsP21-322-2′F, orthotopic bladder tumor models were established by intravesical instillation of KU-7 cells which expressed luciferase and GFP [17]. After confirming tumor presence in the bladder of 29 mice by bioluminescent imaging, they were randomly divided into three groups: iLNP-dsP21-322-2′F (n=10), iLNP-dsCon-2′F (n=10), and PBS (n=9). Treatments were delivered intravesically through catheterization starting on day 4 post tumor cell implantation and repeated every 3 days for a total of 14 doses [17]. Tumor growth in the bladder was monitored weekly by both in vivo bladder ultrasound and bioluminescent imaging, and animal survival was the primary end point. The iLNP-dsP21-322-2′F group had a median survival time of 45 days which was significantly longer than the PBS (9 days) and iLNP-dsCon-2′F (13 days) groups. At the end of the eight-week study, five mice were alive in the iLNP-dsP21-322-2′F, four of whom showed gradual reduction in tumor signal in the course of treatment and three eventually became tumor free. However, the PBS and the iLNP-dsCon-2′F groups had only one and no surviving animal respectively at the end of the study.

Histologic evaluation revealed that iLNP-dsP21-322-2′F treatment resulted in increased p21 expression and cleaved capspase-3, indicative of apoptosis, in harvested bladder tumors [17]. These results demonstrated that iLNP-dsP21-322-2′F has antitumor activity in orthotopic models of bladder cancer by activating p21 expression in vivo.

16.7 Conclusions and Perspectives

RNAa represents one of the few available technologies that could be translated into the clinic to treat disease by stimulating the production of endogenous genes. As a key negative regulator of the cell cycle rarely affected by genetic alteration in cancer, the p21 gene seems to be an ideal druggable target for saRNA therapeutics. The p21 saRNA dsP21-322 has undergone extensive scrutiny by worldwide groups and can be considered as a candidate for clinical development. Further clinical trial is warranted to test its safety properties and pharmacologic effects in human.

References

1. Abbas T, Dutta A (2009) p21 in cancer: intricate networks and multiple activities. Nat Rev Cancer 9(6):400–414. doi:nrc2657 [pii] 10.1038/nrc2657
2. Akinc A, Zumbuehl A, Goldberg M, Leshchiner ES, Busini V, Hossain N, Bacallado SA, Nguyen DN, Fuller J, Alvarez R, Borodovsky A, Borland T, Constien R, de Fougerolles A, Dorkin JR, Narayanannair Jayaprakash K, Jayaraman M, John M, Koteliansky V, Manoharan M, Nechev L, Qin J, Racie T, Raitcheva D, Rajeev KG, Sah DW, Soutschek J, Toudjarska I, Vornlocher HP, Zimmermann TS, Langer R, Anderson DG (2008) A combinatorial library of lipid-like materials for delivery of RNAi therapeutics. Nat Biotechnol 26 (5):561–569. doi:10.1038/nbt1402
3. Akinc A, Goldberg M, Qin J, Dorkin JR, Gamba-Vitalo C, Maier M, Jayaprakash KN, Jayaraman M, Rajeev KG, Manoharan M, Koteliansky V, Rohl I, Leshchiner ES, Langer R, Anderson DG (2009) Development of lipidoid-siRNA formulations for systemic delivery to the liver. Mol Ther 17(5):872–879. doi:10.1038/mt.2009.36
4. Akinc A, Querbes W, De S, Qin J, Frank-Kamenetsky M, Jayaprakash KN, Jayaraman M, Rajeev KG, Cantley WL, Dorkin JR, Butler JS, Qin L, Racie T, Sprague A, Fava E, Zeigerer A, Hope MJ, Zerial M, Sah DW, Fitzgerald K, Tracy MA, Manoharan M, Koteliansky V, Fougerolles A, Maier MA (2010) Targeted delivery of RNAi therapeutics with endogenous and exogenous ligand-based mechanisms. Mol Ther 18(7):1357–1364. doi:10.1038/mt.2010.85
5. Chen Z, Place RF, Jia ZJ, Pookot D, Dahiya R, Li LC (2008) Antitumor effect of dsRNA-induced p21(WAF1/CIP1) gene activation in human bladder cancer cells. Mol Cancer Ther 7 (3):698–703. doi:10.1158/1535-7163.MCT-07-2312
6. Clasen S, Schulz WA, Gerharz CD, Grimm MO, Christoph F, Schmitz-Drager BJ (1998) Frequent and heterogeneous expression of cyclin-dependent kinase inhibitor WAF1/p21 protein and mRNA in urothelial carcinoma. Br J Cancer 77(4):515–521
7. Coelho T, Adams D, Silva A, Lozeron P, Hawkins PN, Mant T, Perez J, Chiesa J, Warrington S, Tranter E, Munisamy M, Falzone R, Harrop J, Cehelsky J, Bettencourt BR, Geissler M, Butler JS, Sehgal A, Meyers RE, Chen Q, Borland T, Hutabarat RM, Clausen VA,

Alvarez R, Fitzgerald K, Gamba-Vitalo C, Nochur SV, Vaishnaw AK, Sah DW, Gollob JA, Suhr OB (2013) Safety and efficacy of RNAi therapy for transthyretin amyloidosis. N Engl J Med 369(9):819–829. doi:10.1056/NEJMoa1208760
8. De Paula D, Bentley MV, Mahato RI (2007) Hydrophobization and bioconjugation for enhanced siRNA delivery and targeting. RNA 13(4):431–456. doi:10.1261/rna.459807
9. el-Deiry WS, Harper JW, O'Connor PM, Velculescu VE, Canman CE, Jackman J, Pietenpol JA, Burrell M, Hill DE, Wang Y et al (1994) WAF1/CIP1 is induced in p53-mediated G1 arrest and apoptosis. Cancer Res 54(5):1169–1174
10. Dong Z, Dang Y, Chen Y (2014) Small double-stranded RNA mediates the anticancer effects of p21WAF1/ClP1 transcriptional activation in a human glioma cell line. Yonsei Med J 55 (2):324–330. doi:10.3349/ymj.2014.55.2.324201403324 [pii]
11. Eastham JA, Hall SJ, Sehgal I, Wang J, Timme TL, Yang G, Connell-Crowley L, Elledge SJ, Zhang WW, Harper JW et al (1995) In vivo gene therapy with p53 or p21 adenovirus for prostate cancer. Cancer Res 55(22):5151–5155
12. Falsini S, Ciani L, Ristori S, Fortunato A, Arcangeli A (2014) Advances in lipid-based platforms for RNAi therapeutics. J Med Chem 57(4):1138–1146. doi:10.1021/jm400791q
13. Harper JW, Adami GR, Wei N, Keyomarsi K, Elledge SJ (1993) The p21 Cdk-interacting protein Cip1 is a potent inhibitor of G_1 cyclin-dependent kinases. Cell 75(4):805–816
14. Hu J, Chen Z, Xia D, Wu J, Xu H, Ye ZQ (2012) Promoter-associated small double-stranded RNA interacts with heterogeneous nuclear ribonucleoprotein A2/B1 to induce transcriptional activation. Biochem J 447(3):407–416. doi:10.1042/BJ20120256BJ20120256 [pii]
15. Huang V, Qin Y, Wang J, Wang X, Place RF, Lin G, Lue TF, Li LC (2010) RNAa is conserved in mammalian cells. PLoS One 5(1):e8848. doi:10.1371/journal.pone.0008848
16. Janowski BA, Younger ST, Hardy DB, Ram R, Huffman KE, Corey DR (2007) Activating gene expression in mammalian cells with promoter-targeted duplex RNAs. Nat Chem Biol 3 (3):166–173. doi:nchembio860 [pii] 10.1038/nchembio860
17. Kang MR, Yang G, Place RF, Charisse K, Epstein-Barash H, Manoharan M, Li LC (2012) Intravesical delivery of small activating RNA formulated into lipid nanoparticles inhibits orthotopic bladder tumor growth. Cancer Res 72(19):5069–5079. 0008-5472.CAN-12-1871 [pii] 10.1158/0008-5472.CAN-12-1871
18. Kosaka M, Kang MR, Yang G, Li LC (2012) Targeted p21WAF1/CIP1 activation by RNAa inhibits hepatocellular carcinoma cells. Nucleic Acid Ther 22(5):335–343. doi:10.1089/nat.2012.0354
19. Li LC, Okino ST, Zhao H, Pookot D, Place RF, Urakami S, Enokida H, Dahiya R (2006) Small dsRNAs induce transcriptional activation in human cells. Proc Natl Acad Sci U S A 103 (46):17337–17342
20. Li C, Jiang W, Hu Q, Li LC, Dong L, Chen R, Zhang Y, Tang Y, Thrasher JB, Liu CB, Li B (2016) Enhancing *DPYSL3* gene expression via a promoter-targeted small activating RNA approach suppresses cancer cell motility and metastasis. Oncotarget 7(16):22893–22910. doi:10.18632/oncotarget.82908290 [pii]
21. Place RF, Li LC, Pookot D, Noonan EJ, Dahiya R (2008) MicroRNA-373 induces expression of genes with complementary promoter sequences. Proc Natl Acad Sci U S A 105 (5):1608–1613. doi:0707594105 [pii] 10.1073/pnas.0707594105
22. Place RF, Noonan EJ, Foldes-Papp Z, Li LC (2010) Defining features and exploring chemical modifications to manipulate RNAa activity. Curr Pharm Biotechnol 11(5):518–526. doi:BSP/CPB/E-Pub/0094-11-6aaa [pii]
23. Place RF, Wang J, Noonan EJ, Meyers R, Manoharan M, Charisse K, Duncan R, Huang V, Wang X, Li L-C (2012) Formulation of small activating RNA into lipidoid nanoparticles inhibits xenograft prostate tumor growth by inducing p21 expression. Mol Ther Nucleic Acids 1:e15. doi:10.1038/mtna.2012.5
24. Portnoy V, Lin SH, Li KH, Burlingame A, Hu ZH, Li H, Li LC (2016) saRNA-guided Ago2 targets the RITA complex to promoters to stimulate transcription. Cell Res 26(3):320–335. doi:10.1038/cr.2016.22cr201622 [pii]

25. Semple SC, Akinc A, Chen J, Sandhu AP, Mui BL, Cho CK, Sah DW, Stebbing D, Crosley EJ, Yaworski E, Hafez IM, Dorkin JR, Qin J, Lam K, Rajeev KG, Wong KF, Jeffs LB, Nechev L, Eisenhardt ML, Jayaraman M, Kazem M, Maier MA, Srinivasulu M, Weinstein MJ, Chen Q, Alvarez R, Barros SA, De S, Klimuk SK, Borland T, Kosovrasti V, Cantley WL, Tam YK, Manoharan M, Ciufolini MA, Tracy MA, de Fougerolles A, MacLachlan I, Cullis PR, Madden TD, Hope MJ (2010) Rational design of cationic lipids for siRNA delivery. Nat Biotechnol 28 (2):172–176. doi:10.1038/nbt.1602nbt.1602 [pii]
26. Shen Z, Shen T, Wientjes MG, O'Donnell MA, Au JL (2008) Intravesical treatments of bladder cancer: review. Pharm Res 25(7):1500–1510
27. Slaton JW, Benedict WF, Dinney CP (2001) P53 in bladder cancer: mechanism of action, prognostic value, and target for therapy. Urology 57(5):852–859
28. Spruck CH 3rd, Ohneseit PF, Gonzalez-Zulueta M, Esrig D, Miyao N, Tsai YC, Lerner SP, Schmutte C, Yang AS, Cote R et al (1994) Two molecular pathways to transitional cell carcinoma of the bladder. Cancer Res 54(3):784–788
29. Stein JP, Ginsberg DA, Grossfeld GD, Chatterjee SJ, Esrig D, Dickinson MG, Groshen S, Taylor CR, Jones PA, Skinner DG, Cote RJ (1998) Effect of p21WAF1/CIP1 expression on tumor progression in bladder cancer. J Natl Cancer Inst 90(14):1072–1079
30. Tam YY, Chen S, Cullis PR (2013) Advances in lipid nanoparticles for siRNA delivery. Pharmaceutics 5(3):498–507. doi:10.3390/pharmaceutics5030498
31. Teraishi F, Kadowaki Y, Tango Y, Kawashima T, Umeoka T, Kagawa S, Tanaka N, Fujiwara T (2003) Ectopic p21sdi1 gene transfer induces retinoic acid receptor beta expression and sensitizes human cancer cells to retinoid treatment. Int J Cancer 103(6):833–839
32. Vaishnaw AK, Gollob J, Gamba-Vitalo C, Hutabarat R, Sah D, Meyers R, de Fougerolles T, Maraganore J (2010) A status report on RNAi therapeutics. Silence 1(1):14. doi:1758-907X-1-14 [pii] 10.1186/1758-907X-1-14
33. Voutila J, Saetrom P, Mintz P, Sun G, Alluin J, Rossi JJ, Habib NA, Kasahara N (2012) Gene expression profile changes after short-activating RNA-mediated induction of endogenous pluripotency factors in human mesenchymal stem cells. Mol Ther Nucleic Acids 1:e35. doi:10.1038/mtna.2012.20mtna201220 [pii]
34. Waga S, Hannon GJ, Beach D, Stillman B (1994) The p21 inhibitor of cyclin-dependent kinases controls DNA replication by interaction with PCNA. Nature 369(6481):574–578
35. Warfel NA, El-Deiry WS (2013) p21WAF1 and tumorigenesis: 20 years after. Curr Opin Oncol 25(1):52–58. doi:10.1097/CCO.0b013e32835b639e
36. Watts JK, Yu D, Charisse K, Montaillier C, Potier P, Manoharan M, Corey DR (2010) Effect of chemical modifications on modulation of gene expression by duplex antigene RNAs that are complementary to noncoding transcripts at gene promoters. Nucleic Acids Res 38:5242–5259. doi:gkq258 [pii] 10.1093/nar/gkq258
37. Wei J, Zhao J, Long M, Han Y, Wang X, Lin F, Ren J, He T, Zhang H (2010) p21WAF1/CIP1 gene transcriptional activation exerts cell growth inhibition and enhances chemosensitivity to cisplatin in lung carcinoma cell. BMC Cancer 10:632. doi:1471-2407-10-632 [pii] 10.1186/1471-2407-10-632
38. Whitson JM, Noonan EJ, Pookot D, Place RF, Dahiya R (2009) Double stranded-RNA-mediated activation of P21 gene induced apoptosis and cell cycle arrest in renal cell carcinoma. Int J Cancer 125(2):446–452. doi:10.1002/ijc.24370
39. Wu M, Bellas RE, Shen J, Sonenshein GE (1998) Roles of the tumor suppressor p53 and the cyclin-dependent kinase inhibitor p21WAF1/CIP1 in receptor-mediated apoptosis of WEHI 231 B lymphoma cells. J Exp Med 187(10):1671–1679
40. Wu ZM, Dai C, Huang Y, Zheng CF, Dong QZ, Wang G, Li XW, Zhang XF, Li B, Chen G (2011) Anti-cancer effects of p21WAF1/CIP1 transcriptional activation induced by dsRNAs in human hepatocellular carcinoma cell lines. Acta Pharmacol Sin 32(7):939–946. doi: aps201128 [pii] 10.1038/aps.2011.28
41. Yang K, Zheng XY, Qin J, Wang YB, Bai Y, Mao QQ, Wan Q, Wu ZM, Xie LP (2008) Up-regulation of p21(WAF1/Cip1) by saRNA induces G_1-phase arrest and apoptosis in T24 human bladder cancer cells. Cancer Lett 265(2):206–214

Index

L.-C. Li (ed.), *RNA Activation*, Advances in Experimental Medicine and Biology 983, DOI 10.1007/978-981-10-4310-9

MIX
Papier aus verantwortungsvollen Quellen
Paper from responsible sources
FSC® C105338

If you have any concerns about our products,
you can contact us on
ProductSafety@springernature.com

In case Publisher is established outside the EU,
the EU authorized representative is:
Springer Nature Customer Service Center GmbH
Europaplatz 3, 69115 Heidelberg, Germany

Printed by Libri Plureos GmbH
in Hamburg, Germany